Gluten Related Disorders: Coeliac Disease and Beyond

Gluten Related Disorders: Coeliac Disease and Beyond

Editors

Isabel Comino
Carolina Sousa

MDPI • Basel • Beijing • Wuhan • Barcelona • Belgrade • Manchester • Tokyo • Cluj • Tianjin

Editors
Isabel Comino
Universidad de Sevilla
Spain

Carolina Sousa
Universidad de Sevilla
Spain

Editorial Office
MDPI
St. Alban-Anlage 66
4052 Basel, Switzerland

This is a reprint of articles from the Special Issue published online in the open access journal *Nutrients* (ISSN 2072-6643) (available at: https://www.mdpi.com/journal/nutrients/special_issues/ Gluten-Related_Disorders_Coeliac_Disease_and_Beyond).

For citation purposes, cite each article independently as indicated on the article page online and as indicated below:

LastName, A.A.; LastName, B.B.; LastName, C.C. Article Title. *Journal Name* **Year**, *Volume Number*, Page Range.

ISBN 978-3-0365-5103-6 (Hbk)
ISBN 978-3-0365-5104-3 (PDF)

Contents

Editorial

Advances in Celiac Disease and Gluten-Free Diet

Isabel Comino * and Carolina Sousa

Department of Microbiology and Parasitology, Faculty of Pharmacy, University of Seville, 41012 Seville, Spain; csoumar@us.es

* Correspondence: icomino@us.es

Citation: Comino, I.; Sousa, C. Advances in Celiac Disease and Gluten-Free Diet. *Nutrients* **2022**, *14*, 570. https://doi.org/10.3390/nu14030570

Received: 27 December 2021

Accepted: 11 January 2022

Published: 28 January 2022

Publisher's Note: MDPI stays neutral with regard to jurisdictional claims in published maps and institutional affiliations.

Celiac disease (CD) is a systemic disease that causes chronic enteropathy of the small intestine and develops through an inadequate immune response to gluten in genetically predisposed individuals [1,2]. To date, the only effective treatment is a gluten-free diet (GFD) that essentially relies on the consumption of naturally gluten-free foods and gluten-free dietary products that may not contain more than 20 mg/kg of gluten according to Codex Alimentarius (Codex Standard 118-19792). Research on CD is changing rapidly due to a steady increase in knowledge that addresses its pathophysiology, diagnosis, follow-up, and therapeutic options. For this reason, this Special Issue includes 12 peer-reviewed articles reporting on the latest research findings and evidence related to CD and a GFD.

CD is characterized by a heterogeneous clinical presentation, affecting any organ or tissue with gastrointestinal, extraintestinal, seronegative, or nonresponsive manifestations. A common, and sometimes the only, clinical finding in untreated patients is anemia, which is generally caused by damage of the duodenal mucosa and the resulting iron malabsorption. However, a poor correlation has been found between the presence of anemia, an abnormal expression of duodenal iron transport proteins, and the severity of histological damage [3]. In other cases, the onset of EC is represented by subclinical manifestations, some of which can be found in the mouth. In this Special Issue, Nota et al. [4] report significant associations between the clinical characteristics of CD and the prevalence of caries and dentin sensitivity. Additionally, an inappropriate GFD was associated with oral manifestations.

Regarding extraintestinal manifestations, CD has been associated with IgA nephropathy (IgAN), the most common primary chronic glomerular disease worldwide. Furthermore, it is well known that IgA-class tissue transglutaminase (tTG) autoantibodies are deposited in the small intestine mucosa and extraintestinal organs. Nurmi et al. [5] have identified IgA deposits targeted for tTG in kidney biopsies of gluten-consuming IgAN patients with or without known CD.

The diagnosis of CD is based on several criteria, including positive serology, a spectrum of duodenal damage, clinical symptoms and/or risk conditions, and response to a GFD in patients with HLA-DQ2 or DQ8 genotypes. In the absence of some of these criteria, especially when serology is negative or duodenal atrophy is incomplete, the diagnosis of CD becomes challenging. Ruiz-Ramírez et al. [6] have confirmed the high diagnostic accuracy of the intraepithelial lymphocyte cytometric pattern as a tool in the diagnosis of CD regardless of the degree of mucosal damage and age.

As Viitasalo et al. [7] proposed, close relatives of patients with CD, with partially shared living environments and genetic factors, could have increased seroreactivity to microbial markers. They studied the seropositivity and levels of anti-*Saccharomyces cerevisiae* (ASCA), *Pseudomonas fluorescens*-associated sequence (anti-I2), and *Bacteroides caccae* TonB-linked outer membrane protein (anti-OmpW) in first-degree relatives and patients with CD. The results showed an increase in seroreactivity to serum microbial markers, particularly ASCA and anti-I2, in relatives of patients with CD, even in the absence of disease-specific autoantibodies or other signs of active CD. This observation was not explained by the presence or absence of predisposing HLA haplotypes, suggesting the role of other genetic and environmental factors.

Since the only effective treatment for CD is a GFD, several patients have difficulty controlling their diet, especially those who live in a rural setting [8] and, therefore, regularly consume sufficient gluten to trigger symptoms. In their reviews, Weiser et al. [9,10] discuss gluten contamination and adherence to a GFD. The available evidence on the degree of adherence to a GFD, barriers to its implementation [9,10], and methods to assess it [11] were examined. Despite the availability of diverse traditional GFD adherence markers, such as diet tests or serology, none of them are an accurate diet evaluation method. Thus, the use of gluten immunogenic peptides (GIP) detection in urine has been developed as a direct test for GFD monitoring, contrary to classical methods. Coto et al. provide new knowledge on gluten metabolism and GIP excretion in urine [12].

Segura et al. [13] discuss emerging therapeutic options for CD based on the removal of toxic gluten peptides, the modulation of intestinal permeability, or the restoration of the gut microbiota. These treatment options have shown encouraging preliminary results in phase II and III clinical trials. If successful, these novel approaches raise the possibility of reintroducing gluten, in amounts to be determined, into the diets of patients with CD. However, a GFD is the mainstay of CD therapy for the immediate future, pending FDA and/or EMA approval of any of these treatment options.

Finally, a GFD has been evaluated in other gastrointestinal pathologies. Several trials have evaluated the 'bottom-up' approach of a GFD in irritable bowel syndrome (IBS), with a response rate between 34 and 71%. Fernández-Bañares et al. [14] studied the effect of a GFD in patients with functional bowel disorders (FBD) and evaluated the role of both the low-grade celiac score and the celiac lymphogram in the probability of responding to a GFD. Their study shows that a GFD is effective in the long-term treatment of patients with previously unexplained chronic watery diarrhea or dominant bloating symptoms that meet the criteria for FBD. The response rate was much higher in the subgroup of patients defined by the presence of a positive low-grade celiac score and celiac lymphogram.

In conclusion, these valuable studies provide a deeper understanding of the diagnosis, follow-up of patients with EC, and the effect of a GFD. We thank all the authors for their contributions, the reviewers for their constructive comments, and the Nutrients Publishing Team for their professional assistance in the development of this Special Issue.

Author Contributions: Conceptualization, I.C. and C.S.; writing—review and editing, I.C. and C.S. All authors have read and agreed to the published version of the manuscript.

Funding: This research received no external funding.

Conflicts of Interest: The authors declare no conflict of interest.

References

1. Lebwohl, B.; Sanders, D.S.; Green, P.H.R. Coeliac disease. *Lancet* **2018**, *391*, 70–81. [CrossRef]
2. Lebwohl, B.; Rubio-Tapia, A. Epidemiology, presentation, and diagnosis of celiac disease. *Gastroenterology* **2021**, *160*, 63–75. [CrossRef] [PubMed]
3. Repo, M.; Hannula, M.; Taavela, J.; Hyttinen, J.; Isola, J.; Hiltunen, P.; Popp, A.; Kaukinen, K.; Kurppa, K.; Lindfors, K. Iro-transporter protein expressions in children with celiac disease. *Nutrients* **2021**, *13*, 776. [CrossRef] [PubMed]
4. Nota, A.; Abati, S.; Bosco, F.; Rota, I.; Polizzi, E.; Tecco, S. General health, systemic diseases and oral status in adult patients with coeliac disease. *Nutrients* **2020**, *12*, 3836. [CrossRef] [PubMed]
5. Nurmi, R.; Korponay-Szabó, I.; Laurila, K.; Huhtala, H.; Niemelä, O.; Mustonen, J.; Mäkelä, S.; Kaukinen, K.; Lindfors, K. Celiac disease-type tissue transglutaminase autoantibody deposits in kidney biopsies of patients with IgA nephropathy. *Nutrients* **2021**, *13*, 1594. [CrossRef]
6. Ruiz-Ramírez, P.; Carreras, G.; Fajardo, I.; Tristán, E.; Carrasco, A.; Salvador, I.; Zabana, Y.; Andújar, X.; Ferrer, C.; Horta, D.; et al. Intraepithelial lymphocyte cytometric pattern is a useful diagnostic tool for coeliac disease diagnosis irrespective of degree of mucosal damage and age—A validation cohort. *Nutrients* **2021**, *13*, 1684. [CrossRef] [PubMed]
7. Viitasalo, L.; Iltanen, S.; Huhtala, H.; Saavalainen, P.; Kaukinen, K.; Lindfors, K.; Kurppa, K. First-degree Relatives of Celiac Disease patients have increased seroreactivity to serum microbial markers. *Nutrients* **2020**, *12*, 1073. [CrossRef] [PubMed]
8. Lee, R.; Crowley, E.T.; Baines, S.K.; Heaney, S.; Brown, L.J. Patient Perspectives of living with coeliac disease and accessing dietetic services in rural Australia: A qualitative study. *Nutrients* **2021**, *13*, 2074. [CrossRef] [PubMed]

9. Wieser, H.; Ruiz-Carnicer, Á.; Segura, V.; Comino, I.; Sousa, C. Challenges of monitoring the gluten-free diet adherence in the management and follow-up of patients with celiac disease. *Nutrients* **2021**, *13*, 2274. [CrossRef] [PubMed]

10. Wieser, H.; Segura, V.; Ruiz-Carnicer, Á.; Sousa, C.; Comino, I. Food safety and cross-contamination of gluten-free products: A narrative review. *Nutrients* **2021**, *13*, 2244. [CrossRef] [PubMed]

11. Rodríguez-Herrera, A.; Reyes-Andrade, J.; Rubio-Escudero, C. Rationale for timing of follow-up visits to assess gluten-free diet in celiac disease patients based on data mining. *Nutrients* **2021**, *13*, 357. [CrossRef] [PubMed]

12. Coto, L.; Sousa, C.; Cebolla, A. Dynamics and considerations in the determination of the excretion of gluten immunogenic peptides in urine: Individual variability at low gluten intake. *Nutrients* **2021**, *13*, 2624. [CrossRef] [PubMed]

13. Segura, V.; Ruiz-Carnicer, Á.; Sousa, C.; Moreno, M.d.L. New insights into non-dietary treatment in celiac disease: Emerging therapeutic options. *Nutrients* **2021**, *13*, 2146. [CrossRef] [PubMed]

14. Fernández-Bañares, F.; Arau, B.; Raga, A.; Aceituno, M.; Tristán, E.; Carrasco, A.; Ruiz, L.; Martín-Cardona, A.; Ruiz-Ramírez, P.; Esteve, M. Long-term effect of a gluten-free diet on diarrhoea- or bloating-predominant functional bowel disease: Role of the 'low-grade coeliac score' and the 'coeliac lymphogram' in the response rate to the diet. *Nutrients* **2021**, *13*, 1812. [CrossRef] [PubMed]

Article

Iron Transporter Protein Expressions in Children with Celiac Disease

Marleena Repo [1,2], **Markus Hannula** [3], **Juha Taavela** [4], **Jari Hyttinen** [3], **Jorma Isola** [5,6], **Pauliina Hiltunen** [1], **Alina Popp** [7], **Katri Kaukinen** [2,8], **Kalle Kurppa** [1,9] and **Katri Lindfors** [2,*]

[1] Tampere Centre for Child Health Research, Tampere University and Tampere University Hospital, 33014 Tampere, Finland; marleena.repo@tuni.fi (M.R.); pauliina.hiltunen@pshp.fi (P.H.); kalle.kurppa@tuni.fi (K.K.)

[2] Celiac Disease Research Center, Faculty of Medicine and Health Technology, Tampere University, 33520 Tampere, Finland; katri.kaukinen@tuni.fi

[3] Faculty of Medicine and Health Technology and BioMediTech Institute, Tampere University, 33520 Tampere, Finland; markus.hannula@tuni.fi (M.H.); jari.hyttinen@tuni.fi (J.H.)

[4] Central Finland Central Hospital, 40620 Jyväskylä, Finland; juha.taavela@tuni.fi

[5] Laboratory of Cancer Biology, Faculty of Medicine and Health Technology, Tampere University, 33520 Tampere, Finland; jorma.isola@tuni.fi

[6] Jilab Inc, 33520 Tampere, Finland

[7] National Institute for Mother and Child Health, Carol Davila University of Medicine and Pharmacy, 050474 Bucharest, Romania; alina.popp@tuni.fi

[8] Department of Internal Medicine, Tampere University Hospital, 33521 Tampere, Finland

[9] Department of Pediatrics, Seinäjoki Central Hospital and University Consortium of Seinäjoki, 60320 Seinäjoki, Finland

* Correspondence: katri.lindfors@tuni.fi; Tel.: +358-456786374

Citation: Repo, M.; Hannula, M.; Taavela, J.; Hyttinen, J.; Isola, J.; Hiltunen, P.; Popp, A.; Kaukinen, K.; Kurppa, K.; Lindfors, K. Iron Transporter Protein Expressions in Children with Celiac Disease. *Nutrients* **2021**, *13*, 776. https://doi.org/10.3390/nu13030776

Academic Editor: Isabel Comino

Received: 6 February 2021
Accepted: 25 February 2021
Published: 27 February 2021

Publisher's Note: MDPI stays neutral with regard to jurisdictional claims in published maps and institutional affiliations.

Abstract: Anemia is a frequent finding in children with celiac disease but the detailed pathophysiological mechanisms in the intestine remain obscure. One possible explanation could be an abnormal expression of duodenal iron transport proteins. However, the results have so far been inconsistent. We investigated this issue by comparing immunohistochemical stainings of duodenal cytochrome B (DCYTB), divalent metal transporter 1 (DMT1), ferroportin, hephaestin and transferrin receptor 1 (TfR1) in duodenal biopsies between 27 children with celiac disease and duodenal atrophy, 10 celiac autoantibody-positive children with potential celiac disease and six autoantibody-negative control children. Twenty out of these 43 subjects had anemia. The expressions of the iron proteins were investigated with regard to saturation and the percentage of the stained area or stained membrane length of the enterocytes. The results showed the stained area of ferroportin to be increased and the saturation of hephaestin to be decreased in celiac disease patients compared with controls. There were no differences in the transporter protein expressions between anemic and non-anemic patients. The present results suggest an iron status-independent alteration of ferroportin and hephaestin proteins in children with histologically confirmed celiac disease.

Keywords: celiac disease; anemia; iron transporter

1. Introduction

Celiac disease is an immune-mediated disorder driven by ingested gluten [1]. A frequent and sometimes the only clinical finding in untreated patients is anemia, generally considered to be caused by damaged duodenal mucosa and the resulting malabsorption of iron [2,3]. Nevertheless, there is a poor correlation between the presence of anemia and the severity of histological damage [2,4,5]. Moreover, duodenal absorption of only about 10% of the dietary iron fulfills the daily needs [6], indicating that the reduced mucosal surface area is not the sole explanation for anemia. In fact, it may be present in so-called potential celiac disease, referring to subjects with endomysial (EmA) or transglutaminase 2 (TGA) celiac autoantibodies but with a normal small bowel morphology [5,7–9], suggesting that the pathophysiologic mechanisms behind iron deficiency and anemia are more complex than previously thought.

In healthy conditions, iron is absorbed from the gut by a sophisticated and tightly regulated process [6,10]. In the apical membrane of enterocytes, the duodenal cytochrome B (DCYTB) reduces iron to a ferrous form. A divalent metal transporter (DMT1) transfers ferrous iron into the enterocyte where it is either utilized in mitochondria, stored as ferritin or transported to the circulation via basolateral ferroportin. Before being able to bind to the plasma iron carrier transferrin, iron must be reconverted into a ferric form by basolateral hephaestin. The enterocytes may also reuptake iron for their own metabolic functions through transferrin receptor 1 (TfR1). A key regulator of iron absorption and metabolism is hepcidin, which reduces the iron uptake in enterocytes and its release from body storages [11,12]. The details of this regulation, however, are not fully understood [13–16].

It has been suggested that the abnormal expression of the iron transporter proteins could provide an explanation for anemia in celiac disease. So far only a few studies have tested this hypothesis with inconsistent findings [17–20]. We therefore aimed to investigate possible altered transporter protein expression by staining the DMT, DCYTB, ferroportin, hephaestin and TfR1 in duodenal biopsies of children with histologically confirmed or potential celiac disease and autoantibody-negative controls.

2. Materials and Methods

2.1. Patients and the Study Design

The study was conducted at Tampere University Hospital, Tampere, Finland and the National Institute for Mother and Child Health, Bucharest, Romania. Twenty-seven children (age < 17 years) with EmA and/or TGA and a duodenal lesion comprised the celiac disease group. Ten children with positive EmA and TGA but a non-diagnostic histology comprised the potential celiac disease group. Six children who were endoscopied due to unexplained gastrointestinal symptoms but who had normal duodenal villi and negative EmA/TGA were used as controls. All 43 children were further divided into those with or without anemia.

The study was conducted according to the Helsinki Declaration. The study protocol and patient recruitment were approved by the Ethics Committee of the Pirkanmaa Hospital District, Finland and the Ethics Committees of the University of Medicine and Pharmacy "Carol Davila" and the National Institute for Mother and Child Health "Alessandrescu-Rusescu", Romania. Written informed consent was obtained from all study participants and their guardians.

2.2. Celiac Disease Serology and Small Bowel Mucosal Morphology

EmA titers were measured by an indirect immunofluorescence method using a human umbilical cord as a substrate [21]. A dilution of 1:5 was considered positive and positive sera were further diluted 1:50, 1:100, 1:200, 1:500, 1:1000, 1:2000 and 1:4000. The EliA Celikey test (Phadia, Uppsala, Sweden) was used to determine TGA. The cut-off for seropositivity was set at >7.0 U/L according to the manufacturer's instructions.

A minimum of four representative forceps biopsies were taken from the duodenum. The paraffin-embedded specimens were cut, stained with hematoxylin and eosin and evaluated for celiac disease diagnosis by an experienced pathologist. Only correctly oriented histological sections were accepted for the histological analyses [22]. Subjects with crypt hyperplasia and a villous atrophy in the duodenal mucosa (Corazza–Villanacci B1-B2) were diagnosed with celiac disease whereas children with a non-diagnostic histology (Corazza–Villanacci A) formed the potential celiac disease and control groups [23,24].

2.3. Laboratory Parameters and Hepcidin

The following associated laboratory parameters were measured by standard methods: hemoglobin (reference value (Rf) from 100–141 to 130–160 g/L depending on age and sex [25]), plasma soluble transferrin receptor (sTfR, Rf from 1.6–5.2 mg/L to 2.0–6.8 mg/L), mean corpuscular volume (MCV; Rf from 72–88 to 87–146 fl [25]), serum total iron (Fe, Rf 6–25 mmol/L), plasma ferritin (Rf > 10 mg/L), transferrin iron saturation (Rf 15–50%), serum folate (Rf 10.4–42.4 nmol/L) and serum vitamin B12 (Rf 140–490 pmol/L). In addition, serum bioactive hepcidin (hepcidin-25) levels were measured using a commercial solid-phase enzyme-linked immunosorbent assay (EIA-5258, DRG Diagnostics, Marburg, Germany) according to the manufacturer's instructions [5].

2.4. Immunohistochemistry

For the immunohistochemistry, 5 μm-thick sections were cut from the formalin-fixed, paraffin-embedded duodenal specimens. After deparaffination and rehydration antigens were exposed by heat-induced epitope retrieval. Thereafter, a non-specific staining was blocked followed by overnight incubation with primary antibodies (Supplementary Table S1). After washing the primary antibodies, the specimens were incubated overnight with a secondary antibody prior to the blocking of the endogenic peroxidase and a visualization of the staining with either ImmPRESS or VECTASTAIN Elite ABC reagent (Vector Laboratories Inc, Peterborough, UK). Finally, sections were counterstained with hematoxylin.

2.5. Digital Analysis of the Stained Sections

All slides were scanned as whole-slide images using a SlideStrider scanner at a resolution of 0.16 μm per pixel (Jilab Inc., Tampere, Finland). The images were stored as JPX files and viewed with a JVSview program from where they were exported to a Fiji Image J program for further analysis [26]. Of the DCYTB sections, both the entire visible epithelial apical membrane and the DCYTB stained membrane were drawn and measured. The stained membrane length was divided by the whole membrane length to assess the percentage of the apical membrane covered with the protein. Thereafter, from DMT1, ferroportin, hephaestin and TfR1 stained sections of the epithelium were selected, other parts cut out and the images consisting of only the epithelium were stored as TIF files (Supplementary Figures S1 and S2). Subsequently, the files were transferred to a Matlab program (The MathWorks Inc. Natick, Massachusetts) where they were transformed from RGB to HSV images to access the color saturation independently of the lightness. To measure only the primary antibody staining, a red color was chosen from the hue channel within values 0–0.1 and 0.9–1. The saturation channel was then thresholded according to all sections in each stained protein series using Otsu's method [27]. Finally, the value of the mean saturation of each section divided by the maximum saturation of the protein series and percentage of the stained area were measured for each section.

2.6. Statistical Analysis

All statistical analyses were performed using IBM SPSS Statistics version 26.0 (IBM Corp. Armonk, NY). The clinical characteristics and prevalence of anemia are presented as percentage distributions. The skewness of the quantitative data was assessed by the Shapiro–Wilk method and most of the variables were not normally distributed. For simplicity, all data are thus expressed as medians with quartiles except for age, which is given with a median and a range. Staining results as mean/maximum saturation and the stained area were compared between groups using a non-parametric Mann–Whitney U test. Correlations between hepcidin, plasma transferrin receptor 1, serum ferritin and the DCYTB stained apical border percent and in other proteins' mean/maximum saturations and stained areas were calculated using Spearman's rank (rS) correlation. p values < 0.05 were considered significant.

3. Results

There was no significant difference between children with celiac disease and potential celiac disease in age, gender or median hepcidin values or, despite a non-significant trend, in the frequency of anemia or low MCV (Table 1). The former group nevertheless had a higher frequency of increased sTfR values and lower ferritin (Table 1) as well as a higher median EmA (1:1000 vs 1:50, $p < 0.001$) and TGA (120 U/l vs. 17 U/l, $p = 0.001$). The controls (two boys, two girls, 50% anemia) were slightly older (median 10.6 (range 3.3, 15.3) years) than the celiac and potential celiac patients.

Table 1. Clinical characteristics and laboratory values of 37 children with celiac disease (CD) and potential CD.

Variable	CD, $n = 27$		Potential CD, $n=10$		p Value
	n	%	n	%	
Girls	18	67	8	80	0.431
Anemia	14	52	3	30	0.236
High sTfR	12	46	1	10	0.043
Low MCV	10	35	1	10	0.140
	Median	Q_1, Q_3	Median	Q_1, Q_3	
Age, yrs (range)	6.8	2.7, 14.4	6.1	4.1, 16.9	0.555
Ferritin, mg/L	7.0	4.8, 15.5	20.5	11.3, 29.8	0.017
Hepcidin, ng/mL	13.7	12.6, 15.2	15.4	13.2, 18.2	0.286

MCV, mean corpuscular volume; Q1 and Q3, lower and upper quartiles; sTfR, soluble transferrin receptor. Data was available from all cases except 126 and 217. [1]26 and [2]27

The stained area of ferroportin was increased in the celiac disease patients compared with the controls and a similar although non-significant trend was observed in the saturation of the staining (Table 2). In hephaestin the saturation was significantly decreased in celiac disease compared with the controls with a similar trend in the stained area. No significant differences between the study groups were observed in either saturation or the stained area of the other iron transporters (Table 2), nor were there any differences in either the saturation or the stained area of any of the iron transporters between children with or without anemia (Table 3).

Table 2. Iron transporter protein saturations and the stained areas of enterocytes in the duodenal biopsies of the study subjects.

Iron Transporter Protein	CD $N = 27$		Potential CD $N = 10$		Controls $N = 6$		CD vs. Potential CD	CD vs. Controls	Potential CD vs. Controls
	Median	Q_1, Q_3	Median	Q_1, Q_3	Median	Q_1, Q_3	p Value	p Value	p Value
DCYTB									
Stained apical border, %	54	36, 76	50	24, 79	50	33, 73	0.679	0.751	0.662
DMT1									
Mean/max saturation, %	42	36, 51	43	35, 52	37	33, 50	0.999	0.342	0.828
Stained area, %	59	56, 62	60	49, 67	57	48, 65	0.827	0.653	0.745
Ferroportin									
Mean/max saturation, %	64	62, 66	64	59, 69	61	59, 63	0.827	0.072	0.329
Stained area, %	66	54, 75	68	40, 78	45	22, 57	0.999	0.024	0.129
Hephaestin									
Mean/max saturation, %	27	25, 29	28	26, 31	31	27, 37	0.234	0.028	0.195
Stained area, %	1	0, 22	4	1, 21	16	8, 38	0.266	0.080	0.195
TfR1									
Mean/max saturation, %	52	48, 54	50	49, 55	53	51, 62	0.821	0.325	0.233
Stained area, %	59	49, 69	42	33, 68	64	47, 73	0.257	0.437	0.233

CD, celiac disease; DCYTB, duodenal cytochrome B; DMT1, divalent metal transporter 1; TfR1, transferrin receptor 1. Q_1, Q_3 upper and lower quartiles. Data available in each analysis were from at least 90% of the patients.

There was a positive correlation between ferritin values and TfR1 saturations (r_S 0.594, $p = 0.015$) and the stained area (r_S 0.761, $p = 0.001$) in children with celiac disease. A moderate negative correlation was also found between sTfR values and hephaestin saturation (r_S –0.349, $p = 0.046$) when evaluated in all study subjects whereas this was not observed when evaluated separately in celiac disease patients. No other correlations between the hepcidin, ferritin or sTfR values and the stainings of the iron transporter were detected (data not shown).

Table 3. Iron transporter protein saturations and the stained areas of enterocytes in the duodenal biopsies of children with and without anemia.

Iron Transporter Protein	All Study Children, $n = 43$					Children With CD, $n = 27$				
	Anemia, $n = 20$		No Anemia, $n = 23$		p Value	Anemia, $n = 14$		No Anemia, $n = 13$		p Value
	Median	Q_1, Q_3	Median	Q_1, Q_3		Median	Q_1, Q_3	Median	Q_1, Q_3	
DCYTB										
Stained apical border, %	54	13, 78	56	37, 73	0.999	53	10, 79	63	42, 70	0.689
DMT1										
Mean/max saturation, %	43	37, 51	39	36, 54	0.582	43	39, 51	39	37, 53	0.446
Stained area, %	59	56, 62	59	54, 66	0.388	59	56, 61	59	57, 63	0.744
Ferroportin										
Mean/max saturation, %	64	59, 65	64	60, 68	0.372	64	62, 65	65	62, 69	0.128
Stained area, %	65	46, 74	65	44, 77	0.875	65	55, 74	66	51, 77	0.624
Hephaestin										
Mean/max saturation, %	27 [1]	26, 29	28	25, 32	0.594	27 [2]	25, 29	27	25, 31	0.663
Stained area, %	5 [1]	1, 22	3	0, 23	0.795	3 [2]	0, 19	1	0, 25	0.744
TfR1										
Mean/max saturation, %	50 [1]	49, 54	52	49, 55	0.452	50 [3]	49, 54	53	48, 55	0.750
Stained area, %	55 [1]	42, 62	61	43, 70	0.292	55 [3]	55, 64	61	50, 70	0.469

CD, celiac disease; DCYTB, duodenal cytochrome B; DMT1, divalent metal transporter 1; TfR1, transferrin receptor 1. Q_1, Q_3 upper and lower quartiles. Data available in each analysis were from at least 90% of the patients except [1] 17, [2] 12 and [3] 11 patients.

4. Discussion

The main finding of the present study was an increased expression of ferroportin and a decreased expression of hephaestin in children with histologically confirmed celiac disease compared with the non-celiac controls. There were no other significant differences between the study groups in the expression of iron transporter proteins. In addition, no differences in any of these proteins were detected when anemic and non-anemic children were evaluated separately.

The expression of the iron transporter proteins and/or their coding mRNAs in celiac disease have previously been reported in three studies comprising adult patients and in one pediatric study [17–20]. In line with our results, Sharma et al. showed an iron status-independent increase in protein levels of ferroportin but also of DMT1 in untreated adult celiac disease [17]. Additionally, they found increased DMT1 and ferroportin mRNAs in iron deficient celiac disease patients and also in anemic non-celiac controls. Tolone et al. later reported that DMT1 mRNA was increased in celiac disease children with mild but not with severe atrophy compared with controls with normal duodenal mucosa [20]. However, they included both potential celiac disease patients and suspected gastroesophageal reflux disease patients in the control group. Additionally, Matysiak-Budnik reported an upregulation of TfR1 protein levels in adults with untreated celiac disease [19]. Barisani et al. reported increased mRNAs and protein levels of DMT1, ferroportin, hephaestin and TfR1 in adult celiac disease patients but, in contrast to the protein levels in ours and Sharma's studies, these findings were iron status-dependent [18]. However, unlike others, Barisani et al. included both untreated patients and patients on a gluten-free diet in the celiac disease group. No earlier studies have reported the decreased hephaestin expression observed here.

These partially inconsistent results between the studies may be attributable to the differences in the number and clinical characteristics of the participants and/or by the variable use of primary antibodies and staining protocols. On the other hand, there may in fact be significant differences between children and adults in intestinal iron transporter protein expression [28]. As our results lacked major outliers and were also consistent within and between the study groups, we believe the present findings to reflect the true state of iron transporter protein expression in the duodenal mucosa of children with untreated celiac disease.

Our findings would suggest that changes in ferroportin and hephaestin expression do not explain the intestinal pathophysiology of anemia in celiac disease but may rather reflect the immaturity of the epithelium [29] of the atrophic duodenal mucosa. Interestingly, Tolone et al. found a distinct polymorphism in the DMT1 gene to be significantly more frequent in anemic than in non-anemic children with celiac disease; in fact, the polymorphism conferred a four-fold risk for the development of anemia [20]. Furthermore, a polymorphism in the transmembrane serine protease 6 gene can be overrepresented in celiac disease patients and its presence predicts an inadequate response to iron supplementation [30,31] whereas polymorphisms in the human hemochromatosis protein gene may provide protection against anemia in celiac disease [31–33]. Thus, genetic variants affecting iron metabolism may at least partially determine a predisposition to anemia in celiac disease.

As an additional novel finding of the present study, we observed a moderately positive correlation between the TfR1 saturation and stained area and the serum ferritin levels in children with celiac disease. Additionally, a negative correlation between the saturation of hephaestin and sTfR levels was shown among all of the children although this was not seen in celiac disease patients when evaluated separately. As sTfR usually increases and ferritin decreases in subjects with iron deficiency, an opposite correlation pointing towards a compensatory increase of intestinal iron absorption would have been expected [34]. However, both the origin and function of circulating ferritin and sTfR are currently unknown [10] and thus their connection with the duodenal iron transporters needs to be further studied.

5. Conclusions

To conclude, the iron status-independent changes observed here in ferroportin and hephaestin in children with histologically confirmed celiac disease likely reflect the immature nature of the epithelium in the atrophic disease state and do not explain the intestinal pathophysiology of anemia in children with celiac disease. Further investigations with a larger number of study subjects and in both children and adults are needed to understand the complex mechanisms of abnormal iron metabolism leading to anemia in celiac disease.

Supplementary Materials: The following are available online at https://www.mdpi.com/2072-6643/13/3/776/s1. Supplementary Table S1. Specific characteristics of the staining procedures, Supplementary Figure S1. The process for measuring the saturation and stained area of iron transporter proteins in enterocytes as exemplified by ferroportin staining in a patient with a subtotal villous atrophy and anemia. Supplementary Figure S2. The process for measuring the saturation and stained area of iron transporter proteins in enterocytes as exemplified by ferroportin staining in a control child with anemia.

Author Contributions: Conceptualization, K.L. and K.K. (Kalle Kurppa); methodology, M.R., M.H., J.H. and J.T.; software, M.H., J.I. and J.H.; investigation, M.R.; resources, K.K. (Kalle Kurppa), K.K. (Katri Kaukinen) and K.L.; data curation, K.K. (Kalle Kurppa), P.H., A.P.; writing—original draft preparation, M.R.; writing—review and editing, K.L. and K.K. (Kalle Kurppa); visualization, M.H., J.I. and M.R.; supervision, K.L. and K.K. (Kalle Kurppa); funding acquisition, K.K. (Kalle Kurppa), K.L. and M.R. All authors have read and agreed to the published version of the manuscript.

Funding: The study was supported by the Academy of Finland, the Sigrid Juselius Foundation, The Competitive State Research Financing of the Expert Area of Tampere University Hospital, the Finnish Cultural Foundation, the Foundation for Pediatric Research, the Finnish Celiac Society, the Finnish Medical Foundation and the Päivikki and Sakari Sohlberg Foundation.

Institutional Review Board Statement: The study was conducted according to the Helsinki Declaration. The study protocol and patient recruitment were approved by the Ethics Committee of the Pirkanmaa Hospital District, Finland and the Ethics Committees of the University of Medicine and Pharmacy "Carol Davila" and the National Institute for Mother and Child Health "Alessandrescu-Rusescu", Romania.

Informed Consent Statement: Written informed consent was obtained from all study participants and their guardians.

Data Availability Statement: Due to the protection of patient privacy, the original data used to support the findings of this study cannot be shared.

Conflicts of Interest: The authors declare that they have no conflict of interest.

References

1. Ludvigsson, J.F.; Leffler, D.A.; Bai, J.C.; Biagi, F.; Fasano, A.; Green, P.H.R.; Hadjivassiliou, M.; Kaukinen, K.; Kelly, C.P.; Leonard, J.N.; et al. The Oslo definitions for coeliac disease and related terms. *Gut* **2013**, *62*, 43–52. [CrossRef] [PubMed]
2. Harper, J.W.; Holleran, S.F.; Ramakrishnan, R.; Bhagat, G.; Green, P.H.R. Anemia in celiac disease is multifactorial in etiology. *Am. J. Hematol.* **2007**, *82*, 996–1000. [CrossRef]
3. Freeman, H.J. Iron deficiency anemia in celiac disease. *World J. Gastroenterol.* **2015**, *21*, 9233–9238. [CrossRef] [PubMed]
4. Rajalahti, T.; Repo, M.; Kivelä, L.; Huhtala, H.; Mäki, M.; Kaukinen, K.; Lindfors, K.; Kurppa, K. Anemia in Pediatric Celiac Disease: Association with Clinical and Histological Features and Response to Gluten-free Diet. *J. Pediatr. Gastroenterol. Nutr.* **2016**, *64*, e1–e6. [CrossRef]
5. Repo, M.; Lindfors, K.; Mäki, M.; Huhtala, H.; Laurila, K.; Lähdeaho, M.-L.; Saavalainen, P.; Kaukinen, K.; Kurppa, K. Anemia and Iron Deficiency in Children with Potential Celiac Disease. *J. Pediatr. Gastroenterol. Nutr.* **2017**, *64*, 56–62. [CrossRef]
6. Evstatiev, R.; Gasche, C. Iron sensing and signalling. *Gut* **2012**, *61*, 933–952. [CrossRef]
7. Kondala, R.; Puri, A.S.; Banka, A.K.; Sachdeva, S.; Sakhuja, P. Short-term prognosis of potential celiac disease in Indian patients. *United Eur. Gastroenterol. J.* **2016**, *4*, 275–280. [CrossRef]
8. Shahriari, M.; Honar, N.; Yousefi, A.; Javaherizadeh, H. Association of Potential Celiac Disease and Refractory Iron Deficiency Anemia in Children and Adolescents. *Arq. Gastroenterol.* **2018**, *55*, 78–81. [CrossRef]
9. Tosco, A.; Aitoro, R.; Auricchio, R.; Ponticelli, D.; Miele, E.; Paparo, F.; Greco, L.; Troncone, R.; Maglio, M. Intestinal anti-tissue transglutaminase antibodies in potential coeliac disease. *Clin. Exp. Immunol.* **2013**, *171*, 69–75. [CrossRef] [PubMed]
10. Camaschella, C.; Nai, A.; Silvestri, L. Iron metabolism and iron disorders revisited in the hepcidin era. *Haematologica* **2020**, *105*, 260–272. [CrossRef] [PubMed]
11. Kautz, L.; Jung, G.; Valore, E.V.; Rivella, S.; Nemeth, E.; Ganz, T. Identification of erythroferrone as an erythroid regulator of iron metabolism. *Nat. Genet.* **2014**, *46*, 678–684. [CrossRef]
12. Aschemeyer, S.; Qiao, B.; Stefanova, D.; Valore, E.V.; Sek, A.C.; Alex Ruwe, T.; Vieth, K.R.; Jung, G.; Casu, C.; Rivella, S.; et al. Structure-function analysis of ferroportin defines the binding site and an alternative mechanism of action of hepcidin. *Blood* **2018**, *131*, 899–910. [CrossRef]
13. Brasse-Lagnel, C.; Karim, Z.; Letteron, P.; Bekri, S.; Bado, A.; Beaumont, C. Intestinal DMT1 cotransporter is down-regulated by hepcidin via proteasome internalization and degradation. *Gastroenterology* **2011**, *140*, 1261–1271. [CrossRef]
14. Chaston, T.; Chung, B.; Mascarenhas, M.; Marks, J.; Patel, B.; Srai, S.K.; Sharp, P. Evidence for differential effects of hepcidin in macrophages and intestinal epithelial cells. *Gut* **2008**, *57*, 374. [CrossRef]
15. Chung, B.; Chaston, T.; Marks, J.; Srai, S.K.; Sharp, P.A. Hepcidin Decreases Iron Transporter Expression in Vivo in Mouse Duodenum and Spleen and in Vitro in THP-1 Macrophages and Intestinal Caco-2 Cells. *J. Nutr.* **2009**, *139*, 1457–1462. [CrossRef] [PubMed]
16. Mena, N.P.; Esparza, A.; Tapia, V.; Valdés, P.; Núñez, M.T. Hepcidin inhibits apical iron uptake in intestinal cells. *Am. J. Physiol. Gastrointest. Liver Physiol.* **2008**, *294*. [CrossRef] [PubMed]
17. Sharma, N.; Begum, J.; Eksteen, B.; Elagib, A.; Brookes, M.; Cooper, B.T.; Tselepis, C.; Iqbal, T.H. Differential ferritin expression is associated with iron deficiency in coeliac disease. *Eur. J. Gastroenterol. Hepatol.* **2009**, *21*, 794–804. [CrossRef] [PubMed]
18. Barisani, D.; Parafioriti, A.; Bardella, M.T.; Zoller, H.; Conte, D.; Armiraglio, E.; Trovato, C.; Koch, R.O.; Weiss, G. Adaptive changes of duodenal iron transport proteins in celiac disease. *Physiol. Genom.* **2004**, *17*, 316–325. [CrossRef]
19. Matysiak-Budnik, T.; Moura, I.C.; Arcos-Fajardo, M.; Lebreton, C.; Menard, S.; Candalh, C.; Ben-Khalifa, K.; Dugave, C.; Tamouza, H.; van Niel, G.; et al. Secretory IgA mediates retrotranscytosis of intact gliadin peptides via the transferrin receptor in celiac disease. *J. Exp. Med.* **2008**, *205*, 143–154. [CrossRef]
20. Tolone, C.; Bellini, G.; Punzo, F.; Papparella, A.; Miele, E.; Vitale, A.; Nobili, B.; Strisciuglio, C.; Rossi, F. The DMT1 IVS4+44C>A polymorphism and the risk of iron deficiency anemia in children with celiac disease. *PLoS ONE* **2017**, *12*, 1–13. [CrossRef] [PubMed]
21. Ladinser, B.; Rossipal, E.; Pittschieler, K. Endomysium antibodies in coeliac disease: An improved method. *Gut* **1994**, *35*, 776–778. [CrossRef] [PubMed]
22. Taavela, J.; Koskinen, O.; Huhtala, H.; Lähdeaho, M.-L.; Popp, A.; Laurila, K.; Collin, P.; Kaukinen, K.; Kurppa, K.; Mäki, M. Validation of Morphometric Analyses of Small-Intestinal Biopsy Readouts in Celiac Disease. *PLoS ONE* **2013**, *8*, 1–8. [CrossRef] [PubMed]
23. Corazza, G.; Villanacci, V. Coeliac disease. *J. Clin. Pathol.* **2005**, *58*, 573–574. [CrossRef]
24. Oberhuber, G.; Granditsch, G.; Vogelsang, H. The histopathology of coeliac disease: Time for a standardized report scheme for pathologists. *Eur. J. Gastroenterol. Hepatol.* **1999**, *11*, 1185–1194. [CrossRef]
25. Repo, M.; Rajalahti, T.; Hiltunen, P.; Sotka, A.; Kivelä, L.; Huhtala, H.; Kaukinen, K.; Lindfors, K.; Kurppa, K. Diagnostic findings and long-term prognosis in children with anemia undergoing GI endoscopies. *Gastrointest. Endosc.* **2020**, *91*, 1272–1281.e2. [CrossRef]

26. Schindelin, J.; Arganda-Carreras, I.; Frise, E.; Kaynig, V.; Longair, M.; Pietzsch, T.; Preibisch, S.; Rueden, C.; Saalfeld, S.; Schmid, B.; et al. Fiji: An open-source platform for biological-image analysis. *Nat. Methods* **2012**, *9*, 676–682. [CrossRef] [PubMed]
27. Otsu, N. A Threshold Selection Method from Gray-Level Histograms. *IEEE Trans. Syst. Man Cybern.* **1979**, *9*, 62–66. [CrossRef]
28. Lönnerdal, B. Development of iron homeostasis in infants and young children. *Am. J. Clin. Nutr.* **2017**, *106*, 1575S–1580S. [CrossRef] [PubMed]
29. Oittinen, M.; Popp, A.; Kurppa, K.; Lindfors, K.; Mäki, M.; Kaikkonen, M.U.; Viiri, K. Polycomb Repressive Complex 2 Enacts Wnt Signaling in Intestinal Homeostasis and Contributes to the Instigation of Stemness in Diseases Entailing Epithelial Hyperplasia or Neoplasia. *Stem Cells* **2017**, *35*, 445–457. [CrossRef] [PubMed]
30. Elli, L.; Poggiali, E.; Tomba, C.; Andreozzi, F.; Nava, I.; MT, B.; Campostrini, N.; Girelli, D.; Conte, D.; MD, C. Does TMPRSS6 RS855791 polymorphism contribute to iron deficiency in treated celiac disease? *Am. J. Gastroenterol.* **2015**, *110*, 200–202. [CrossRef]
31. De Falco, L.; Tortora, R.; Imperatore, N.; Bruno, M.; Capasso, M.; Girelli, D.; Castagna, A.; Caporaso, N.; Iolascon, A.; Rispo, A. The role of TMPRSS6 and HFE variants in iron deficiency anemia in celiac disease. *Am. J. Hematol.* **2018**, *93*, 383–393. [CrossRef] [PubMed]
32. Butterworth, J.R.; Cooper, B.T.; Rosenberg, W.M.C.; Purkiss, M.; Jobson, S.; Hathaway, M.; Briggs, D.; Howell, W.M.; Wood, G.M.; Adams, D.H.; et al. The role of hemochromatosis susceptibility gene mutations in protecting against iron deficiency in celiac disease. *Gastroenterology* **2002**, *123*, 444–449. [CrossRef] [PubMed]
33. Zanella, I.; Caimi, L.; Biasiotto, G. About TMPRSS6 rs855791 polymorphism, iron metabolism and celiac disease. *Am. J. Gastroenterol.* **2015**, *110*, 1240. [CrossRef] [PubMed]
34. Pootrakul, P.; Kitcharoen, K.; Yansukon, P.; Wasi, P.; Fucharoen, S.; Charoenlarp, P.; Brittenham, G.; Pippard, M.J.; Finch, C.A. The effect of erythroid hyperplasia on iron balance. *Blood* **1988**, *71*, 1124–1129. [CrossRef] [PubMed]

Article

General Health, Systemic Diseases and Oral Status in Adult Patients with Coeliac Disease

Alessandro Nota, Silvio Abati, Floriana Bosco, Isabella Rota, Elisabetta Polizzi and Simona Tecco *

Dental School, Vita-Salute San Raffaele University and IRCCS San Raffaele Hospital, 20132 Milano, Italy; nota.alessandro@hsr.it (A.N.); abati.silvio@hsr.it (S.A.); boscofloriana@gmail.com (F.B.); rota.isabella@hsr.it (I.R.); polizzi.elisabetta@hsr.it (E.P.)
* Correspondence: tecco.simona@hsr.it; Tel.: +39-3755565708

Received: 16 November 2020; Accepted: 14 December 2020; Published: 15 December 2020

Abstract: The prevalence of coeliac disease in the general population is 0.5–1%; however, most patients remain undiagnosed until adult age. In some cases, the onset is represented by sub-clinical signs, some of which can be found in the mouth. The aim of this research was to identify any associations between the clinical characteristics of coeliac disease and oral manifestations. A structured questionnaire was administered to a group of 237 individuals with coeliac disease. 100% of the subjects fully completed the questionnaire. Among them, 182 (76.7%) were female, 64 patients (27%) were aged 15 to 24 years, 159 (67%) were aged 25 to 55 and 14 (6%) were aged 56 and over. Significant associations were observed in caries prevalence and dentin sensitivity; in addition, an inappropriate diet was related to oral manifestations; following a gluten-free diet could be important to control the gingival bleeding levels and to manage oral symptoms associated to coeliac disease. In general, the presence of inflammatory symptoms in the mouth seems to be associated with general symptoms of inflammation related to coeliac disease.

Keywords: coeliac disease; oral diseases; oral prevention; gingival bleeding; sleep-related breathing disorders; oral health; enamel defects; interceptive orthodontics

---

1. Introduction

Coeliac disease is an immune-mediated disease, typical of genetically predisposed individuals; it is caused by gluten [1]. A gluten-free diet is the only treatment strategy accepted in these patients [2]. The prevalence of coeliac disease in the general population is 0.5–1%; however, most patients remain undiagnosed until adult age. There are several clinical onsets of the disease, the most common caused by malabsorption (iron deficiency anemia, hypovitaminosis etc.) [3]; in some cases, the onset is represented by sub-clinical signs, some of which are found in the mouth (herpes-like lesions, recurrent aphthous stomatitis, hypoplasia and dyschromia of the enamel, etc.) [4]. It is characterized by a variety of symptoms, both intestinal and extra-intestinal, including oral manifestations [5]. Coeliac disease can develop at any age after people start consuming food that contains gluten. If untreated, coeliac condition can lead to a worsening of health, for example the development of other autoimmune disorders, such as type I diabetes, multiple sclerosis, dermatitis herpetiformis, anemia, osteoporosis, infertility and miscarriage, neurological conditions such as epilepsy and migraine, short stature and intestinal tumors [6]. The aim of this research is to identify, through the analysis of a questionnaire distributed to coeliac adult patients, any association between the clinical systemic characteristics of coeliac disease and oral manifestations, with the ultimate purpose of improving the clinical dental management of this category of patients. The primary outcome was the association of the clinical systemic signs and symptoms associated with coeliac disease with the lifestyle of patients with intra-oral manifestations.

2. Subjects and Methods

This is an observational study based on the data derived from an anonymous structured questionnaire distributed to coeliac adult subjects. Data were recorded during the period from April to September 2018. Initially, on the basis of a literature review of the medical conditions potentially associated with coeliac disease, a specific questionnaire was developed to identify clinical systemic signs and symptoms, signs concerning the oral cavity (halitosis, reflux, oral lesions, caries, etc.), and the lifestyle of the enrolled patients. Then, the questionnaire was proposed to a sample of coeliac adult patients. These subjects were informed about this project through connection platforms (Facebook) and/or through advertisements in restaurants or supermarkets specializing in coeliac customers. A group of 237 patients participated and completed the anonymous questionnaire on-line via Google Forms.

The participants did not have to specify any demographic data (for example name, surname, or telephone number), as the questionnaire remained completely anonymous. They were invited to complete the questionnaire sincerely, and to give full importance to the project. Data management and analyses were organized to guarantee the participants' full privacy.

2.1. The Questionnaire

First, useful information in order to understand the clinical history of the coeliac patient was gathered. Then, the questionnaire attempted to identify the main systemic diseases that affected the patients. The subjects were asked to select all items of interest in order to identify the pathologies potentially related to coeliac disease (for example, autoimmune thyroiditis, rheumatoid arthritis, type 1 diabetes, autoimmune hepatitis, osteoporosis, fertility disorders, dermatitis herpetiformis, selective IgA deficiency, Sjorgen syndrome, primary biliary cirrhosis, Crohn's disease, Down's syndrome, Ullrich-Turner syndrome, Williams syndrome, epilepsy). Then, a series of questions were asked in order to identify all the symptoms that may have characterized the patients' clinical history. As coeliac disease is a multifactorial disease that can occur in different forms, it is important to identify correlations between symptoms; for this reason a series of questions concerned general symptoms (short stature, weight loss, puberty delay, asthenia, apathy, malaise, edema); gastrointestinal symptoms (dyspepsia, nausea, diarrhea, vomiting, constipation, abdominal distention, flatulence); neurological-psychiatric symptoms (peripheral neuropathy, ataxia, epilepsy, paresthesia, anxiety, depression, irritability); hematological problems (anemia, iron deficiency, folate deficiency, bleeding, ecchymosis); osteoarticular and muscular disorders (arthritis, osteoporosis, osteo-malacia, cramps, myopathy).Then, the main eating habits of the subjects were collected, followed by questions about their daily life.

A gluten-free diet is essential for the coeliac patient, but not all patients follow it with dedication. So some questions about gluten-free diet were included. In addition, it was requested if any other food had been removed from the patients' diets for medical reasons: for example, lactose, sugars, red meats, white meats, sausages, or fat. Finally, a group of questions concerned oro-dental characteristics that may be congenital (glossitis, agenesis of teeth, etc.), or acquired, e.g., oral ulcer. Clinical data were requested before and after beginning an appropriate diet, in order to understand any changes in the acquired pathologies (bad breath, gastroesophageal reflux, xerostomia, humming, tension-type headache, joint clicks). In addition, questions about the home oral hygiene and prevention programs of the patient were addressed (for example, the frequency of dental appointments, the occurrence of recurring caries, and/or gingivitis signs and symptoms, dentin sensitivity, bleeding after brushing).

2.2. Statistical Analysis

Once a considerable number of responses were collected on time, the anonymous data were entered into an Excel data collection file, so that each response could be analyzed correctly. This file was then transferred to a statistical analysis program (SPSS—Hong Kong—Ltd., Rm 1804, 18/F, Westlands

Road, Quarry Bay, Hong Kong, China). Data were dichotomized as presence or absence of a sign (or a symptom), or yes or no for a particular habit (concerning lifestyle). The results are shown as percentages of subjects answering yes or not, present or absent. Cross-tabulations and Odds Ratio (with confidence interval) are showed when statistically significant. The p value was set at 0.5%.

3. Results

237 individuals were recruited and participated to the project, answering the anonymous questionnaire. 100% of the subjects fully completed the questionnaire. Among them, 182 (76.7%) were female. 64 patients (27%) aged 15 to 24 years, 159 (67%) aged 25 to 55 and 14 (6%) aged 56 and over. 211 individuals (89%) reported adherence to a gluten-free diet, while 25 (10%) reported not adhering at all. One subject (0.42%) reported not following any diet.

Some differences in oral manifestations were observed between males and females. Regarding caries prevalence, for example, out of 182 (76.79%) coeliac females, 43 (18.14%) had caries, while out of 55 (23.21%) coeliac males, only 6 (12.24%) had caries. It can be observed that coeliac males predisposition to caries is about two times (OR: 0.40; CI: 0.16–0.99) lower than coeliac females. In addition, regarding dentin sensitivity, out of 182 (76.79%) coeliac females, 89 (37.55%) suffered from dentin sensitivity, whereas out of 55 (23.21%) coeliac males, only 14 (5.91%) suffered from dentin sensitivity. It can be observed that coeliac males are predisposed to suffer from dentin sensitivity almost 2. 5 times (OR: 0.36; CI: 0.18–0.70) less than females.

In addition, some oral manifestations resulted in significant correlation with the age of the subjects, as evidenced in Table 1. For example, the percentage of coeliac patients with gingivitis signs and symptoms increased significantly as the age group advanced. Similarly, the percentage of coeliac patients with dentin sensitivity increased significantly with increasing age.

Table 1. Age range of subjects included in the study, and oral manifestations. Statistically significant differences.

Oral Manifestations		Age Range			*p* Value
		0–25 Years (64 Subjects)	26–55 Years (159 Subjects)	>55 Years (14 Subjects)	
Modifications in the oral manifestations after gluten-free diet	NO	49 (76.56%)	97 (61.01%)	5 (35.71%)	0.007
	YES	15 (23.44%)	62 (38.99%)	9 (64.29%)	
Gingivitis signs and symptoms	NO	51 (100%)	122 (89.71%)	10 (71.43%)	0.003
	YES	0 (0%)	14 (10.29%)	4 (28.57%)	
Dentin sensitivity	NO	47 (73.44%)	81 (50.94%)	6 (42.86%)	0.005
	YES	17 (26.56%)	78 (49.06%)	8 (57.14%)	

In general, among coeliac patients with particular general health conditions correlated to coeliac disease (for example, illness, weight loss, short stature, puberty delay), the oral manifestations appeared evenly distributed (Table 2). But it can be noted that coeliac patients who didn't manifest any particular general health condition (a sub-group of 45 coeliac subjects, 18.99% of the whole sample, described in Table 2) showed—for the most part (40 subjects, 88.89% of the sub-group)—no significant oral manifestation after gluten-free diet assumption (Table 2). In addition, 42 subjects out of 45 (93.33%) did not present halitosis. 39 subjects out of 45 (86.67%) did not suffer from nocturnal snoring (Table 2).

Table 2. Associations between general health symptoms and oral manifestations.

Oral Manifestations		General Health Particular Condition Correlated To Coeliac Diases						*p* Value
		None (45 Subjects)	Illness (85 Subjects)	Puberty Delay (2 Subjects)	Weight Loss (19 Subjects)	Short Stature (9 Subjects)	Combined Symptoms (77 Subjects)	
Modifications in the oral cavity after gluten-free diet	NO	40 (88.89%)	51 (60%)	1 (50%)	12 (63.16%)	4 (44.44%)	43 (55.84%)	0.006
	YES	5 (11.11%)	34 (40%)	1 (50%)	7 (36.84%)	5 (55.56%)	34 (44.16%)	
Halitosis	NO	42 (93.33%)	67 (78.82%)	0 (0%)	18 (94.74%)	6 (88.67%)	63 (81.82%)	0.004
	YES	3 (6.67%)	18 (21.18%)	2 (100%)	1 (5.26%)	3 (33.33%)	14 (18.18%)	
Gastroesophageal reflux	NO	41 (91.11%)	56 (65.88%)	1 (50%)	13 (68.42%)	8 (88.89%)	45 (58.44%)	0.005
	YES	4 (8.86%)	29 (34.12%)	1 (50%)	6 (31.58%)	1 (11.11%)	32 (41.56%)	
Nocturnal snoring	NO	39 (86.67%)	64 (75.29%)	1 (50%)	16 (84.21%)	9 (100%)	70 (90.91%)	0.048
	YES	6 (13.33%)	21 (24.71%)	1 (50%)	3 (15.79%)	0 (0%)	7 (9.09%)	

Some differences were observed between coeliac patients without gastrointestinal symptoms (a sub-group of 49 subjects) and those with these symptoms (a sub-group of 188 subjects) (Table 3). In the sub-group of 49 coeliac patients who did not report gastrointestinal symptoms, 40 subjects found no changes in their oral cavity after gluten-free therapy, whereas out of 188 patients with such problems, oral cavity changes were found by 77 (32.49%) patients. It can therefore be said that coeliac patients with associated gastrointestinal symptoms showed a risk of presenting oral manifestations about three times higher (OR: 3. 08; CI: 1.41–6.72) than those without them.

Table 3. Associations between gastrointestinal symptoms and oral manifestations.

Oral Manifestations		Gastrointestinal Symptoms		OR [1]	CI [2]	*p* Value
		NO (49 Subjects)	YES (188 Subjects)			
Changes in the oral cavity after gluten-free diet	NO	40 (81.63%)	111 (59.04%)	3.08	1.41–6.72	0.003
	YES	9 (18.37%)	77 (40.96%)			

[1] Confidence Interval; [2] Odds Ratio.

Some statistically significant differences were also observed between coeliac patients suffering from neuropsychiatric disorders (a sub-group of 79 subjects) and those without these disorders (a sub-group of 147 subjects) (Table 4). Out of the 79 patients suffering from psychiatric disorders, 38 subjects reported sporadic gingival bleeding, while 15 reported frequent bleeding. In the other cases, the bleeding was equally distributed.

In addition, some correlations were observed in the sub-group of coeliac patients reporting associated hematological problems (161 subjects), as 68 subjects out of 161 reported specific oral characteristics (Table 5). Among the 76 coeliac patients without hematological problems, 58 subjects found no changes in their oral cavity. Therefore, coeliac patients with hematological problems showed a risk of presenting changes in the oral cavity more than two times higher (OR: 2. 36; CI: 1.27–4.36) than those who did not have them. More specifically, among 76 subjects without hematological problems, 66 patients did not show any enamel defects, while among 161 patients with hematological problems, 43 subjects reported dental enamel defects. Therefore, coeliac patients with hematological problems can be said to have a 2.5 times higher risk of showing enamel defects (OR: 2.67; CI: 1.25–5.82) than patients who do not have them. Also, coeliac patients with hematological problems have an almost four times higher risk of presenting xerostomia (OR: 3. 80; CI 1.42–10.16) than those without hematological problems (out of 76 coeliac patients without hematological problems, 71 did not have xerostomia, while among 161 patients with hematological problems, 34 subjects had xerostomia). The same was found for dentin sensitivity. It can be observed that coeliac patients with hematological problems have a risk of experiencing dental sensitivity more than two times higher (OR: 2. 28; CI: 1.28–4.06) than those who do not.

Table 4. Associations between neuropsychiatric disorders and oral manifestations.

Oral Manifestations		Neuropshychiatric Disorders				*p* Value
		No (147 Subjects)	Psychiatric (79 Subjects)	Neurological (5 Subjects)	Combined (6 Subjects)	
Modifications in the oral cavity after gluten-free diet	NO	111 (75.51%)	38 (48.10%)	2 (40%)	0 (0%)	<0.001
	YES	36 (24.29%)	41 (51.60%)	3 (60%)	6 (100%)	
Gastroesophageal reflux	NO	114 (77.55%)	45 (56.96%)	45 (56.96%)	2 (33.33%)	
	YES	33 (22.45%)	34 (43.04%)	34 (43.04%)	4 (66.67%)	
Enamel defects	NO	120 (81.63%)	59 (75.64%)	2 (40%)	3 (50%)	0.04
	YES	27 (18.37%)	19 (24.36%)	3 (60%)	3 (50%)	
Gingival bleeding	NO	76 (51.70%)	26 (32.91%)	3 (60%)	3 (50%)	0.003
	SOMETIMES	64 (43.54%)	38 (48.10%)	2 (40%)	1 (16.67%)	
	YES	7 (4.76%)	15 (18.99%)	0 (0%)	2 (33.33%)	
Aphthous stomatitis	NO	110 (74.83%)	45 (56.96%)	1 (20%)	4 (66.67%)	<0.001
	PRE-DIET	19 (12.93%)	12 (15.19%)	4 (80%)	1 (16.67%)	
	YES	18 (12.24%)	22 (27.85%)	0 (0%)	1 (16.67%)	
Xerostomia	NO	129 (87.76%)	59 (74.68%)	5 (100%)	5 (83.33%)	0.06
	YES	18 (12.24%)	20 (25.32%)	0 (0%)	1 (16.67%)	
Articular clicks	NO	110 (74.83%)	47 (59.49%)	5 (100%)	3 (50%)	0.029
	YES	37 (25.17%)	32 (40.51%)	0 (0%)	3 (50%)	
Tension-type headache	NO	96 (65.31%)	40 (50.63%)	1 (20%)	3 (50%)	0.045
	YES	51 (34.69%)	39 (49.37%)	4 (80%)	3 (50%)	
Dentin sensitivity	NO	95 (64.63%)	34 (43.04%)	3 (60%)	2 (33.33%)	0.011
	YES	52 (35.37%)	45 (56.96%)	2 (40%)	4 (66.67%)	

Table 5. Associations between hematological disorders and oral manifestations.

Oral Manifestations		Hematological Disorders		OR [1]	CI [2]	*p* Value
		NO (76 Subjects)	YES (161 Subjects)			
Modifications in the oral cavity after gluten free diet	NO	58 (76.32%)	93 (57.76%)	2.36	1.27–4.36	0.006
	YES	18 (23.68%)	68 (42.24%)			
Enamel defects	NO	66 (88.00%)	118 (73.29%)	2.67	1.23–5.82	0.011
	YES	9 (12.00%)	43 (26.71%)			
Xerostomia	NO	71 (93.42%)	127 (78.88%)	3.8	1.42–10.16	0.005
	YES	5 (6.58%)	34 (21.12%)			
Dentin sensitivity	NO	53 (69.74%)	81 (50.31%)	2.28	1.28–4.06	0.005
	YES	23 (30.26%)	80 (49.69%)			

[1] Confidence Interval; [2] Odds Ratio.

In addition, some statistically significant correlations were observed in the sub-group of coeliac patients reporting muscular disorders (a sub-group of 60 subjects) (Table 6). Among 177 coeliac patients without muscular disorders, 121 reported no changes in their oral cavity following gluten-free therapy, while out of 60 coeliac patients with such problems, oral changes after diet were reported by 30 subjects. Therefore, it can be stated that coeliac patients with muscular disorders have a risk of presenting changes at the level of the oral cavity after the adoption of a gluten-free diet that is twice as high (OR:2.16; CI:1.19–3.93) compared to those who do not have these problems. In addition, coeliac patients with concurrent muscular disorders showed a risk (OR: 1.98; CI 1.07–3.64) of suffering from TMJ clicks twice as high than those who do not suffer from muscular disorders. Finally, it can be said that coeliac patients with muscular disorders have a risk of presenting dentin sensitivity approximately 2.5 (OR: 2.46; CI: 1. 35–4. 48) times higher than those who do not suffer from muscle disorders.

Table 6. Associations between muscular disorders and oral manifestations.

Oral Manifestations		Muscular Disorders		OR [1]	CI [2]	*p* Value
		NO (177 Subjects)	YES (60 Subjects)			
Modifications in the oral cavity after gluten free diet	NO	121 (68.36%)	30 (50.00%)	2.16	1.19–3.93	0.011
	YES	56 (31.64%)	30 (50.00%)			
Gastroesophageal reflux	NO	132 (74.58%)	32 (53.33%)	2.57	1.40–4.72	0.002
	YES	45 (25.42%)	28 (46.67%)			
Xerostomia	NO	156 (88.14%)	42 (70.00%)	3.18	1.56–6.51	0.001
	YES	21 (11.86%)	18 (30.00%)			
Gingivitis signs and symptoms	NO	143 (93.46%)	40 (83.33%)	2.86	1.06–7.72	0.032
	YES	10 (6.54%)	8 (16.67%)			
Articular clicks	NO	130 (73.45%)	35 (58.33%)	1.98	1.07–3.64	0.028
	YES	47 (26.55%)	25 (41.67%)			
Dentin sensitivity	NO	110 (62.15%)	24 (40.00%)	2.46	1.35–4.48	0.003
	YES	67 (37.85%)	36 (60.00%)			

[1] Confidence Interval; [2] Odds Ratio.

Other correlations were observed between the total and partial gluten-free diet (Table 7). The number of patients with gingival bleeding and partial gluten-free diet was 117 out of 211, while the number of patients on a totally gluten-free diet and gingival bleeding was 11 out of 25. In addition, the number of patients with nocturnal snoring among subjects with a partial gluten-free diet was 35 out of 211, whereas among patients with a totally gluten-free diet, nocturnal snoring resulted in 2 out of 25 patients.

Table 7. Associations between partial/total gluten free diet and oral manifestations.

Oral Manifestations		Diet			*p* Value
		None (1 Subject)	Total Gluten Free Diet (211 Subjects)	Partial Gluten Free Diet (25 Subjects)	
Bleeding	NO	0 (0%)	94 (44.55%)	14 (56%)	0.039
	SOMETIMES	0 (0%)	96 (45.50%)	9 (36%)	
	YES	1 (100%)	21 (9.95%)	2 (8%)	
Nocturnal snoring	NO	0 (0%)	176 (83.41%)	23 (92%)	0.039
	YES	1 (100%)	35 (16.59%)	2 (8%)	

Finally, some correlations were observed with the removal of particular foods (Table 8). Out of 123 patients who did not give up any food, 92 subjects did not report any changes in the oral cavity with the introduction of the gluten-free diet. But out of only five patients who did not eat meat, four reported changes in their oral cavity with the introduction of the gluten-free diet. Out of 123 (51.90%) patients who did not give up any food, only three suffered from glossitis. But out of five (2.11%) patients who did not eat meat, two subjects suffered from glossitis. In addition, out of 123 patients who did not give up any food, only 14 reported the disappearance of aphthous lesions after the introduction of the gluten-free diet. Of the eight patients who do not eat fat foods, three reported the total disappearance of mouth ulcers with the introduction of a gluten-free diet.

Table 8. Associations between foods consumption and oral manifestations.

Oral Manifestations		Other Food						*p* Value
		None (123 Subjects)	Lactose (45 Subjects)	Sugar (11 Subjects)	Meat (5 Subjects)	Fat (8 Subjects)	Combined (45 Subjects)	
Modifications in the oral cavity after gluten-free diet	NO	92 (74.80%)	23 (51.11%)	8 (72.73%)	1 (20%)	4 (50%)	23 (51.11%)	0.003
	YES	31 (25.20%)	22 (48.89%)	3 (27.27%)	4 (80%)	4 (50%)	22 (48.89%)	
Gastroesophageal reflux	NO	95 (77.24%)	27 (60%)	7 (63.64%)	0 (0%)	8 (100%)	27 (60%)	<0.001
	YES	28 (22.76%)	18 (40%)	4 (36.36%)	5 (100%)	0 (0%)	18 (40%)	
Symptoms of Glossitis	NO	120 (97.56%)	43 (95.56%)	10 (90.91%)	3 (60%)	7 (87.50%)	42 (93.33%)	0.007
	YES	3 (2.44%)	2 (4.44%)	1 (9.09%)	2 (40%)	1 (12.50%)	3 (6.67%)	
Aphthous lesions	NO	92 (74.80%)	23 (51.11%)	9 (81.82%)	1 (20%)	5 (62.50%)	30 (66.67%)	0.002
	PRE-DIET	14 (11.38%)	11 (24.44%)	1 (9.09%)	0 (0%)	3 (37.50%)	7 (15.56%)	
	YES	17 (13.82%)	11 (24.44%)	1 (9.09%)	4 (80%)	0 (0%)	8 (17.78%)	

4. Discussion

In this study, 237 individuals suffering from coeliac disease were recruited through connection platforms such as Facebook and in restaurants and supermarkets specializing in coeliac clients. According to the Annual Report to the Parliament on Coeliac Disease, at the end of 2016 there were 198,427 coeliac patients in Italy, instead of the approximately 600,000 expected. Coeliac females (138,902 in 2016) were more than twice the number of males (59,525).

From the present findings, some differences in oral manifestations were observed between males and females. Regarding caries prevalence, for example, coeliac females were almost twice as susceptible to caries than men. This result is not clearly confirmed in the literature as far as the healthy population is concerned, but there are data (relating to the non-coeliac population) that suggest that there could be such a difference in the predisposition to caries between the two genders. For example, the study by Galvao-Moreira et al. [7] shows that female salivary pH is more acidic than male pH, and the same was suggested by the study by Eliasson et al. [8] in which lower rates of buccal and labial salivation were found in females, as well as lower levels of IgA.

In addition, some statistically significant correlations were found in relation to the age of the subjects. For example, with progressing age, the prevalence of dentin sensitivity and TMJ clicks increased, with a peak prevalence in subjects aged between 45 and 64. The analysis of the data on the prevalence of TMJ clicks in the various age groups is confirmed by previous literature [9,10], and this ensures the reliability of the data collected.

In the literature it has been demonstrated that coeliac disease presents a heterogeneity in the manifestation of symptoms, and therefore there are cases with manifestations of "minor" entity, and clinical cases in which the symptoms are more evident and heterogeneous. From the present study, it appears evident that clinical cases with a more evident manifestation of general coeliac characteristics (i.e., those cases in which the coeliac patients also present general related characteristics) are the same with a greater symptomatology at the level of the oral cavity as well. This aspect has been observed, for example, for related pathologies such as hematological systemic disorders, for symptoms of muscle disorders, and for neuro-psychiatric symptoms.

From the present survey, examples of oral cavity interest in coeliac patients with related pathologies are the observed correlation between tension-type headache and dentin sensitivity, and between general symptoms in coeliac patients (such as gastroesophageal reflux) and halitosis, that are often also associated in non-coeliac subjects [11,12]. Dentin sensitivity was also found to be associated with coexistence of other related diseases in coeliac patients, such as xerostomia. This result is most likely to be related to the fact that coeliac sufferers often present enamel defects [13–15] which could lead to dentin sensitivity. The literature partly seems to confirm these associations. For example, with regard to the association between xerostomia and coeliac disease, the study by van Gils et al. [16] reports the evaluation of 740 patients with coeliac disease and 270 control subjects, showing that oral health problems are more commonly experienced in adult patients with coeliac disease than in the comparison group. Regarding the association between halitosis and coeliac disease, in the literature the study by Tsai et al. [6], based on a sample of children, states that there is a correlation. Finally, regarding the association between tension-type headache and coeliac disease, the literature seems to confirm the present observation. The study by Zis et al. [4] states that the average aggregate prevalence of tension-type headache among coeliac patients is 26% in adult populations. The study concludes by inviting patients with headache of unknown origin to be screened for coeliac disease, as these patients could benefit from a gluten-free diet.

According to the present findings following a gluten-free diet appears crucial to managing oral diseases associated to coeliac pathology. In fact, gingival bleeding levels increase with a high systemic inflammatory rate (i.e., the detection of a whole series of inflammatory symptoms related to coeliac disease), in coeliac patients who do not follow a gluten-free diet carefully. In addition to gingivitis management with chlorhexidine [17], a general program of prevention in these subjects should also be recommended. A previous NHANES (National Health and Nutrition Examination Survey) study, apparently in disagreement with the present findings, showed the absence of association of coeliac disease with periodontal disease and the absence of difference between subjects with diagnosed and undiagnosed coeliac disease, but the sample size of the coeliac disease group was low and the severity of the general and oral manifestations of the disease was not considered as a possible confounding factor [18].

Another example of the potential role of the gluten-free diet was observed on nocturnal snoring, a condition which is now increasingly managed in the dental field, as the data of the present study reveal that a worsening in nocturnal snoring is observed when the coeliac patient does not strictly follow the correct diet.

Even if previous studies have also reported the manifestations of coeliac disease on oral health [13,14], to the authors' knowledge this is the first study that evidences the potential importance of following a strictly gluten-free diet in controlling gingival bleeding levels and nocturnal snoring in patients with coeliac disease. Thus, clinical studies should be encouraged to confirm these results and estimate the impact of the gluten-free diet on periodontal indices and sleep-related breathing disorders of subjects affected by coeliac disease.

In general, it can be said that the results of the present study showed that an adult coeliac patient with associated systemic diseases could also present a significant prevalence of diseases and symptoms at the level of the oral cavity. Patients with fewer "systemic" symptoms, on the other hand, show more modest oral symptoms. The present data suggest that it is essential to monitor frequently, over time, the general health of the coeliac patient, as well as his/her oral health, as the presence of symptoms of inflammation at the level of the oral cavity is often associated with systemic symptoms [19–21] that can be also related to coeliac disease. Therefore, the prevention of inflammatory diseases in the oral cavity inevitably includes a prevention program that invests in the general health of the patient and dentists are also called upon to take part because of the particular role they play in care of the oral cavity.

Nowadays, it is increasingly clear that the figure of the dentist not only has the task of monitoring oral diseases but is often identified as an educator regarding a healthy lifestyle and proper nutrition.

5. Conclusions

The results of the present study suggest that the dentist should implement a specific clinical protocol for coeliac patients, due to the peculiar and heterogeneous clinical situation that they may present. This protocol should include frequent follow-ups with monitoring of "general" health, in addition to oral health, including several recommendations for compliance with the gluten-free diet. In fact, following a gluten-free diet could be important to control gingival bleeding levels and to manage oral symptoms associated with coeliac disease.

In addition to monitoring the appearance of specific symptoms and signs in the mouth, the dentist should encourage the patient to perform other general health checks as well.

Author Contributions: Conceptualization, A.N. and S.T.; methodology, A.N., S.A. and S.T.; validation, A.N., E.P. and S.T.; formal analysis, A.N., I.R. and S.T.; investigation, A.N. and S.T.; resources, I.R., E.P.; data curation, A.N. and S.T.; writing—original draft preparation, A.N., I.R., F.B., S.A. and S.T.; supervision, S.T.; project administration, S.T. All authors have read and agreed to the published version of the manuscript.

Funding: This research received no external funding.

Conflicts of Interest: The authors declare no conflict of interest.

References

1. Macho, V.M.P.; Coelho, A.S.; Veloso e Silva, D.M.; de Andrade, D.J.C. Oral manifestations in pediatric patients with coeliac disease—A review article. *Open Dent. J.* **2017**, *11*, 539–545. [CrossRef]
2. Walker, M.M.; Ludvigsson, J.F.; Sanders, D.S. Coeliac disease: Review of diagnosis and management. *Med. J. Aust.* **2017**, *207*, 173–178. [CrossRef] [PubMed]
3. Cichewicz, A.B.; Mearns, E.S.; Taylor, A.; Boulanger, T.; Gerber, M.; Leffler, D.A.; Drahos, J.; Sanders, D.S.; Thomas Craig, K.J.; Lebwohl, B. Diagnosis and treatment patterns in celiac disease. *Dig. Dis. Sci.* **2019**, *64*, 2095–2106. [CrossRef] [PubMed]
4. Zis, P.; Hadjivassiliou, M. Treatment of neurological manifestations of gluten sensitivity and coeliac disease. *Curr. Treat. Options Neurol.* **2019**, *21*, 10. [CrossRef] [PubMed]
5. de Carvalho, F.K.; de Queiroz, A.M.; Bezerra da Silva, R.A.; Sawamura, R.; Bachmann, L.; Bezerra da Silva, L.A.; Nelson-Filho, P. Oral aspects in celiac disease children: Clinical and dental enamel chemical evaluation. *Oral Surg. Oral Med. Oral Pathol. Oral Radiol.* **2015**, *119*, 636–643. [CrossRef]
6. Tsai, C.-C.; Chou, H.-H.; Wu, T.-L.; Yang, Y.-H.; Ho, K.-Y.; Wu, Y.-M.; Ho, Y.-P. The levels of volatile sulfur compounds in mouth air from patients with chronic periodontitis. *J. Periodontal Res.* **2008**, *43*, 186–193. [CrossRef]
7. Galvão-Moreira, L.V.; de Andrade, C.M.; de Oliveira, J.F.F.; Bomfim, M.R.Q.; Figueiredo, P.d.M.S.; Branco-de-Almeida, L.S. Sex differences in salivary parameters of caries susceptibility in healthy individuals. *Oral Health Prev. Dent.* **2018**, *16*, 71–77.
8. Eliasson, L.; Birkhed, D.; Österberg, T.; Carlén, A. Minor salivary gland secretion rates and immunoglobulin A in adults and the elderly. *Eur. J. Oral Sci.* **2006**, *114*, 494–499. [CrossRef]
9. Tecco, S.; Festa, F.; Salini, V.; Epifania, E.; D'Attilio, M. Treatment of joint pain and joint noises associated with a recent TMJ internal derangement: A comparison of an anterior repositioning splint, a full-arch maxillary stabilization splint, and an untreated control group. *Cranio* **2004**, *22*, 209–219. [CrossRef]
10. Tecco, S.; Mummolo, S.; Marchetti, E.; Tetè, S.; Campanella, V.; Gatto, R.; Gallusi, G.; Tagliabue, A.; Marzo, G. sEMG activity of masticatory, neck, and trunk muscles during the treatment of scoliosis with functional braces. A longitudinal controlled study. *J. Electromyogr. Kinesiol.* **2011**, *21*, 885–892. [CrossRef]
11. Bernardi, S.; Zeka, K.; Mummolo, S.; Marzo, G.; Continenza, M. Development of a new protocol: A macroscopic study of the tongue dorsal surface (Note). *Ital. J. Anat. Embriol.* **2013**, *118*, 24.

12. Pennazza, G.; Marchetti, E.; Santonico, M.; Mantini, G.; Mummolo, S.; Marzo, G.; Paolesse, R.; D'Amico, A.; Di Natale, C. Application of a quartz microbalance based gas sensor array for the study of halitosis. *J. Breath Res.* **2008**, *2*, 017009. [CrossRef] [PubMed]

13. Pastore, L.; Carroccio, A.; Compilato, D.; Panzarella, V.; Serpico, R.; Muzio, L.L. Oral manifestations of celiac disease. *J. Clin. Gastroenterol.* **2008**, *42*, 224–232. [CrossRef] [PubMed]

14. Cheng, J.; Malahias, T.; Brar, P.; Minaya, M.T.; Green, P.H.R. The association between celiac disease, dental enamel defects, and aphthous ulcers in a United States Cohort. *J. Clin. Gastroenterol.* **2010**, *44*, 191–194. [CrossRef]

15. Nota, A.; Palumbo, L.; Pantaleo, G.; Gherlone, E.F.; Tecco, S. Developmental Enamel Defects (DDE) and their association with oral health, preventive procedures, and children's psychosocial attitudes towards home oral hygiene: A cross-sectional study. *Int. J. Environ. Res. Public Health* **2020**, *17*, 4025. [CrossRef]

16. van Gils, T.; Bouma, G.; Bontkes, H.J.; Mulder, C.J.J.; Brand, H.S. Self-reported oral health and xerostomia in adult patients with celiac disease versus a comparison group. *Oral Surg. Oral Med. Oral Pathol. Oral Radiol.* **2017**, *124*, 152–156. [CrossRef]

17. Mummolo, S.; Severino, M.; Campanella, V.; Barlattani, A.; Quinzi, V.; Marchetti, E. Chlorhexidine gel used as antiseptic in periodontal pockets. *J. Biol. Regul. Homeost. Agents* **2019**, *33*, 83–88.

18. Spinell, T.; DeMayo, F.; Cato, M.; Thai, A.; Helmerhorst, E.J.; Green, P.H.R.; Lebwohl, B.; Demmer, R.T. The association between coeliac disease and periodontitis: Results from NHANES 2009–2012. *J. Clin. Periodontol.* **2018**, *45*, 303–310. [CrossRef]

19. Mummolo, S.; Severino, M.; Campanella, V.; Barlattani, A.; Quinzi, V.; Marchetti, E. Periodontal disease in subjects suffering from coronary heart disease. *J. Biol. Regul. Homeost. Agents* **2019**, *33*, 73–82.

20. Baldini, A.; Nota, A.; Cioffi, C.; Ballanti, F.; Cozza, P. Infrared thermographic analysis of craniofacial muscles in military pilots affected by bruxism. *Aerosp. Med. Hum. Perform.* **2015**, *86*, 374–378. [CrossRef]

21. Tecco, S.; Sciara, S.; Pantaleo, G.; Nota, A.; Visone, A.; Germani, S.; Polizzi, E.; Gherlone, E.F. The association between minor recurrent aphthous stomatitis (RAS), children's poor oral condition, and underlying negative psychosocial habits and attitudes towards oral hygiene. *BMC Pediatr.* **2018**, *18*, 136. [CrossRef] [PubMed]

Article

Celiac Disease-Type Tissue Transglutaminase Autoantibody Deposits in Kidney Biopsies of Patients with IgA Nephropathy

Rakel Nurmi [1], Ilma Korponay-Szabó [1,2,3], Kaija Laurila [1], Heini Huhtala [4], Onni Niemelä [5,6], Jukka Mustonen [6,7], Satu Mäkelä [6,7], Katri Kaukinen [1,7] and Katri Lindfors [1,*]

1   Celiac Disease Research Center, Faculty of Medicine and Health Technology, Tampere University, 33520 Tampere, Finland; rakel.nurmi@fimnet.fi (R.N.); ilma.korponay-szabo@tuni.fi (I.K.-S.); kaija.laurila@tuni.fi (K.L.); katri.kaukinen@tuni.fi (K.K.)
2   Celiac Disease Center, Heim Pál National Pediatric Institute, 1089 Budapest, Hungary
3   Department of Pediatrics, Faculty of Medicine and Clinical Center, University of Debrecen, 4032 Debrecen, Hungary
4   Faculty of Social Sciences, Tampere University, 33520 Tampere, Finland; heini.huhtala@tuni.fi
5   Medical Research Unit, Seinäjoki Central Hospital, 60220 Seinäjoki, Finland; onni.niemela@epshp.fi
6   Faculty of Medicine and Health Technology, Tampere University, 33520 Tampere, Finland; jukka.mustonen@tuni.fi (J.M.); satu.m.makela@pshp.fi (S.M.)
7   Department of Internal Medicine, Tampere University Hospital, 33521 Tampere, Finland
*   Correspondence: katri.lindfors@tuni.fi; Tel.: +358-50-3186306

Citation: Nurmi, R.; Korponay-Szabó, I.; Laurila, K.; Huhtala, H.; Niemelä, O.; Mustonen, J.; Mäkelä, S.; Kaukinen, K.; Lindfors, K. Celiac Disease-Type Tissue Transglutaminase Autoantibody Deposits in Kidney Biopsies of Patients with IgA Nephropathy. *Nutrients* **2021**, *13*, 1594. https://doi.org/10.3390/nu13051594

Academic Editors: Vassilios Liakopoulos and Stefano Guandalini

Received: 26 March 2021
Accepted: 8 May 2021
Published: 11 May 2021

Publisher's Note: MDPI stays neutral with regard to jurisdictional claims in published maps and institutional affiliations.

Abstract: An association between celiac disease and IgA nephropathy (IgAN) has been suggested. In celiac disease, in addition to circulating in serum, IgA-class tissue transglutaminase (tTG) autoantibodies are deposited in the small bowel mucosa and extraintestinal organs. In this case series of IgAN patients with or without celiac disease, we studied whether celiac disease-type IgA-tTG deposits occur in kidney biopsies. The study included nine IgAN patients, four of them with celiac disease. At the time of the diagnostic kidney biopsy serum tTG autoantibodies were measured and colocalization of IgA and tTG was investigated in the frozen kidney biopsies. Three IgAN patients with celiac disease had IgA-tTG deposits in the kidney even though in two of these the celiac disease diagnosis had been set years later. These deposits were not found in a patient with already diagnosed celiac disease following a gluten-free diet. Of the five non-celiac IgAN patients, three had IgA-tTG deposits in the kidney. We conclude that tTG-targeted IgA deposits can be found in the kidney biopsies of gluten-consuming IgAN patients but their specificity to celiac disease seems limited.

Keywords: IgA nephropathy; celiac disease; tissue transglutaminase autoantibody; tissue transglutaminase-targeted IgA deposits

1. Introduction

Celiac disease, an immune-mediated enteropathy, is driven by the ingestion of cereals, wheat, rye, and barley containing gluten and characterized by a disease-specific autoantibody response targeting tissue transglutaminase (tTG) [1]. During gluten consumption, IgA-class tTG autoantibodies circulate in serum but are also deposited in the small bowel mucosa, where they are bound to their antigen tTG around mucosal capillaries and on the basement membrane below the mucosal epithelium [2,3]. Interestingly, these small intestinal IgA deposits may be present even before the development of small bowel mucosal villous atrophy or the detection of the tTG autoantibodies in the circulation, and they may predict forthcoming manifest celiac disease [3–5]. Moreover, IgA deposits in the gut have also been found in celiac disease patients with negative serum tTG autoantibodies [6,7]. Upon introduction of a gluten-free diet (GFD), the gold standard treatment for celiac disease, these small intestinal mucosal deposits disappear along with serum tTG autoantibodies [4,8]. IgA-class tTG-targeted autoantibody deposits can be also found in several other tissues, including liver, muscle, and brain, often coinciding with extraintestinal manifestations of celiac disease affecting the organ in question (e.g., hepatitis, muscle weakness, and ataxia) [3,7]. However, little is known about the occurrence of such extraintestinal IgA-tTG deposits long before diagnosis of celiac disease or their dependence on the presence of gluten in the diet.

It has been proposed that celiac disease may be associated with IgA nephropathy (IgAN), globally the most common primary chronic glomerular disease [9,10]. The diagnostic hallmark of IgAN is the predominance of hypo-galactosylated IgA1 deposits in the mesangium of the glomeruli [11]. In IgAN, the interaction between environmental antigens, dysregulation of IgA immune responses and pathogenic circulating IgA complexes eventually leads to IgA1 deposits in the kidney [12]. Although celiac disease and IgAN target different organs, they have a great deal in common. As mentioned above, aberrant IgA response is involved in both diseases [13]. tTG is known to play a decisive role in the pathogenesis of celiac disease by modifying wheat gluten-derived gliadin into a more immunogenic form [14]. Although the role of tTG in IgAN is less clear, studies conducted in both human patients and mice suggest that tTG is needed for the development of IgAN-type mesangial IgA deposits and the impairment of the clinical course of IgAN [15,16]. The ingestion of dietary gluten is required for the development of celiac disease, and may also promote the development of IgAN, at least in mice [14,17,18]. In addition, GFD being the only treatment for celiac disease, it is intriguing that some reports have described clinical improvement of IgAN with the same diet [19–21].

The association between celiac disease and IgAN is still being actively researched. In this case series we investigated whether celiac disease-type tTG-targeted IgA autoantibody deposits occur in the kidney biopsies of IgAN patients with or without concomitant or subsequent celiac disease.

2. Materials and Methods

2.1. Patients and Clinical Data

This retrospective study reports a case series of nine adult IgAN patients on whom diagnostic kidney biopsy was performed during the period 1981–1987 at Tampere University Hospital, Finland. IgAN was defined as glomerulonephritis with typical light microscopy features and IgA as the sole or main glomerular immunofluorescence finding [9]. Four of the IgAN patients also had diagnosed celiac disease. Five non-celiac patients who had received their IgAN diagnoses during the same period were selected as controls. The patients were followed-up until recent available laboratory results or death. Clinical data at the time of the IgAN diagnoses and at follow-up were collected from the medical records and included demographic data, creatinine values, and data on celiac disease diagnosis. The outcomes (chronic dialysis, kidney transplant, and death) were documented. Creatinine values and the CKD-EPI (Chronic Kidney Disease Epidemiology Collaboration) equation were used to determine estimated glomerular filtration rate (eGFR). eGFR $\geq$90 indicates normal kidney function, eGFR 30–59 moderate renal impairment, eGFR 15–29 severe renal impairment, and eGFR <15 renal failure [22].

2.2. Determination of Serum and Tissue tTG-Targeted IgA Autoantibodies

Stored frozen (-80 °C) serum samples taken at the time of kidney biopsy were used to identify IgA-class tTG autoantibodies using the ELIA Celikey assay (Celikey®, Phadia, GmbH, Freiburg, Germany) according to manufacturers' instructions. Values higher than 7.0 U/mL were regarded as positive.

Celiac disease-type tissue deposits with colocalization of IgA and tTG were determined in frozen biopsies by evaluators blinded to the clinical data. These included kidney biopsies for diagnosing IgAN and small bowel mucosal biopsies of two celiac disease patients taken at the time of the celiac disease diagnoses. The deposits were detected using the technique described earlier by Korponay-Szabó et al. [3] from snap frozen biopsies embedded in optimal cutting temperature compound (OCT, Tissue Tek, Sakura Finetek

Europe B.V., AJ Alphen aan den Rijn, The Netherlands). The frozen sections of 5 μm thickness were stained with fluorescein isothiocyanate (FITC)-labeled rabbit antibodies against human IgA (Dako AS, Glostrup, Denmark) at a dilution of 1:40 in phosphate buffered saline and with monoclonal mouse antibodies against tTG (CUB7402, NeoMarkers, Fremont, CA, USA), which were detected with rhodamine-conjugated anti-mouse immunoglobulin rabbit antibodies (Dako) diluted 1:200 in phosphate buffered saline.

The kidney samples were investigated for the presence of tTG and IgA. IgA deposits around the basement membrane of extraglomerular blood vessels, of the parietal layer of Bowman's capsules and of the proximal or distal tubuli in colocalization with tTG were regarded as celiac disease-type deposits. Although the glomerular capillaries also contain tTG around their basement membrane, IgA deposition on tTG within the glomeruli was not clearly discernible because all patients had extensive mesangial IgA deposits related to IgAN itself. Therefore, glomerular IgA was not taken into account for celiac disease-type deposit evaluation.

In the small bowel mucosal biopsies, subepithelial IgA deposits found on the basement membrane below the villous and crypt epithelium and around the mucosal blood vessels were regarded as celiac disease-type IgA deposits, as in non-celiac subjects small-bowel mucosa IgA is detected only inside the plasma and epithelial cells [3,6]. Colocalization of IgA and tTG was regarded as celiac disease-type deposits.

2.3. Ethical Considerations

Informed consent for the study was obtained from the patients. The research protocol (E99105, R20056) was approved by the Regional Ethics Committee of Pirkanmaa Hospital District. The study protocol follows the ethical guidelines of the Declaration of Helsinki.

3. Results

By definition, all nine IgAN patients in our case series had glomerular IgA deposits characteristic of IgAN (Figure 1a). Four of the subjects were female (Table 1). At the time of the kidney biopsy and IgAN diagnosis their median age was 34 years (range 20–50 years) and eight of them were on a normal gluten-containing diet (Table 1). One patient (9M) with previously diagnosed celiac disease was on GFD. One patient (1F) was diagnosed with celiac disease during the same treatment episode as the diagnosis of IgAN. During follow-up two additional subjects (2M and 3M) received celiac disease diagnoses after eight and ten years respectively. At the time of celiac disease diagnoses, both of these cases had tTG-targeted autoantibodies in serum and IgA-tTG deposits in the small bowel mucosal biopsies (Figure 2).

In the kidney biopsies, celiac disease-type deposits characterized by colocalization of IgA and tTG were detected in all three patients with both IgAN and celiac disease on a normal gluten-containing diet (Figure 1b–d, Table 1). It is noteworthy that in two patients (2M and 3M) diagnosed with celiac disease during follow-up, the celiac disease-type IgA-tTG autoantibody deposits were already present in the renal tissue at the time of the kidney biopsies taken eight and ten years prior to celiac disease diagnosis. In all three patients with both IgAN and celiac disease, connective tissue IgA deposits were found around both proximal and distal tubuli. Moreover, in two of these patients (1F and 3M) deposits were also seen in the periglomerular region around Bowman's capsule. Interestingly, tTG autoantibody levels in the serum taken at the time of kidney biopsy and determined retrospectively from stored samples were already elevated in all these celiac disease patients, including those subsequently diagnosed with celiac disease during follow-up (Table 1).

Of the five IgAN patients without celiac disease, three had celiac disease-type IgA-tTG deposits around the proximal and distal tubuli in the kidney (4F, 5F, 6M) without having elevated levels of serum tTG autoantibodies at the cross-sectional serologic evaluation at diagnosis (Table 1). Of the remaining two IgAN patients without celiac disease (7F, 8M), neither had celiac disease-type IgA-tTG deposits in the kidney. Of these patients, 7F had no serum tTG autoantibodies, while 8M had no serum sample available for analysis. Patient 9M on GFD due to earlier diagnosed celiac disease, had no celiac disease-type tTG-targeted IgA deposits in the diagnostic kidney biopsy, although serum tTG autoantibody levels were still slightly elevated.

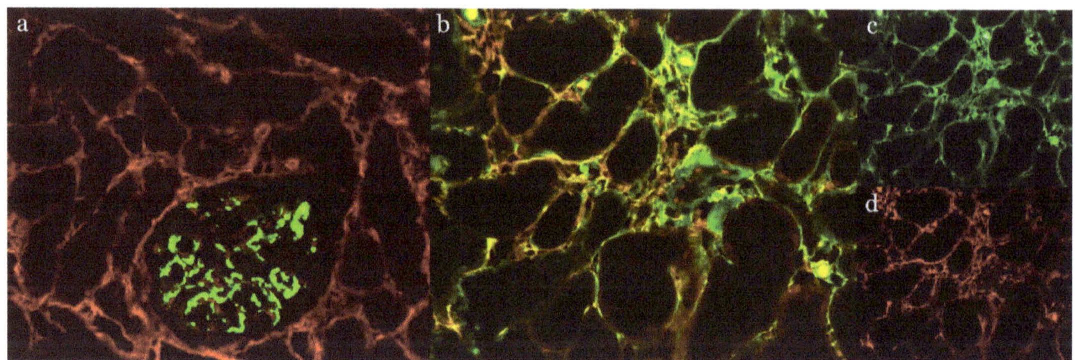

Figure 1. Immunofluorescent staining for immunoglobulin A (IgA) (green) and tissue transglutaminase (tTG) (red), and their colocalization (yellow) in the renal biopsies taken at the time of the IgA nephropathy (IgAN) diagnosis. (**a**) In a patient with IgAN without celiac disease, the IgA is only found in the glomerular mesangium. No colocalization of IgA and tTG is detected; (**b**) representative figure demonstrating colocalization of IgA and tTG (yellow) in the extracellular matrix around the renal tubuli in patients with both IgAN and celiac disease; (**c**) IgA; (**d**) tTG staining in the same specimen. Magnification 20x.

Table 1. Background and follow-up data and biopsy findings among nine patients with IgA nephropathy (IgAN).

Patient/Sex	Diagnosis	At IgAN Diagnosis			At CD Diagnosis			Follow-Up [1], Years	Disease Progression
		Age	IgA-tTG Deposits in the Kidney	Serum tTG Autoantibody Levels (U/mL)	Age	Gastrointestinal Symptoms and Signs	Duodenal IgA-tTG Deposits		
	On gluten containing diet								
1F	IgAN + CD	28	Yes	>100	28	Malabsorption	No data	35	eGFR 17
2M	IgAN + CD	35	Yes	36	42	Diarrhea, malabsorption	Yes	28	Dialysis and death
3M	IgAN + CD	41	Yes	>100	51	No symptoms [2]	Yes	29	eGFR 23
4F	IgAN	20	Yes	1.8	-	-		20	eGFR 110
5F	IgAN	50	Yes	1.4	-	-		35	eGFR 43
6M	IgAN	32	Yes	0.9	-	-		15	eGFR 71
7F	IgAN	28	No	0.7	-	-		10	eGFR 82
8M	IgAN	39	No	No data	-	-		32	eGFR 58
	On a gluten-free diet								
9M	IgAN + CD [3]	34	No	17	<34 [4]	No data	No data	5	Death

IgAN, IgA nephropathy; CD, celiac disease; tTG, tissue transglutaminase; eGFR, estimated glomerular filtration rate. tTG autoantibody levels higher than 7.0 U/mL were regarded as positive. eGFR $\geq$ 90 indicates normal kidney function, eGFR 60–89 mild loss of kidney function, eGFR 30–59 moderate renal impairment, eGFR 15–29 severe renal impairment and eGFR < 15 renal failure. [1] Started from kidney biopsy. [2] Risk-group screening. [3] The diagnosis of CD was made before the diagnosis of IgAN and the patient followed gluten-free diet at the time of kidney biopsy. Initial serum tTG autoantibody level was not known. [4] Exact time of diagnosis of CD was not known.

The median follow-up for all patients was 28 (range 5–35) years (mean 23, standard deviation 11 for comparison). Four patients with IgAN and celiac disease were followed-up for a median 28 years (range 5–35) (mean 24, standard deviation 13 for comparison). Two of these had died and two suffered from severe loss of kidney function (eGFR 17 and 23 respectively). Five IgAN patients without celiac disease were followed-up for a median 20 years (range 10–35) (mean 22, standard deviation 11 for comparison). One had normal kidney function, two mild loss of kidney function, and two moderate renal impairment (Table 1). None of these patients required dialysis treatment. No follow-up data on serum tTG autoantibody levels were available.

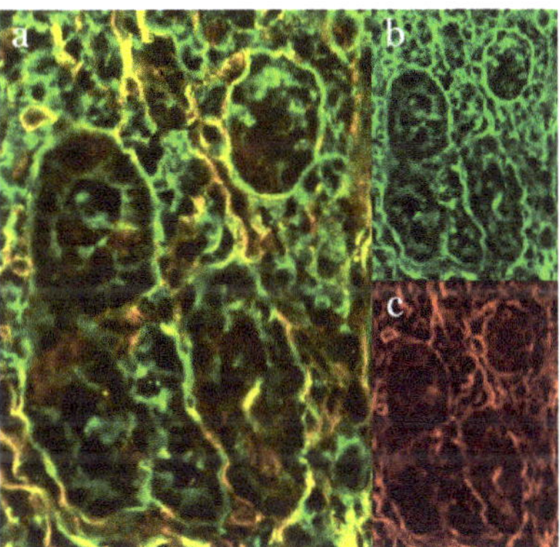

Figure 2. Small bowel mucosal biopsy of a celiac disease patient at the time of diagnosis of celiac disease stained for celiac disease-type IgA-tTG deposits. (**a**) Immunofluorescent staining demonstrating the colocalization of IgA and tTG (yellow); (**b**) staining for IgA (green), and (**c**) tissue transglutaminase (tTG) (red) in the same specimen. Magnification 20x.

4. Discussion

In this case series, we found that, during ingestion of gluten, celiac disease-type tTG-targeted IgA deposits were present in the kidney biopsies of all patients with both IgAN and celiac disease and thus our results are in line with the case report by Costa et al. [23]. Two of the IgAN patients with celiac disease (2M, 3M) were found to have IgA-tTG deposits in the kidney years before the diagnosis of celiac disease. Parallel deposits were also detected in the small bowel mucosal biopsies of these two patients (2M, 3M) at the time of celiac disease diagnosis, but unfortunately no small bowel mucosal biopsies taken at the time of IgAN diagnosis were available for the determination of IgA-tTG deposits. However, as small bowel mucosal IgA-tTG deposits may be present prior to small bowel mucosal damage diagnostic for celiac disease [4], the finding of celiac disease-type IgA-tTG deposits in the kidney is interesting and suggests that such deposits may also occur prior to celiac disease diagnosis. It must be noted, however, that both patients (2M, 3M) with IgA-tTG deposits in the kidney also had tTG autoantibodies in serum at the time of IgAN diagnosis. This would suggest that they already had celiac disease at this point, even though the clinical diagnosis of celiac disease came only later. In fact, one of these patients (3M) had tTG autoantibody levels 10 times above the cut-off and this would have been sufficient to warrant a diagnosis of celiac disease under the current guidelines [24]. In any case, as patient 9M with diagnosed celiac disease and already on a GFD at the time of kidney biopsy did not have celiac disease-type IgA-tTG deposits in the kidney, this would suggest that these deposits may be gluten-dependent. The fact that this patient had circulating tTG autoantibodies in low concentrations casts doubt on the strictness of the diet in this patient but may also suggest that these celiac disease-type deposits may disappear from the extraintestinal organs even before the complete clearance of autoantibodies from circulation.

In our study the IgA-tTG deposits in the kidney were not specific to celiac disease patients since they were also found in three non-celiac IgAN patients without serum tTG autoantibodies. Although no data is available on the presence of extraintestinal IgA-tTG deposits in seronegative celiac disease, such patients have been reported to have such deposits in the small bowel mucosa [6]. Unfortunately, no small bowel mucosal specimens taken at the time of IgAN diagnosis were available for determination of mucosal morphology and IgA-tTG deposits, and therefore we cannot be certain whether these IgAN patients with IgA-tTG deposits in the kidney were indeed antibody-negative celiac disease patients. Moreover, serum samples or small bowel mucosal biopsies during follow-up were not collected and thus it cannot be ascertained whether these individuals developed celiac disease later. However, it is interesting that our earlier studies describe small bowel inflammation and stress in IgAN in the absence of celiac disease [25,26]. Furthermore, increased immune reactivity to dietary antigens, including gluten, has been suggested among IgAN patients even without overt dietary intolerance [20,27]. Intestinal IgA-tTG deposits have been reported in patients without celiac disease and also in a patient with gluten ataxia, in whom the deposits were also found in the brain [7,28]. Hence the renal IgA-tTG deposits in IgAN patients without celiac disease could indicate a similar phenomenon. It is possible that the colocalization of IgA-tTG deposits in the kidney among IgAN patients with celiac disease is a familiar antigen-antibody interaction, while that among patients with IgAN reflects another type of molecular phenomenon related, for instance, to increased expression of tTG in the renal biopsies of IgAN patients [16].

In our study the patients with IgAN and celiac disease seemed to have worse outcomes of IgAN than did the patients with IgAN only. This finding is interesting in the light of our earlier study and the Swedish study by Rehnberg and co-workers showing that prognosis in IgAN may be poorer with concomitant comorbid bowel disease [29,30]. However, given the small number of patients and the fact that several factors affected the outcome of IgAN patients [31], conclusions concerning the renal survival of IgAN patients with celiac disease cannot be drawn on the basis of our data. Additionally, the follow-up of IgAN patients with celiac disease was longer, which may likewise affect this finding. In any case, the impact of the celiac disease-type IgA-tTG deposits in the kidney seemed not to be related to renal function or outcome, as also suggested by earlier findings of IgA in the kidneys of celiac disease patients without any renal problems [32].

To conclude, tTG-targeted IgA deposits were found in the kidney biopsies of gluten-consuming IgAN patients with or without known celiac disease. The significance of this interesting finding remains open and therefore in the future larger studies, preferably with more data on small bowel histochemistry and regular follow-up serology for tTG antibodies, are needed to investigate the association between celiac disease and IgAN in greater detail.

Author Contributions: Conceptualization and methodology, R.N., I.K.-S., J.M., S.M., K.K. and K.L. (Katri Lindfors); formal analysis and investigation, R.N., I.K.-S., K.L.(Kaija Laurila), H.H., O.N., K.K. and K.L. (Katri Lindfors); writing—original draft preparation, R.N. and K.L. (Katri Lindfors); writing—review and editing, R.N., I.K.-S., K.L. (Kaija Laurila), H.H., O.N., J.M., S.M., K.K. and K.L. (Katri Lindfors); supervision, J.M., S.M., K.K. and K.L. (Katri Lindfors); funding acquisition, R.N., K.K. and K.L. (Katri Lindfors). All authors have read and agreed to the published version of the manuscript.

Funding: This study was supported by the Competitive State Research Financing of the Expert Responsibility Area of Tampere University Hospital, the Academy of Finland, the Sigrid Jusélius Foundation, the Finnish Celiac Society, Tampere University Doctoral School, the Finnish Kidney and Liver Association, research grants NKFI 120392 from the Hungarian National Research, Development and Innovation Fund and the GINOP-2.3.2-15-2016-00015 and Interreg Danube DTP571 CD SKILLS projects, co-financed by the European Union and the Hungarian State.

Institutional Review Board Statement: The study was conducted according to the guidelines of the Declaration of Helsinki, and approved by the Ethics Committee of Pirkanmaa Hospital District (protocol codes E99105, R20056).

Informed Consent Statement: Informed consent was obtained from the subjects involved in the study or the Ethics Committee of Pirkanmaa Hospital District in cases of deceased patients.

Data Availability Statement: The data are not publicly available due to ethical and privacy reasons.

Acknowledgments: The valuable work of Ole Wirta (MD, PhD), Suvi Kalliokoski (PhD), Immo Rantala (PhD), and Johannes Valkonen is greatly appreciated.

Conflicts of Interest: The authors declare no conflict of interest. The funders had no role in the design of the study; in the collection, analyses, or interpretation of data; in the writing of the manuscript, or in the decision to publish the results.

References

1. Lindfors, K.; Ciacci, C.; Kurppa, K.; Lundin, K.E.A.; Makharia, G.K.; Mearin, M.L.; Murray, J.A.; Verdu, E.F.; Kaukinen, K. Coeliac disease. *Nat. Rev. Dis. Primers* **2019**, *5*, 3. [CrossRef]
2. Dieterich, W.; Ehnis, T.; Bauer, M.; Donner, P.; Volta, U.; Riecken, E.O.; Schuppan, D. Identification of tissue transglutaminase as the autoantigen of celiac disease. *Nat. Med.* **1997**, *3*, 797–801. [CrossRef] [PubMed]
3. Korponay-Szabó, I.R.; Halttunen, T.; Szalai, Z.; Laurila, K.; Király, R.; Kovács, J.B.; Fésüs, L.; Mäki, M. In vivo targeting of intestinal and extraintestinal transglutaminase 2 by coeliac autoantibodies. *Gut* **2004**, *53*, 641–648. [CrossRef]
4. Koskinen, O.; Collin, P.; Korponay-Szabo, I.; Salmi, T.; Iltanen, S.; Haimila, K.; Partanen, J.; Mäki, M.; Kaukinen, K. Gluten-dependent small bowel mucosal transglutaminase 2–specific IgA deposits in overt and mild enteropathy coeliac disease. *J. Pediatr. Gastroenterol. Nutr.* **2008**, *47*, 436–442. [CrossRef] [PubMed]
5. Kaukinen, K.; Peräaho, M.; Collin, P.; Partanen, J.; Woolley, N.; Kaartinen, T.; Nuutinen, T.; Halttunen, T.; Mäki, M.; Korponay-Szabo, I. Small-bowel mucosal transglutaminase 2-specific IgA deposits in coeliac disease without villous atrophy: A prospective and randomized clinical study. *Scand. J. Gastroenterol.* **2005**, *40*, 564–572. [CrossRef] [PubMed]
6. Salmi, T.T.; Collin, P.; Korponay-Szabó, I.R.; Laurila, K.; Partanen, J.; Huhtala, H.; Király, R.; Lorand, L.; Reunala, T.; Mäki, M.; et al. Endomysial antibody-negative coeliac disease: Clinical characteristics and intestinal autoantibody deposits. *Gut* **2006**, *55*, 1746–1753. [CrossRef]
7. Hadjivassiliou, M.; Mäki, M.; Sanders, D.S.; Williamson, C.A.; Grünewald, R.A.; Woodroofe, N.M.; Korponay-Szabó, I.R. Autoantibody targeting of brain and intestinal transglutaminase in gluten ataxia. *Neurology* **2006**, *66*, 373–377. [CrossRef] [PubMed]
8. Agardh, D.; Lynch, K.; Brundin, C.; Ivarsson, S.A.; Lernmark, Å.; Cilio, C.M. Reduction of tissue transglutaminase autoantibody levels by gluten-free diet is associated with changes in subsets of peripheral blood lymphocytes in children with newly diagnosed coeliac disease. *Clin. Exp. Immunol.* **2006**, *144*, 67–75. [CrossRef] [PubMed]
9. Collin, P.; Syrjänen, J.; Partanen, J.; Pasternack, A.; Kaukinen, K.; Mustonen, J. Celiac disease and HLA DQ in patients with IgA nephropathy. *Am. J. Gastroenterol.* **2002**, *97*, 2572–2576. [CrossRef]
10. Welander, A.; Sundelin, B.; Fored, M.; Ludvigsson, J.F. Increased risk of IgA nephropathy among individuals with celiac disease. *J. Clin. Gastroenterol.* **2013**, *47*, 678–683. [CrossRef]
11. Wyatt, R.J.; Julian, B.A. IgA nephropathy. *N. Engl. J. Med.* **2013**, *368*, 2402–2414. [CrossRef] [PubMed]
12. Pouria, S.; Barratt, J. Secondary IgA nephropathy. *Semin. Nephrol.* **2008**, *28*, 27–37. [CrossRef] [PubMed]
13. Papista, C.; Berthelot, L.; Monteiro, R.C. Dysfunctions of the Iga system: A common link between intestinal and renal diseases. *Cell. Mol. Immunol.* **2011**, *8*, 126–134. [CrossRef]
14. Di Sabatino, A.; Vanoli, A.; Giuffrida, P.; Luinetti, O.; Solcia, E.; Corazza, G.R. The function of tissue transglutaminase in celiac disease. *Autoimmun. Rev.* **2012**, *11*, 746–753. [CrossRef]
15. Berthelot, L.; Papista, C.; Maciel, T.T.; Biarnes-Pelicot, M.; Tissandie, E.; Wang, P.H.M.; Tamouza, H.; Jamin, A.; Bex-Coudrat, J.; Gestin, A.; et al. Transglutaminase is essential for IgA nephropathy development acting through IgA receptors. *J. Exp. Med.* **2012**, *209*, 793–806. [CrossRef]
16. Ikee, R.; Kobayashi, S.; Hemmi, N.; Saigusa, T.; Namikoshi, T.; Yamada, M.; Imakiire, T.; Kikuchi, Y.; Suzuki, S.; Miura, S. Involvement of transglutaminase-2 in pathological changes in renal disease. *Nephron Clin. Pract.* **2007**, *105*, c139–c146. [CrossRef] [PubMed]
17. Coppo, R.; Mazzucco, G.; Martina, G.; Roccatello, D.; Amore, A.; Novara, R.; Bargoni, A.; Piccoli, G.; Sena, L.M. Gluten-induced experimental IgA glomerulopathy. *Lab. Investig.* **1989**, *60*, 499–506.
18. Papista, C.; Lechner, S.; Ben Mkaddem, S.; LeStang, M.B.; Abbad, L.; Bex-Coudrat, J.; Pillebout, E.; Chemouny, J.M.; Jablonski, M.; Flamant, M.; et al. Gluten exacerbates IgA nephropathy in humanized mice through gliadin–CD89 interaction. *Kidney Int.* **2015**, *88*, 276–285. [CrossRef]
19. Ludvigsson, J.F.; Bai, J.C.; Biagi, F.; Card, T.R.; Ciacci, C.; Ciclitira, P.J.; Green, P.H.R.; Hadjivassiliou, M.; Holdoway, A.; van Heel, D.A.; et al. Diagnosis and management of adult coeliac disease: Guidelines from the British Society of Gastroenterology. *Gut* **2014**, *63*, 1210–1228. [CrossRef]
20. Coppo, R.; Amore, A.; Roccatello, D. Dietary antigens and primary immunoglobulin A nephropathy. *J. Am. Soc. Nephrol.* **1992**, *2*, S173–S180. [CrossRef]
21. Koivuviita, N.; Tertti, R.; Heiro, M.; Metsärinne, K. A case report: A patient with IgA nephropathy and coeliac disease. Complete clinical remission following gluten-free diet. *NDT Plus* **2009**, *2*, 161–163. [CrossRef] [PubMed]
22. Levin, A.; Stevens, P.E.; Bilous, R.W.; Coresh, J.; de Francisco, A.L.M.; de Jong, P.E.; Griffith, K.E.; Hemmelgarn, B.R.; Iseki, K.; Lamb, E.J.; et al. Kidney Disease: Improving Global Outcomes (KDIGO) CKD Work Group. KDIGO 2012 Clinical Practice Guideline for the Evaluation and Management of Chronic Kidney Disease. *Kidney Int. Suppl.* **2013**, *3*, 1–150. [CrossRef]
23. Costa, S.; Currò, G.; Pellegrino, S.; Lucanto, M.C.; Tuccari, G.; Ieni, A.; Visalli, G.; Magazzù, G.; Santoro, D. Case report on pathogenetic link between gluten and IgA nephropathy. *BMC Gastroenterol.* **2018**, *18*, 64. [CrossRef] [PubMed]
24. Husby, S.; Koletzko, S.; Korponay-Szabó, I.; Kurppa, K.; Mearin, M.L.; Ribes-Koninckx, C.; Shamir, R.; Troncone, R.; Auricchio, R.; Castillejo, G.; et al. European Society Paediatric Gastroenterology, Hepatology and Nutrition Guidelines for Diagnosing Coeliac Disease 2020. *J. Pediatr. Gastroenterol. Nutr.* **2020**, *70*, 141–156. [CrossRef]

25. Rantala, I.; Collin, P.; Holm, K.; Kainulainen, H.; Mustonen, J.; Mäki, M. Small bowel T cells, HLA class II antigen DR, and GroEL stress protein in IgA nephropathy. *Kidney Int.* **1999**, *55*, 2274–2280. [CrossRef] [PubMed]
26. Honkanen, T.; Mustonen, J.; Kainulainen, H.; Myllymäki, J.; Collin, P.; Hurme, M.; Rantala, I. Small bowel cyclooxygenase 2 (COX-2) expression in patients with IgA nephropathy. *Kidney Int.* **2005**, *67*, 2187–2195. [CrossRef]
27. Floege, J.; Feehally, J. The mucosa–kidney axis in IgA nephropathy. *Nat. Rev. Nephrol.* **2016**, *12*, 147–156. [CrossRef] [PubMed]
28. Maglio, M.; Ziberna, F.; Aitoro, R.; Discepolo, V.; Lania, G.; Bassi, V.; Miele, E.; Not, T.; Troncone, R.; Auricchio, R. Intestinal Production of Anti-Tissue Transglutaminase 2 Antibodies in Patients with Diagnosis Other Than Celiac Disease. *Nutrients* **2017**, *9*, 1050. [CrossRef] [PubMed]
29. Nurmi, R.; Metso, M.; Pörsti, I.; Niemelä, O.; Huhtala, H.; Mustonen, J.; Kaukinen, K.; Mäkelä, S. Celiac disease or positive tissue transglutaminase antibodies in patients undergoing renal biopsies. *Dig. Liver Dis.* **2018**, *50*, 27–31. [CrossRef] [PubMed]
30. Rehnberg, J.; Symreng, A.; Ludvigsson, J.F.; Emilsson, L. Inflammatory Bowel Disease Is More Common in Patients with IgA Nephropathy and Predicts Progression of ESKD: A Swedish Population-Based Cohort Study. *J. Am. Soc. Nephrol.* **2021**, *32*, 411–423. [CrossRef]
31. Barbour, S.; Reich, H. An update on predicting renal progression in IgA nephropathy. *Curr. Opin. Nephrol. Hypertens.* **2018**, *27*, 214–220. [CrossRef] [PubMed]
32. Pasternack, A.; Collin, P.; Mustonen, J.; Reunala, T.; Rantala, I.; Laurila, K.; Teppo, A.M. Glomerular IgA deposits in patients with celiac disease. *Clin. Nephrol.* **1990**, *34*, 56–60. [PubMed]

Article

Intraepithelial Lymphocyte Cytometric Pattern Is a Useful Diagnostic Tool for Coeliac Disease Diagnosis Irrespective of Degree of Mucosal Damage and Age—A Validation Cohort

Pablo Ruiz-Ramírez [1], Gerard Carreras [1], Ingrid Fajardo [1], Eva Tristán [1,2], Anna Carrasco [1,2], Isabel Salvador [1], Yamile Zabana [1,2], Xavier Andújar [1,2], Carme Ferrer [3], Diana Horta [1], Carme Loras [1,2], Roger García-Puig [4], Fernando Fernández-Bañares [1,2] and Maria Esteve [1,2,*]

[1] Department of Gastroenterology, Hospital Universitari Mútua Terrassa, Universitat de Barcelona, 08221 Barcelona, Spain; pruiz@mutuaterrassa.es (P.R.-R.); gcarreras@mutuaterrassa.cat (G.C.); ifajardo@mutuaterrassa.cat (I.F.); etristan@mutuaterrassa.cat (E.T.); acarrasco@mutuaterrassa.es (A.C.); isalvador@mutuaterrassa.cat (I.S.); yzabana@mutuaterrassa.es (Y.Z.); xandujar@mutuaterrassa.es (X.A.); dhorta@mutuaterrassa.cat (D.H.); cloras@mutuaterrassa.cat (C.L.); ffbanares@mutuaterrassa.es (F.F.-B.)
[2] Centro de Investigación Biomédica en Red de Enfermedades Hepáticas y Digestivas (CIBERehd), Instituto de Salud Carlos III, 28029 Madrid, Spain
[3] Department of Pathology, Hospital Universitari Mútua Terrassa, Universitat de Barcelona, 08221 Barcelona, Spain; carmeferrer@mutuaterrassa.es
[4] Department of Pediatrics, Hospital Universitari Mútua Terrassa, Universitat de Barcelona, 08221 Barcelona, Spain; rgarcia@mutuaterrassa.cat
* Correspondence: mariaesteve@mutuaterrassa.cat; Tel.: +34-937365050

Citation: Ruiz-Ramírez, P.; Carreras, G.; Fajardo, I.; Tristán, E.; Carrasco, A.; Salvador, I.; Zabana, Y.; Andújar, X.; Ferrer, C.; Horta, D.; et al. Intraepithelial Lymphocyte Cytometric Pattern Is a Useful Diagnostic Tool for Coeliac Disease Diagnosis Irrespective of Degree of Mucosal Damage and Age—A Validation Cohort. *Nutrients* **2021**, *13*, 1684. https://doi.org/10.3390/nu13051684

Academic Editors: Isabel Comino and Carolina Sousa

Received: 2 April 2021
Accepted: 12 May 2021
Published: 15 May 2021

Abstract: Introduction: The study of intraepithelial lymphocytes (IEL) by flow cytometry is a useful tool in the diagnosis of coeliac disease (CD). Previous data showed that an increase in %TCRγδ+ and decrease of %CD3− IEL constitute a typical CD cytometric pattern with a specificity of 100%. However, there are no data regarding whether there are differences in the %TCRγδ+ related to sex, age, titers of serology, and degree of histological lesion. Study aims: To confirm the high diagnostic accuracy of the coeliac cytometric patterns. To determine if there are differences between sex, age, serology titers, and histological lesion grade. Results: We selected all patients who fulfilled "4 of 5" rule for CD diagnosis ($n = 169$). There were no differences in %TCRγδ+ between sexes ($p = 0.909$), age groups ($p = 0.986$), serology titers ($p = 0.53$) and histological lesion grades ($p = 0.41$). The diagnostic accuracy of complete CD cytometric pattern was: specificity 100%, sensitivity 82%, PPV 100%, NPV 47%. Conclusion: We confirmed, in a validation cohort, the high diagnostic accuracy of complete CD pattern irrespective of sex, age, serology titers, and grade of mucosal lesion.

Keywords: coeliac disease; flow cytometry; age; sex; lesion grade; intraepithelial lymphocytes TCRγδ+

1. Introduction

The diagnosis of coeliac disease (CD) is based on several criteria including positive serology, a spectrum of duodenal damage and clinical symptoms and/or risk conditions, and response to a gluten-free diet (GFD) in patients bearing the HLA-DQ2 or DQ8 genotypes. When some of these criteria are lacking, especially when serology is negative or the duodenal atrophy is not complete, the CD diagnosis is a challenge [1]. In these difficult situations, the study of duodenal intraepithelial lymphocytes (IEL) by flow cytometry is a useful tool for CD diagnosis. It has been shown to be of value in the diagnosis of CD with atrophy [2–5] and refractory CD [6,7]. An increase in %CD3+ TCRγδ+ IEL (%TCRγδ+) with a decrease in %CD3− IEL (%CD3−) has been described as the typical pattern of CD [8].

The diagnosis of CD in the case of mild histological lesions (Marsh 1) can be difficult due to low sensitivity of serology and low specificity of the lymphocytic enteritis [9,10].

However, the diagnosis of Marsh 1 patients with CD is important because they present with similar clinical symptoms to patients with atrophy that reverse with a gluten-free diet (GFD) [11,12]. Previous ESPGHAN guidelines suggest that both an increase in %TCRγδ⁺ count assessed by immunohistochemical analysis of biopsies and the presence of IgA anti-tissue transglutaminase (anti-TG2) deposits increase the chances of a diagnosis of CD [7].

The increase of %TCRγδ⁺ has occasionally been found in some other conditions such as cow's milk intolerance, food allergy, cryptosporidiosis, Giardiasis, Sjögren syndrome, Olmesartan enteropathy, and IgA deficiency. Nevertheless, the increase in %TCRγδ⁺ in these diseases tends to be mild and transient [13]. CD is the only disease in which %TCRγδ⁺ has been found to be systematically and permanently increased, even in patients following a GFD. The concomitant decrease in %CD3⁻ provides increased specificity for the diagnosis [14]. Therefore, this particular cytometric pattern may be used to confirm the CD diagnosis in patients that had already started a GFD before the diagnosis confirmation.

A previous study by our group demonstrated good diagnostic accuracy (sensitivity 85%, specificity 100%, PPV 100%, and NPV 72%) for the typical CD cytometric pattern (increased %TCRγδ⁺ and decreased %CD3⁻) in the diagnosis of CD in patients with positive serology, both Marsh 1 and Marsh 3 [8]. However, these findings should be confirmed with a larger validation cohort.

Another important issue is learning whether the cut-off values established for %TCRγδ⁺ and %CD3⁻ reveal a cytometric CD pattern influenced by age, sex, and degree of histological lesion. In this sense, the information is very limited, but it has been suggested that γδ⁺ IEL decreases with age [6].

The aims of our study were to determine: (1) whether there are differences in the percentage of TCRγδ⁺ IEL in CD patients related to sex, age, degree of histological lesion, levels of serology; and (2) the diagnostic accuracy in a large validation cohort of the typical cytometric CD pattern and of the increase in %TCRγδ⁺ IEL, without the simultaneous decrease in %CD3⁻.

2. Materials and Methods

2.1. Patients and Controls

For the period of January 2013 to December 2019, we prospectively included all patients who fulfilled CD diagnostic criteria based on the rule of '4 of 5' proposed by Catassi and Fasano [1]: typical symptoms of CD, positivity of serum coeliac disease IgA class autoantibodies, HLA-DQ2 or DQ8 genotypes, coeliac enteropathy at the small intestinal biopsy, response to the GFD (at least 4 of 5 diagnostic criteria or 3 of 4 if the HLA Genotype is not performed). The control group consisted of patients referred to the gastroenterology department for endoscopic assessment including duodenal biopsy (histopathology and flow cytometry) because they had digestive symptoms or/and anemia. Digestive symptoms were defined by the chronic or intermittent presence of either diarrhea, dyspepsia, bloating, and/or abdominal pain. Controls were consecutively included based on the following criteria to rule out CD: (1) negative coeliac serology, (2) negative HLA-DQ2.5 and HLA-DQ8, and (3) normal duodenal biopsy. We excluded patients with intake of NSAIDs and Olmesartan, and patients with Crohn's disease, autoimmune disease-associated enteropathy, collagenous sprue associated with collagenous colitis, lymphocytic enteritis due to intestinal parasitosis or *Helicobacter pylori*, and selective IgA deficiency. All CD patients and controls were recorded in a prospective maintained registry.

We performed coeliac serology, HLA genotyping, and duodenal biopsy assessment for histopathology and lymphocyte subpopulations by flow cytometry in all patients and controls.

2.2. Coeliac Serology

Serum IgA-tissue transglutaminase antibody (anti-TG2) and IgA titers were analyzed in serum using a quantitative automated ELISA detection kit (Elia CelikeyTM, Phadia

AB, Freiburg, Germany) with recombinant human TG2 as antigen. A value of anti-TG2 $\geq$8 U/mL was established as the cut-off for normality [15]. Values between 2–8 U/mL were considered as a positive CD serology when titers higher than 1/40 of serum IgA anti-endomisal antibodies (EmA) were also found [16].

2.3. HLA Genotyping

We used a commercial reverse hybridization kit for the determination of CD heterodimers in the HLA genotyping (HLA-DQ2 [A1*0501/0505, B1*0201/*0202], HLA-DQ8 [A1*0301, B1*0301]). HLA-DQ2.5 haplotype is present in 24% of healthy controls and 90% of CD patients in our area [17]. In this study, we considered a positive coeliac genetic when the presence of HLA-DQ2.5, HLA-DQ8 or both was detected [18]. Considering the low frequency of the presence of HLA-DQ2.2 or only one allele of HLA-DQ2 haplotype in CD patients, either DQA1*05 or DQB1*02, the presence of these alleles was allowed in control individuals.

2.4. Duodenal Biopsy Assessment for Histopathology

Four endoscopic biopsies were taken from the second-third portion of the duodenum and one from the duodenal bulb, and these were processed using hematoxylin/eosin staining and CD3 immunophenotyping. Marsh 1 lesion (lymphocytic enteritis) was defined by 25 or more IEL per 100 epithelial nuclei along with normal villous architecture. Two endoscopic biopsies from antrum were also taken to investigate *Helicobacter pylori* infection in all patients. The lymphocyte count was performed as previously described [19,20]. Control group patients were separated into two subgroups according to the percentage of IEL ($\geq$ or < than 18%) since some authors have suggested that a lower cut-off point should be established to redefine lymphocytic enteritis [21].

2.5. Duodenal Biopsy Assessment by Flow Cytometry

We performed IEL flow cytometry in all patients and controls by taking a duodenal sample from the second-third portion of the duodenum. The sample was obtained using a 2.8 mm biopsy forceps (Radial Jaw 4, Boston Scientific®, Marlborough, MA, USA), and immediately processed as previously described [4,8,12].

Briefly, IELs were isolated by gentle rotation in an orbital shaker at 12 rpm for 90 min in a solution of 1 mM DTT and 1 mM EDTA in 10%FBS HBSS, at room temperature. After two washes with HBSS (10 min, 300 g) IEL mixture was immediately stained for 15 min with the antibody mix described in Table 1. Viability (>90%) was assessed by trypan blue exclusion in Neubauer chamber. IELs were acquired in a four-colour FACSCalibur and analyzed with the Cell-Quest Software (BD Biosciences). PMT voltages and compensation values were manually adjusted using single stained samples. Live IELs were gated on CD45 and low scatter basis, and intraepithelial origin was confirmed with CD103 staining. (>90%).

Table 1. Antibodies used for flow cytometry staining

Laser	Fluorochrome	Cell Marker	Antibody Clone	Supplier	Reference	Dilution
488	PerCP	CD3	SK7	BD [1]	345,766	2.5:100
	FITC	CD103	Ber-ACT8	BD	333,155	2.5:100
633	PE	TCR$\gamma\delta$	11F2	BD	333,141	2.5:100
	APC	CD45	2D1	BD	340,910	1.5:100

[1] BD: BD-Biosciences.

Four cytometric patterns were described using the TCR$\gamma\delta^+$ and CD3$^-$ IEL percentages: First cytometric pattern was defined by an increase of %TCR$\gamma\delta^+$ (>8.5%) and a decrease in %CD3$^-$ (<10%) and was labeled as a complete CD IEL flow cytometric pattern (complete FCP). A second cytometric pattern was defined by an isolated increase in %TCR$\gamma\delta^+$ and was labeled an incomplete CD IEL flow cytometric pattern (Incomplete FCP). The third and fourth patterns were defined as non-CD patterns: one of them was defined by an isolated decrease in %CD3$^-$ and the other, labeled normal cytometric pattern, was defined by a TCRγδ$^+$ ≤ 8.5% plus CD3$^-$ > 10%. This corresponds to the normal cut-off established in our laboratory [8,12]. Gating strategy and the four patterns are illustrated in Figure 1.

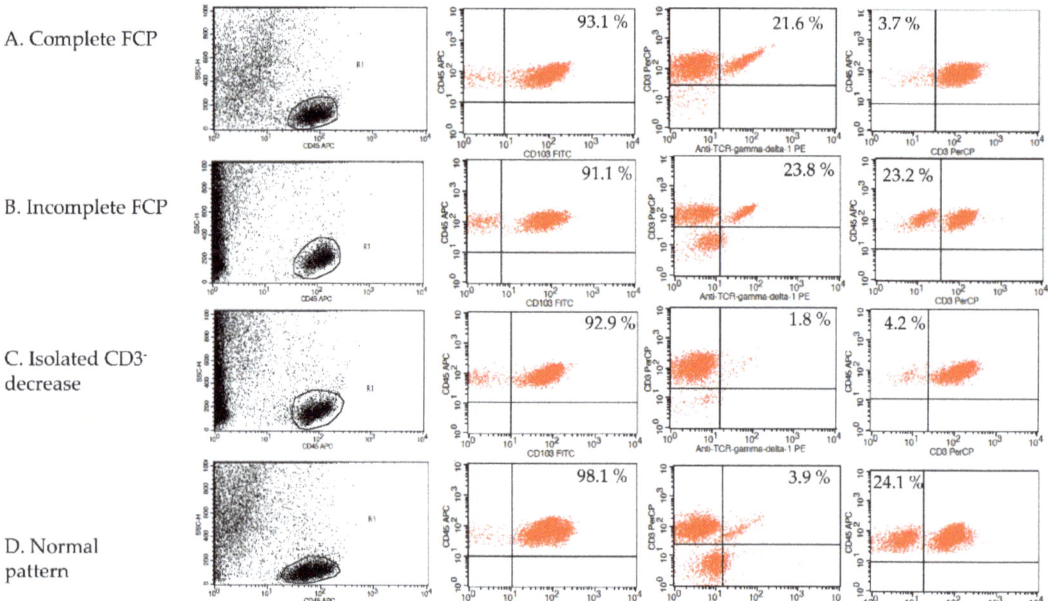

Figure 1. Gating strategy and the four patterns cytometric patterns. Complete and incomplete flow cytometric patterns (FCP) are CD related patterns.

2.6. Statistical Analysis

The results were expressed as mean ± SEM or median (Interquartile range, IQR) or as proportions (and their 95% confidence interval -CI- when appropriate). In order to assess the relationship between age and %TCRγδ$^+$ values, the age was classified in 7 groups (0–10 years, 11–20 years, 21–30 years, 31–40 years, 41–50 years, 51–60 years, ≥61 years). To compare %TCRγδ$^+$ related to anti-TG2 serum titers, three groups were stablished: patients with anti-TG2 titers ≥ 30 U/mL, between 8–30 U/mL and between 2–8 U/mL plus EmA higher than 1/40. We used a student *t* test or ANOVA test for comparing %TCRγδ$^+$ cells related to sex, degree of histological damage, and serology. The non-parametric counterpart (Kruskall–Wallis test) was used to compare the different groups of age because they do not follow a normal distribution assessed by a Kolmogorov–Smirnov test. In addition, we performed a Bonferroni test to assess differences among groups. Sensitivity, specificity, negative predictive value (NPV), and positive predictive value (PPV) for the complete CD pattern and the isolated increase in %TCRγδ$^+$ were calculated using 2 × 2 tables. Statistical analysis was performed using the SPSS for Windows statistical package (SPSS Inc., Chicago, IL, USA).

2.7. Ethical Statements

The study was conducted according to the guidelines of the Declaration of Helsinki. All participants (or their parents in the case of patients less than 16 years old) provided written informed consent. This study is part of a larger registry that prospectively collects all patients who need to be evaluated to rule out CD. This registry was approved by the Ethics Committee of the Hospital Universitari Mútua Terrassa at the start of the registry in 2010 (Code: EO/1011; date: 25 March 2010). Researchers guaranteed strict measures for preserving patient confidentiality.

3. Results

We included 169 patients who fulfilled CD diagnostic criteria (119 women; mean age 18.8 ± 1.5 years, range 1–83 years). One hundred forty-four patients showed villous atrophy (Marsh 3a type, $n = 21$; and 3b-c type, $n = 123$). Twenty-five patients showed architecturally normal small intestinal mucosa with an increase in IEL counts (Marsh type 1 lesion, mean age 36.00 ± 4.48 years, range 4–83 years).

In Table 2 and Figure 2, the percentages of TCR$\gamma\delta^+$ in groups of different degrees of histological lesion, sex, age, and anti-TG2 serum titers are shown. No differences were found relative to any of these variables.

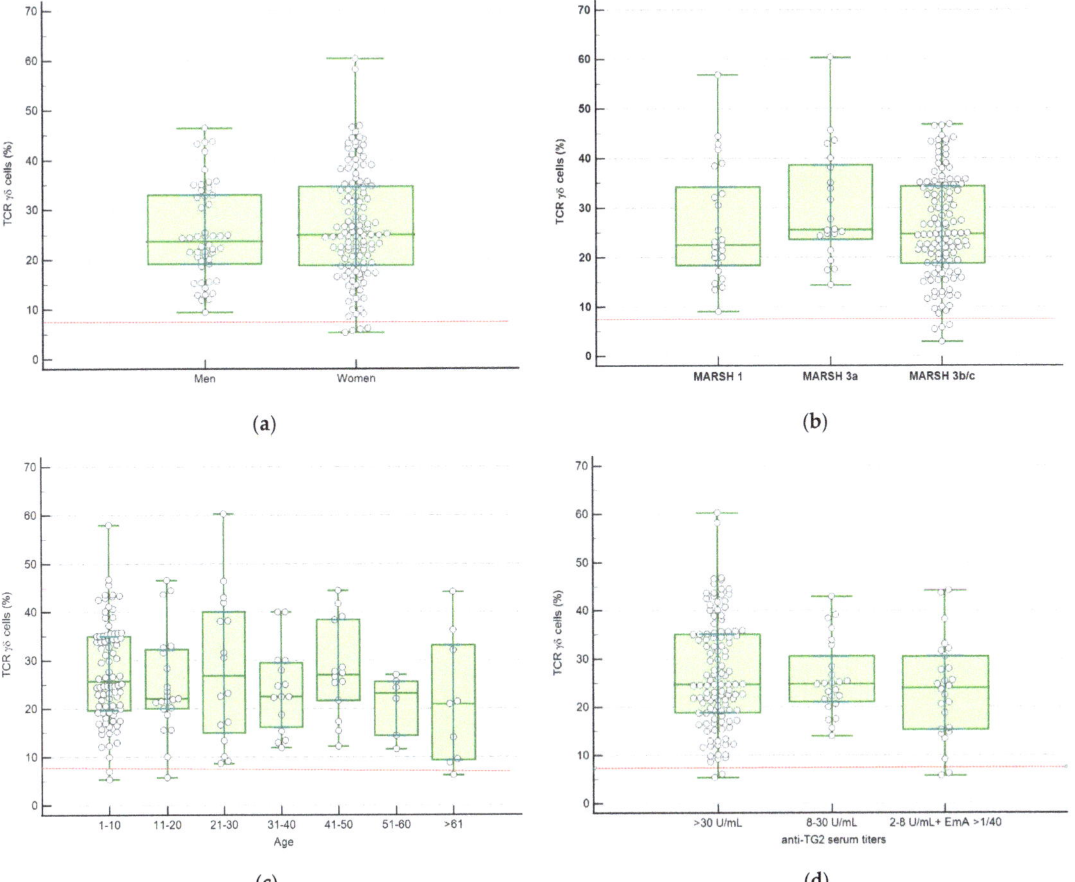

Figure 2. Scatter plot and box-whisker showing the distribution of patients according to sex (**a**), degree of histological lesion (**b**), age (**c**), and anti-TG2 serum titers (**d**). Box-plot rectangle spans the interquartile range, the segment inside the rectangle shows median whereas the whiskers above and below plot, the maximum and the minimum. The dotted red line represents the stablished TCR$\gamma\delta^+$ cut-off (>8.5%).

Table 2. Comparison of %TCR$\gamma\delta^+$ between different groups of sex, age, and degree of histological lesion

	Variable	Median %TCR$\gamma\delta^+$ (IQR)	p
Sex	Male ($n = 50$)	23.70 (18.08–34.00)	0.909
	Female ($n = 119$)	25.40 (18.78–35.31)	
Histology	Marsh 1 ($n = 25$)	22.51 (16.40–35.62)	0.41
	Marsh 3a ($n = 21$)	25.60 (22.85–39.13)	
	Marsh 3b-c ($n = 123$)	24.70 (18.73–34.48)	
Age	0–10 ($n = 86$)	25.03 (19.32–35.04)	0.79
	11–20 ($n = 23$)	22.13 (20.08–32.31)	
	21–30 ($n = 16$)	26.82 (14.98–40.07)	
	31–40 ($n = 15$)	22.53 (16.19–36.38)	
	41–50 ($n = 14$)	26.98 (21.69–38.44)	
	51–60 ($n = 6$)	23.17 (14.38–25.59)	
	$\geq$61 ($n = 9$)	21.47 (12.00–38.28)	
Serology	anti-TG2 $\geq$30 U/mL ($n = 119$)	24.75 (19.20–35.31)	0.53
	anti-TG2 8–30 U/mL ($n = 24$)	24.81 (20.90–33.60)	
	anti-TG2 2–8 U/mL + EmA > 1/40 ($n = 26$)	23.98 (15.18–31.70)	

The control group included 49 subjects (35 women; median age 40.00 (25.00–51.50) years, range 1–67 years). Median value of IEL% was 16.70 (11.50–20.00). Subjects in the control group with IEL count <18% ($n = 27$; 20 women, median age 46.00 (35.00–53.00) years, range 1–67 years) had a median %TCR$\gamma\delta^+$ of 3.36 (2.63–5.64), whereas controls with an IEL count ranging from 18 to 25% ($n = 22$) had a median %TCR$\gamma\delta^+$ of 3.53 (2.59–7.89). Clinical characteristics of the control group are detailed in Table 3.

In Table 4, the four different FCPs found in CD patients and controls are shown. In CD patients, these patterns are provided depending on the degree of histological damage and in controls taking into account whether they had a percentage of IEL <18% or <25%. The majority of patients in the control group had a normal cytometric pattern; only eight of them showed abnormal patterns. Three of them showed an incomplete FCP (isolated increase of %TCR$\gamma\delta^+$ (>8.5%) and the other five showed a selective decrease of %CD3$^-$ (non-coeliac pattern). It must be noted that all three patients with the incomplete FCP had an IEL count between 18–25% and there were no controls showing a complete CD pattern. Therefore, none of the controls with IEL count <18% had a coeliac related FCP.

Among CD patients with atrophy ($n = 144$), 83% had a complete FCP, whereas 13.8% ($n = 20$) had an incomplete FCP and 2.8% ($n = 4$) a normal pattern. A similar picture was found for Marsh 1 CD patients ($n = 25$), with 76% having a complete FCP (76%), 16% ($n = 4$) an incomplete FCP, and 8% ($n = 2$) a selective decrease in %CD3$^-$. Thus, more than 90% of CD patients irrespective of the degree of mucosal damage showed CD related FCP.

Sensitivity, specificity, NPV, and PPV of complete FCP and of the increase of %TCR$\gamma\delta^+$ were calculated considering both control subjects with IEL under 18% ($n = 27$) (Table 5) and all patients in the control group with IEL under 25% ($n = 49$) (Table 6). We found that complete FCP had an 82% sensitivity, 100% specificity, and 100% PPV irrespective of the criteria of IEL normality (below 18% or 25%). By contrast, the more restrictive criteria of IEL normality (<18%) should be adopted only if increased values of %TCR$\gamma\delta^+$ are used as a diagnostic tool, reaching in this case an accuracy close to that obtained with the complete FCP. The largest differences in diagnostic accuracy between the two coeliac FCPs, depending on what we consider normal duodenal mucosa (IEL count < 18% or <25%), were in the NPV. In this sense, the highest probability of not having a CD corresponded to individuals having an IEL count < 25% (non-restrictive criteria of normality) and not having an increased %TCR$\gamma\delta^+$ (NPV 88%).

Table 3. Clinical characteristics of the control group.

Age (years) *	40.00 (25.00–51.50)
Sex (% women)	71.4%
Clinical symptoms [1]	
Diarrhea	19 (36%)
Bloating	10 (20%)
Dyspepsia	10 (20%)
Abdominal pain	4 (8%)
Anaemia	4 (8%)
Autoimmune disease	4 (8%)
HLA Genotyping	
HLA-DQ2.2	16 (32%)
HLA-DQA1 * 05	14 (29%)
HLA-DQB1 * 02	9 (19%)
Without risk alleles	10 (20%)
IEL count (%) *	16.70 (11.50–20.00)
$CD3^+$ $TCR\gamma\delta^+$ IEL (%) *	3.40 (2.63–5.78)
$CD3^-$ IEL (%) *	21.03 (13.79–30.55)
Final diagnosis	
Irritable bowel syndrome	25 (51%)
Fructose malabsorption	8 (17%)
Gastroesophageal reflux disease	6 (12%)
Lactose malabsorption	3 (6%)
Non-coeliac gluten sensitivity	2 (4%)
Autoimmune pancreatitis	1 (2%)
Chronic pancreatitis and exocrine pancreatic insufficiency	1 (2%)
Factitious diarrhea	1 (2%)
Esophageal dysmotility due to systemic sclerosis	1 (2%)
Control biopsy after Helicobacter pylori eradication	1 (2%)

[1] If patients reported more than one symptom, the predominant one was selected. * Median (IQR).

Table 4. Cytometric patterns in CD patients and control patients.

	CD Patients n = 169			Controls (n = 49)	
	Marsh 1 (n= 25)	Marsh 3a (n= 21)	Marsh 3b-c (n= 123)	IEL < 18 (n = 27)	IEL < 25 (n = 49)
Complete FCP	19	19	101	0	0
Incomplete FCP: Isolated increase of %TCRγδ+ IEL	4	2	18	0	3
Isolated decrease of % CD3−	2	0	0	2	5
Increase of %TCRγδ+ IEL [1]	23	21	119	0	3
Normal pattern	0	0	4	25	41

FCP = Flow cytometric pattern. Complete coeliac FCP: $CD3^+$ $TCR\gamma\delta^+$ IEL > 8.5% and $CD3^-$ < 10%. Incomplete coeliac FCP: isolated increase of $CD3^+$ $TCR\gamma\delta^+$ IEL > 8.5%. [1] Total number of patients with increase in %TCRγδ+ (complete + incomplete FCP).

Table 5. Accuracy of coeliac cytometric pattern for the diagnosis of coeliac disease. Control group subjects with IEL count < 18% (*n* = 27).

	Sensitivity % (95% CI)	Specificity % (95% CI)	PPV % (95% CI)	NPV % (95% CI)
Complete FCP	82 (75–88)	100 (84–100)	100 (82–100)	47 (34–61)
Increase of %TCRγδ$^+$ IEL [1]	96 (92–98)	100 (84–100)	100 (97–100)	81 (64–92)

FCP = Flow cytometric pattern. Complete coeliac FCP: TCR CD3$^+$ γδ$^+$ IEL > 8.5% and CD3$^-$ < 10%. [1] Total number of patients with increase in %TCRγδ$^+$ (complete + incomplete FCP).

Table 6. Accuracy of the coeliac cytometric pattern for the diagnosis of coeliac disease. Control group subjects under 25% IEL (*n* = 49).

	Sensitivity % (95% CI)	Specificity % (95% CI)	PPV % (95% CI)	NPV % (95% CI)
Complete FCP	82 (75–88)	100 (91–100)	100 (97–100)	62 (50–73)
Increase of %TCRγδ$^+$ IEL [1]	96 (92–98)	93 (82–98)	98 (93–99)	88 (76–95)

FCP = Flow cytometric pattern. Complete coeliac FCP: TCR CD3$^+$ γδ$^+$ IEL > 8.5% and CD3$^-$ < 10%. [1] Total number of patients with increase in %TCRγδ$^+$ (complete + incomplete FCP).

4. Discussion

The complete IEL cytometric pattern of CD, characterized by an increase of %TCRγδ$^+$ and a concomitant decrease in %CD3$^-$, has been proposed as a in complementary diagnostic tool to reinforce CD diagnosis in doubtful cases, especially when serology is negative [8,22]. This situation may occur in 30% of patients with atrophy and in more than 70% of patients with lymphocytic enteritis or Marsh type 1 CD [23,24].

The most frequent etiology of seronegative duodenal atrophy in Western countries is CD and the percentage increases in patients with positive HLA-DQ2/DQ8 [25]. The diagnosis of CD in cases of lymphocytic enteritis (Marsh 1 lesion) is more challenging since the lesion is much more unspecific than atrophy and other possible etiologies have been proposed [25,26]. As mentioned, only a small percentage of these patients will show a positive coeliac serology and only some patients will progress to villous atrophy after a gluten challenge of eight weeks [27].

The CD diagnosis in seronegative patients is based on the clinical and histological response to a GFD in patients with signs and symptoms of the coeliac spectrum in the presence of a positive HLA-DQ2/-DQ8. This means that the diagnosis of CD is time-consuming and remains uncertain until the effect of a GFD is assessed. In addition, gluten challenge is not well accepted by patients due to the discomfort caused. Nevertheless, it must be considered that this evaluation is sometimes difficult because CD clinical symptoms are quite unspecific and lymphocytic enteritis in non-CD patients may resolve spontaneously [19].

In the present validation cohort, we have confirmed that assessment of the complete FCP is a useful diagnostic tool for CD diagnosis, with a high diagnostic accuracy (82% sensitivity and 100% specificity). In addition, TCRγδ$^+$ IEL subpopulation, which is the main parameter of coeliac lymphogram, is not influenced by age, sex, or the degree of histological damage. Hence, the IEL study through flow cytometry for CD diagnosis can be applied in any situation regardless of the clinical characteristics of the patient. This study also confirms that the normality cut-off previously established for %TCRγδ$^+$ [8] is appropriate in patients bearing the complete coeliac FCP, including elderly patients. However, taking into account that the number of CD patients and controls older than 61 was very small, information focused on this population group is awaited.

Our study was performed in patients with positive serology, to ensure the diagnosis of CD, but it is conceivable that the characteristic behavior of duodenal intraepithelial subpopulations is also maintained in patients with negative serology. In fact, %TCRγδ$^+$ values are not influenced by the levels of serum anti-TG2 titters. Moreover, the results of other studies by our group, showing very high response rates to a GFD in patients with enteropathy of the CD spectrum, negative serology, and coeliac cytometric pattern, lend support to this hypothesis [12,28].

A limitation of studies assessing diagnostic tools in CD is selection of the control group, and this feature of our study merits special mention. The ideal controls should be individuals of the general population who are completely healthy, without digestive symptoms and with negative genetic predisposition and serology. To our knowledge, a study with this type of 'perfect' control group has never been performed. In fact, the cut-off of 18 IEL considered 'normal' in the duodenal mucosa was established in subjects in whom the duodenal mucosa was microscopically assessed due to digestive symptoms [21]. The CD was ruled out with negative serology and negative HLA-DQ2/DQ8. In our study, we also used the same criteria for control group recruitment, excluding all the individuals with a positive HLA-DQ2.5 and DQ8. The recruitment of these controls was consequently very slow because the percentage of individuals in the general population having either HLA-DQ2 or DQ8 exceeds 60% in our area [17], but this makes the diagnosis of CD almost impossible.

Eight subjects in the whole control group had an abnormal FCP. Three of them had an incomplete FCP and the remaining 5 a selective decrease of %CD3$^-$. By contrast, none of the controls with IEL count < 18% had a coeliac FCP and only two of them had a selective decrease of %CD3$^-$, highlighting how this value should be considered the normal cut-off for histopathological analysis. Consequently, we noted a slight decrease in the diagnostic accuracy when we considered the sub-group that presented an IEL count between 18–25% as controls. All these findings are objective data to redefine the cut-off point <18% for considering duodenal mucosa as normal. Also, it is demonstrated that complete FCP is more accurate than an incomplete CD pattern.

In conclusion, the established normality cut-off for %TCR$\gamma\delta^+$ (>8.5%) in IEL flow cytometry study for diagnosis of CD is valid for all age, sex, and histological lesion grade groups. Moreover, we have confirmed the high diagnostic accuracy of the increase in %TCR$\gamma\delta^+$ and the complete FCP for CD diagnosis in a large validation cohort.

Supplementary Materials: The following are available online at https://www.mdpi.com/article/10.3390/nu13051684/s1.

Author Contributions: Guarantor of the article: M.E. Conceptualization, F.F.-B. and M.E.; Data curation, all authors; Formal analysis, P.R.-R., F.F.-B., and M.E.; Investigation, all authors; Methodology, E.T., A.C., C.F., F.F.-B., and M.E.; Project administration, M.E.; Resources, all authors; Supervision, M.E. and F.F.-B.; Validation, M.E.; Visualization, M.E.; Writing—original draft, P.R.-R. and M.E.; Writing—review and editing, all authors. All authors have read and agreed to the published version of the manuscript.

Funding: This research received no external funding.

Institutional Review Board Statement: The following statement is included on page 4 of the manuscript "The study was conducted according to the guidelines of the Declaration of Helsinki and was approved by the Ethics Committee of the Hospital Universitari Mutua Terrassa". Code: EO/1011; date: 25 March 2010.

Informed Consent Statement: Informed consent was obtained from all subjects involved in the study or their parents as noted in the manuscript.

Data Availability Statement: File1.sav, database of the study (supplementary file).

Acknowledgments: Centro de Investigación Biomédica en Red de Enfermedades Hepáticas y Digestivas' (CIBERehd) is an initiative of the Instituto de Salud Carlos III, Madrid, Spain. This institution played no role in the study design, or the acquisition, analysis or interpretation of the data or the report writing.

Conflicts of Interest: The authors declare no conflict of interest.

Writing Assistance: Tom Yohannan.

References

1. Catassi, C.; Fasano, A. Celiac disease diagnosis: Simple rules are better than complicated algorithms. *Am. J. Med.* **2010**, *123*, 691–693. [CrossRef]
2. Eiras, P.; Roldan, E.; Camarero, C.; Olivares, F.; Bootello, A.; Roy, G. Flow cytometry description of a novel CD3−/CD7+ intraepithelial lymphocyte subset in human duodenal biopsies: Potential diagnostic value in celiac disease. *Cytometry* **1998**, *34*, 95–102. [CrossRef]
3. Camarero, C.; Eiras, P.; Leon, F.; Olivares, F.; Escobar, H.; Roy, G. Intraephitelial lymphocytes and celiac disease: Permanent changes in CD3-/CD7+ and T cell γδ receptor subsets studied by flow cytometry. *Acta Paediatr.* **2000**, *89*, 285–290.
4. León, F. Flow cytometry of intestinal intraepithelial lymphocytes in celiac disease. *J. Immunol. Methods* **2011**, *363*, 177–186. [CrossRef]
5. Calleja, S.; Vivas, S.; Santiuste, M.; Arias, L.; Hernando, M.; Nistal, E.; De Morales JG, R. Dynamics of non-conventional intraepithelial lymphocytes-NK, NKT, and γδ T- in celiac disease: Relationship with age, diet, and histopathology. *Dig. Dis. Sci.* **2011**, *56*, 2042–2049. [CrossRef] [PubMed]
6. Järvinen, T.T.; Kaukinen, K.; Laurila, K.; Kyrönpalo, S.; Rasmussen, M.; Mäki, M.; Korhonen, H.; Reunala, T.; Collin, P. Intraepithelial lymphocytes in celiac disease. *Am. J. Gastroenterol.* **2003**, *98*, 1332–1337. [CrossRef] [PubMed]
7. Husby, S.; Koletzko, S.; Korponay-Szabó, I.; Mearin, M.; Phillips, A.; Shamir, R.; Troncone, R.; Giersiepen, K.; Branski, D.; Catassi, C.; et al. European Society for Pediatric Gastroenterology, Hepatology, and Nutrition guidelines for the diagnosis of coeliac disease. *J. Pediatr. Gastroenterol. Nutr.* **2012**, *54*, 136–160. [CrossRef] [PubMed]
8. Fernández-Bañares, F.; Carrasco, A.; García-Puig, R.; Rosinach, M.; González, C.; Alsina, M.; Loras, C.; Salas, A.; Viver, J.M.; Esteve, M. Intestinal Intraepithelial Cytometric pattern is more accurate than subepithelial deposits of anti-tissue transglutaminase Ig A for the diagnosis of Celiac disease in lymphocytic enteritis. *PLoS ONE* **2014**, *9*, e101249. [CrossRef]
9. Vivas, S.; De Morales, J.M.R.; Fernandez, M.; Hernando, M.; Herrero, B.; Casqueiro, J.; Gutierrez, S. Age-related clinical, serological, and histopathological features of celiac disease. *Am. J. Gastroenterol.* **2008**, *103*, 2360–2365. [CrossRef]
10. Dickey, W.; Hughes, D.F.; McMillan, S.A. Disappearance of endomysial antibodies in treated celiac disease does not indicate histological recovery. *Am. J. Gastroenterol.* **2000**, *95*, 712–714. [CrossRef]
11. Esteve, M.; Rosinach, M.; Fernández-Bañares, F.; Farré, C.; Salas, A.; Alsina, M.; Vilar, P.; Abad-Lacruz, A.; Forné, M.; Mariné, M.; et al. Spectrum of gluten-sensitive enteropathy in first-degree relatives of patients with coeliac disease: Clinical relevance of lymphocytic enteritis. *Gut* **2006**, *55*, 1739–1745. [CrossRef] [PubMed]
12. Rosinach, M.; Fernández-Bañares, F.; Carrasco, A.; Ibarra, M.; Temiño, R.; Salas, A.; Esteve, M. Double-blind randomized clinical trial: Gluten versus placebo rechallenge in patients with lymphocytic enteritis and suspected celiac disease. *PLoS ONE* **2016**, *11*, e0157879. [CrossRef]
13. Kokkonen, J.; Holm, K.; Kartunnen, T.J.; Maki, M. Children with untreated food allergy express a relative increment in the density of duodenal gammadelta+ T cells. *Scand. J. Gastroenterol.* **2000**, *35*, 1137. [PubMed]
14. Palomar, P.O.; Ruiz, A.C.; Scapa MA, M.; Prieto, F.L.; Ariño, G.R.; Verge, C.R. Adult celiac disease and intraepithelial lymphocytes. New options for diagnosis? *Gastroenterol. Hepatol.* **2008**, *31*, 555–559.
15. Mariné, M.; Farre, C.; Alsina, M.; Vilar, P.; Cortijo, M.; Salas, A.; Fernández-Bañares, F.; Rosinach, M.; Santaolalla, R.; Loras, C.; et al. The prevalence of coeliac disease is significantly higher in children compared with adults. *Aliment. Pharmacol. Ther.* **2011**, *33*, 477–486. [CrossRef] [PubMed]
16. Chorzelski, T.P.; Beutner, E.H.; Sulej, J.; Tchorzewska, H.; Jablonska, S.; Kumar, V.; Kapuscinska, A. Ig A anti-endomysium antibody. A new immunological marker of dermatitis herpetiformis and coeliac disease. *Br. J. Dermatol.* **1984**, *111*, 395–405. [CrossRef]
17. Farré, C. Enfermedad Celíaca: Marcadores Serológicos y de Predisposición Genética, Aspectos Clínicos y Poblaciones de Riesgo. Ph.D. Thesis, University of Barcelona, Barcelona, Spain, 2002.
18. Megiorni, F.; Pizzuti, A. HLA-DQA1 and HLA-DQB1 in Celiac disease predisposition: Practical implications of the HLA molecular typing. *J. Biomed. Sci.* **2012**, *19*, 88. [CrossRef]
19. Rosinach, M.; Esteve, M.; González, C.; Temiño, R.; Mariné, M.; Monzón, H.; Sainz, E.; Loras, C.; Espinos, J.C.; Forné, M.; et al. Lymphocytic duodenosis: Aetiology and long-term response to specific treatment. *Dig. Liver Dis.* **2012**, *44*, 643–648. [CrossRef]
20. Walker, M.M.; Murray, J.A. An update in the diagnosis of coeliac disease. *Histopathology* **2011**, *59*, 166–179. [CrossRef] [PubMed]
21. Pellegrino, S.; Villanacci, V.; Sansotta, N.; Scarfi, R.; Bassotti, G.; Vieni, G.; Princiotta, A.; Sferlazzas, C.; Magazzù, G.; Tuccari, G. Redefining the intraepithelial lymphocytes threshold to diagnose gluten sensitivity in patients with architecturally normal duodenal histology. *Aliment. Pharmacol. Ther.* **2011**, *33*, 697–706. [CrossRef]
22. Fernández-Bañares, F.; Carrasco, A.; Rosinach, M.; Arau, B.; García-Puig, R.; González, C.; Tristán, E.; Zabana, Y.; Esteve, M. A scoring system for identifying patients likely to be diagnosed with low-grade coeliac enteropathy. *Nutrients* **2019**, *11*, 1050. [CrossRef]
23. Santaolalla, R.; Fernández-Bañares, F.; Rodriguez, R.; Alsina, M.; Rosinach, M.; Marine, M.; Farre, C.; Salas, A.; Forné, M.; Loras, C.; et al. Diagnostic value of duodenal antitissue transglutaminase antibodies in gluten-sensitive enteropathy. *Aliment. Pharmacol. Ther.* **2008**, *27*, 820–829. [CrossRef] [PubMed]

24. Rostami, K.; Kerckhaert, J.; Tiemessen, R.; Von Blomberg, B.M.E.; Meijer, J.W.; Mulder, C.J. Sensitivity of antiendomysium and antigliadin antibodies in untreated celiac disease: Disappointing in clinical practice. *Am. J. Gastroenterol.* **1999**, *94*, 888–894. [CrossRef] [PubMed]
25. Aziz, I.; Evans, K.E.; Hooper, A.D.; Smillie, D.M.; Sanders, D.S. A prospective study into the aetiology of lymphocytic duodenosis. *Aliment. Pharmacol. Ther.* **2010**, *32*, 1392–1397. [CrossRef] [PubMed]
26. Kakar, S.; Nehra, V.; Murray, J.A.; Dayharsh, J.A.; Burgart, R.J. Significance of intraepithelial lymphocytosis in small bowel biopsy samples with normal mucosal architecture. *Am. J. Gastroenterol.* **2003**, *98*, 2027–2033. [CrossRef] [PubMed]
27. Wahab, P.J.; Crusius, B.A.; Meijer, J.W.R.; Mulder, C.J. Gluten challenge in borderline gluten-sensitive enteropathy. *Am. J. Gastroenterol.* **2001**, *96*, 1464–1469. [CrossRef]
28. Fernández-Bañares, F.; Crespo, L.; Núñez, C.; López-Palacios, N.; Tristán, E.; Vivas, S.; Farrais, S.; Arau, B.; Vidal, J.; Roy, G.; et al. Gamma delta[+] intraepithelial lymphocytes and coeliac lymphogram in a diagnostic approach to coeliac disease in patients with seronegative villous atrophy. *Aliment. Pharmacol. Ther.* **2020**, *51*, 699–705. [CrossRef] [PubMed]

Article

First-degree Relatives of Celiac Disease Patients Have Increased Seroreactivity to Serum Microbial Markers

Liisa Viitasalo [1,2], Sari Iltanen [1,3], Heini Huhtala [4], Päivi Saavalainen [5], Katri Kaukinen [6,7], Katri Lindfors [6] and Kalle Kurppa [1,8,*]

[1]   Centre for Child Health Research and Department of Pediatrics, Tampere University Hospital, FI-33521 Tampere, Finland; liisa.viitasalo@tuni.fi (L.V.); sari.iltanen@lshp.fi (S.I.)
[2]   Laboratory of Genetics, HUS Diagnostic Center, Helsinki University Hospital, FI-00029 Helsinki, Finland
[3]   Lapland Central Hospital, FI-96101 Rovaniemi, Finland
[4]   Faculty of Social Sciences, Tampere University, FI-33520 Tampere, Finland; heini.huhtala@tuni.fi
[5]   Research Program Unit, Immunobiology, and Department of Medical and Clinical Genetics, University of Helsinki, FI-00014 Helsinki, Finland; paivi.saavalainen@helsinki.fi
[6]   Celiac Disease Research Centre, Faculty of Medicine and Health Technology, Tampere University, FI-33520 Tampere, Finland; katri.kaukinen@tuni.fi (K.K.); katri.lindfors@tuni.fi (K.L.)
[7]   Department of Internal Medicine, Tampere University Hospital, FI-33521 Tampere, Finland
[8]   Department of Pediatrics, Seinäjoki University Hospital and the University Consortium of Seinäjoki, FI-60320 Seinäjoki, Finland
*   Correspondence: kalle.kurppa@tuni.fi

Received: 23 March 2020; Accepted: 8 April 2020; Published: 13 April 2020

Abstract: Risk of celiac disease (CD) is increased in relatives of CD patients due to genetic and possible environmental factors. We recently reported increased seropositivity to anti-*Saccharomyces cerevisiae* (ASCA), *Pseudomonas fluorescens*-associated sequence (anti-I2) and *Bacteroides caccae* TonB-linked outer membrane protein (anti-OmpW) antibodies in CD. We hypothesized these markers also to be overrepresented in relatives. Seropositivity and levels of ASCA, anti-I2 and anti-OmpW were compared between 463 first-degree relatives, 58 untreated and 55 treated CD patients, and 80 controls. CD-associated human leukocyte antigen (HLA)-haplotypes and transglutaminase (tTGab) and endomysium (EmA) antibodies were determined. One or more of the microbial antibodies was present in 75% of relatives, 97% of untreated and 87% of treated CD patients and 44% of the controls. The relatives had higher median ASCA IgA (9.13 vs. 4.50 U/mL, $p < 0.001$), ASCA IgG (8.91 vs. 5.75 U/mL, $p < 0.001$) and anti-I2 (absorbance 0.74 vs. 0.32, $p < 0.001$) levels than controls. There was a weak, positive correlation between tTGab and ASCA (r = 0.31, $p < 0.001$). Seropositivity was not significantly associated with HLA. To conclude, seropositivity to microbial markers was more common and ASCA and anti-I2 levels higher in relatives of CD patients than controls. These findings were not associated with HLA, suggesting the role of other genetic and environmental factors.

Keywords: celiac disease; relatives; microbiota; *Saccharomyces cerevisiae*; *Pseudomonas fluorescens*; *Bacteroides caccae*

---

1. Introduction

Celiac disease (CD) is an immune-mediated condition characterized by gluten-induced small-bowel enteropathy. Almost all patients carry human leukocyte antigen (HLA) alleles encoding DQ2 or DQ8 heterodimers [1]. These alleles are nevertheless also present in up to 35% of the general population and do not fully explain the genetic risk [2]. Recent genome-wide association studies and immunogenetic studies have identified numerous non-HLA loci and single nucleotide polymorphisms that may modify CD risk [3,4]. Partly due to shared genetic predisposition, the relatives of patients

43

have an increased susceptibility to CD, the average prevalence among first-degree relatives being approximately 8% [5] compared with 1%–2% in the general population [6,7].

However, only a minority of at-risk individuals develop CD, and the concordance even varies between identical twins [8,9], which implicates environmental factors. The prevalence may also vary between adjacent countries with similar genetic backgrounds and gluten consumption [10], and retrospective measurements of stored samples indicate a rise in the true incidence [6,11,12]. As one potentially associated factor, the role of intestinal microbiota in the development of CD has aroused particular interest [13–15]. Previously, we and others observed elevated levels of antibodies to microbial markers *Saccharomyces cerevisiae* (ASCA), *Pseudomonas fluorescens*-associated sequence (anti-I2) and *Bacteroides caccae* TonB-linked outer membrane protein (anti-OmpW) in inflammatory bowel disease [16–18]. We have shown increased seroreactivity to these markers also in overt CD [19] and a decrease of the antibody levels during gluten-free diet (GFD) [20]. Further, these microbial markers are detectable in early stages of the disease even before the presence of villous atrophy and serum CD-specific autoantibodies [21].

We hypothesized that close relatives of CD patients, with partially shared living environments and genetic factors, could have increased seroreactivity to microbial markers. This was investigated by comparing their frequency of seropositivity and levels of microbial antibodies with those in untreated and treated CD patients and in healthy controls.

2. Materials and Methods

2.1. Study Participants

The study was carried out at Tampere University and Tampere University Hospital. Previously diagnosed CD patients were recruited in a nationwide search through newspaper advertisements and via patient societies. Their medical records were obtained with permission, and only subjects with a biopsy-proven diagnosis were included. Relatives of these patients were invited to a screening study comprising personal interviews and measurement of CD serology. Additional blood samples were drawn for research purposes. Exclusion criteria for the relatives were previously diagnosed CD or dermatitis herpetiformis, or otherwise initiated gluten-free diet (GFD). Altogether, 3031 relatives met the inclusion criteria and entered the original screening study. Duodenal biopsy was offered for all relatives with positive CD serology. For the present study, serum samples from 463 first-degree relatives were randomly selected for the measurement of ASCA, anti-I2 and anti-OmpW. The CD control group comprised 58 biopsy-proven patients who underwent measurements of the CD serology and microbial markers at diagnosis and after one year on GFD ($n = 55$). In addition, 80 adult blood donors with negative CD serology served as non-CD controls.

2.2. CD Autoantibodies and Genotyping

Serum immunoglobulin A (IgA) class endomysium autoantibodies (EmA) were tested by an indirect immunofluorescence method using human umbilical cord as substrate [22]. Titers 1: ≥ 5 were deemed positive and diluted up to 1:4000 or until negative. Serum IgA class tissue transglutaminase autoantibodies (tTGab) were measured by an enzyme-linked immunosorbent assay (ELISA, INOVA diagnostics, San Diego, CA) according to the manufacturer's instructions. A cutoff ≥ 30 U/mL was applied for seropositivity. Some of the CD autoantibody-positive relatives declined the biopsy, but, due to the high specificity of EmA/tTGab [23], the vast majority of them are also likely to have CD. They were therefore analyzed as a separate group.

The CD-associated HLA DQ haplotypes (DQ2.5, DQ2.2, DQ8) were determined from the relatives and CD patients with the tagging single nucleotide polymorphism method or with the Olerup SSP DQ low-resolution kit (Olerup SSP AB, Stockholm, Sweden) as described elsewhere [24,25].

2.3. Microbial Antibodies

Serum IgA and IgG class ASCA were measured by a commercial ELISA (Quanta Lite ASCA, INOVA Diagnostics Inc., San Diego, CA) considering levels $\geq$ 25 U/mL positive. *E. coli* XL-1 blue and *E. coli* BL-21 (Stratagene, La Jolla, CA) strains and previously reported antigen purification techniques [26,27] were used to produce I2-GST and OmpW antigens. The serum samples were diluted 1:50, and IgA anti-I2 and anti-OmpW antibodies were measured with an in-house ELISA. For anti-I2, the cutoff level for positivity was set at absorbance 0.5. For anti-OmpW, it was set at 0.6 in children and 1.0 in adults based on our previous studies showing age differences in the normal range [16,19].

2.4. Statistical Analysis

Quantitative data are shown in tables as percentages or as medians with lower and upper quartiles. The data were cross-tabulated in order to ascertain the overlap of seropositivity for microbial antibodies in different study groups. The Kruskal–Wallis test was used to compare the differences in microbial antibody levels between the groups. Correlations between autoantibodies and microbial markers were tested with Spearman's rank correlation coefficient. Associations in the seropositivity to microbial antibodies within and between the families were also tested. The chi-square statistic for the change in the -2 log-likelihood from the constant only model to the model with "family" was used to determine whether the inclusion of "family" contributed significantly to model fit. A p value < 0.05 was considered significant. Statistical analyses were carried out with SPSS Statistics for Windows (IBM Corp., Armonk, NY, USA).

2.5. Ethical Aspects

The study protocol was approved by the Ethics Committee of the Pirkanmaa Hospital District, study identification code ETL R05183. All participants or, in the case of children, their legal guardians gave written informed consent. The paper follows the rules of the Declaration of Helsinki.

3. Results

The gender distribution was fairly equal among the relatives, whereas a majority of CD patients were women, and there were more men in the non-CD control group (Table 1). There were no major differences in the median ages between the groups (Table 1), but 49 (10.6%) of the relatives were <18 years of age, while the other groups comprised only adults.

Table 1. Demographic data on relatives of celiac disease (CD) patients, CD patients and non-celiac controls.

	Seropositive Relatives	Seronegative Relatives *	CD at Diagnosis	CD on GFD	Non-CD Controls
	$n = 49$	$n = 414$	$n = 58$	$n = 55$	$n = 80$
Females, %	42.9	57.2	77.6	76.4	35.0
Age, median (quartiles), y	41 (31–54)	42 (28–59)	45 (36–59)	46 (38-60)	41 (31–56)

* Negative serum endomysium (titer 1: < 5) and tissue transglutaminase (< 30 U/mL) antibodies. GFD, gluten-free diet.

The relatives were divided into CD autoantibody-negative ($n = 414$) and autoantibody-positive ($n = 49$) groups and were analyzed separately (Table 1). Among the autoantibody-negative relatives, seropositivity for at least one of the microbial markers was more common than in the non-CD controls but less frequent than in the CD patients (Figure 1). The most notable difference was seen in ASCA, as 19% of the relatives without CD-autoantibodies and none of the controls were seropositive for ASCA IgA, ASCA IgG, or both. In addition, anti-I2 and anti-OmpW positivity was more common among the autoantibody-negative relatives than controls (61% and 40% vs. 31% and 24%, respectively; Figure 1).

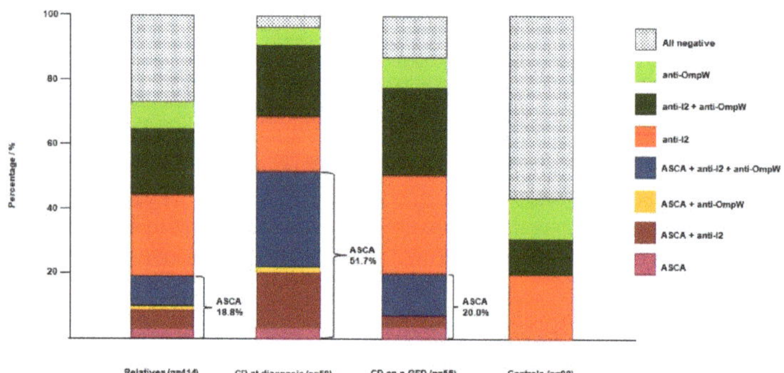

Figure 1. Distribution of seropositivity to antibodies against *Saccharomyces cerevisiae* (ASCA), *Pseudomonas fluorescens*-associated sequence (anti-I2 antibodies) and *Bacteroides caccae* TonB-linked outer membrane protein (anti-OmpW antibodies) among autoantibody-negative relatives of celiac disease (CD) patients, CD patients (at diagnosis and on a GFD) and controls.

The median levels of ASCA IgA, ASCA IgG and anti-I2 were also significantly higher in the autoantibody-negative relatives than those in the control group (Figure 2a–c), whereas anti-OmpW was higher only in untreated and treated CD patients (Figure 2d). ASCA IgG was higher in both untreated and treated CD patients and anti-I2/OmpW in untreated patients when compared with autoantibody-negative relatives (Figure 2b–d).

Altogether, 46 out of the 49 autoantibody-positive relatives had HLA-DQ2 haplotype, DQ8 haplotype, or both. As many as 86% of them showed seroreactivity to at least one microbial marker compared to 73% of the CD antibody-negative relatives, and the median levels of the microbial antibodies were also higher (ASCA IgA 11.1 vs. 8.90 U/mL, $p = 0.019$; ASCA IgG 12.8 vs. 8.37 U/mL, $p = 0.001$; absorbance for anti-I2 0.93 vs. 0.71, $p = 0.320$ and for anti-OmpW 1.00 vs. 0.81, $p = 0.022$, respectively). In contrast to the autoantibody-negative group, anti-OmpW levels were also significantly higher than in the controls (absorbance 0.79, $p = 0.043$).

Adjusting for age and gender or exclusion of children from the comparisons did not affect the results of the prevalence of seropositivity nor median levels of the microbial markers, although the medians were significantly lower in children than in adults (ASCA IgA 6.30 vs. 9.64 U/mL, $p < 0.001$; ASCA IgG 7.13 vs. 9.18 U/mL, $p = 0.070$; absorbance for anti-I2 0.34 vs. 0.79, $p < 0.001$ and for anti-OmpW 0.54 vs. 0.87, $p < 0.001$, respectively).

Seropositivity to anti-I2 and anti-OmpW was significantly more frequent between relatives in the same family than between different families ($p < 0.001$ for anti-I2 and $p = 0.001$ for anti-OmpW, respectively). In ASCA, this was observed only when autoantibody-positive relatives were also included in the analysis ($p = 0.007$).

There were no significant differences in the distribution of seropositivity across microbial markers when the relatives were categorized according to their HLA haplotypes (Table 2).

There was a weak, positive correlation between the values of tTGab and ASCA IgA ($r = 0.31$, $p < 0.001$), whereas correlation coefficients between the other microbial markers and tTGab or EmA were <0.3.

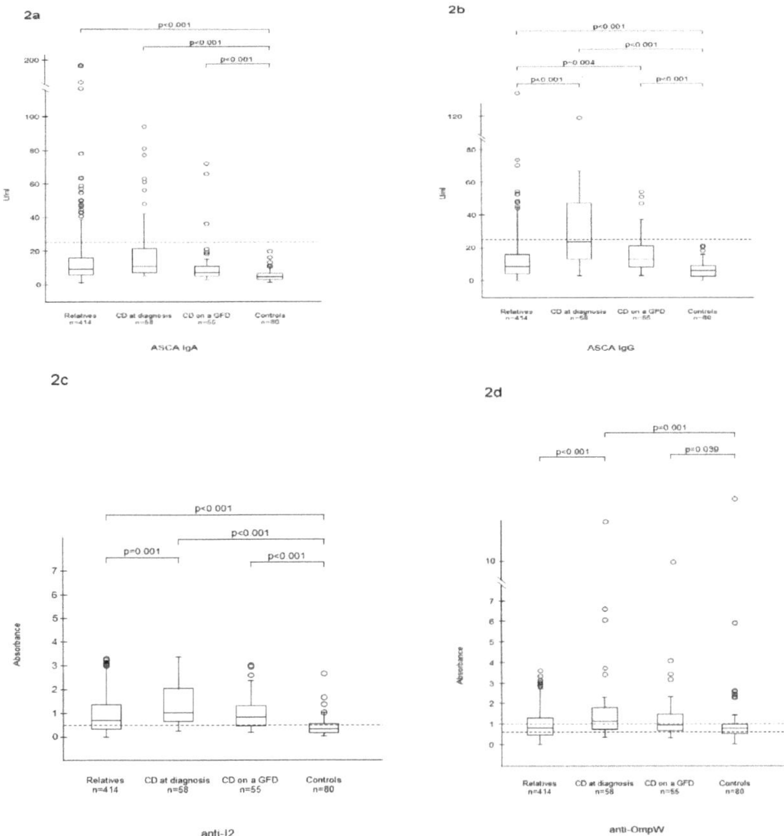

Figure 2. Serum levels of antibodies to *Saccharomyces cerevisiae* (ASCA) in IgA (**a**) and IgG (**b**) classes, *Pseudomonas fluorescens*-associated sequence (anti-I2) (**c**) and *Bacteroides caccae* TonB-linked outer membrane protein (anti-OmpW) (**d**) in autoantibody-negative relatives. Horizontal lines indicate the cutoff level for seropositivity of each antibody.

Table 2. Frequency of seropositivity to microbial markers in autoantibody-negative relatives of celiac disease patients with different human leukocyte antigen (HLA) haplotypes.

	DQ2 $n = 233$	DQ8 $n = 67$	DQ2 + DQ8 $n = 8$	DQ2/8 Negative $n = 89$
	%	%	%	%
ASCA IgA	11.2	10.4	12.5	10.1
ASCA IgG	12.9	13.4	0	14.6
Anti-I2	58.4	61.2	75.0	66.3
Anti-OmpW	39.5	35.8	25.0	43.8

ASCA, Anti-*Saccharomyces cerevisiae* antibodies; anti-I2, antibodies to *Pseudomonas fluorescens*-associated sequence; anti-OmpW, antibodies to *Bacteroides Caccae* TonB-linked outer membrane protein; DQ2, HLA-DQA1*05-DQB1*02 (DQ2.5) or HLA-DQA1*02-DQB1*02 (DQ2.2); DQ8, HLA-DQA1*03-DQB1*0302. There were no statistically significant differences between the groups in the distribution of seropositivity.

4. Discussion

The main finding of the present study was increased seroreactivity to microbial markers in the relatives of CD patients compared with controls even after the exclusion of CD autoantibody-positive individuals. This was observed particularly with ASCA and anti-I2, the median levels of which were also significantly higher than levels in the controls, although they were lower than in CD patients. To the best of our knowledge, the only study to report on this issue so far was a conducted by Da Silva et al., who investigated seropositivity to ASCA in relatives of CD patients [28]. They divided 76 relatives into EmA/tTGab negative and positive groups, while 57 individuals with negative CD autoantibodies and no family risk served as controls. Partly in contrast to us, there was a significantly higher frequency of positivity to ASCA IgA/G only in autoantibody-positive relatives compared with the controls [28]. This discrepancy may, at least in part, be explained by the smaller number of participants since there was a trend toward overrepresentation of ASCA, also among the CD autoantibody-negative relatives. There may also have been methodological differences, as the authors did not report the kits used for the ASCA measurements.

Owing to the high specificity of tTGab and EmA [23], most of the autoantibody-positive relatives were likely CD patients. Therefore, their increased seroreactivity to microbial markers is logically in line with that observed in already-diagnosed CD. By contrast, the increased frequency of seroreactivity to a part of the microbial markers in the autoantibody-negative relatives is not as easily explained. It is to be noted that Setty and colleagues [29] previously reported that tTGab-negative relatives of CD patients had signs of intestinal epithelial stress, demonstrated by ultrastructural alterations of microvilli, and increased expression of heat shock proteins and interleukin-15 along with elevated expression of activating NK receptors on intraepithelial cytotoxic T cells. Thus, even in the absence of CD autoantibodies or characteristic histological damage to the intestine, at least some of the relatives appeared to display proinflammatory responses reminiscent of CD. This raises the question of whether the observed abnormal microbial antibody production could also be implicated in this process.

Setty et al. also speculated about a possible genetic predisposition to epithelial stress [29] and suggested a possible HLA and other as yet-unidentified genetic associations. We observed no significant association between the distribution of ASCA, anti-I2 and anti-OmpW positivity and the CD-related HLA haplotypes, suggesting that at least HLA genetics does not markedly affect the serological response. In line with this, HLA DQ2/8 are not overexpressed in inflammatory bowel disease (IBD) patients [30] who also may have increased seropositivity to microbial markers [16,17]. Genetics may still play a role in microbial antibody production in intestinal diseases, as demonstrated by two studies comparing levels of microbial antibodies between monozygous and dizygous twin pairs with IBD. Amcoff et al. reported that the differences in the anti-I2 antibody levels were smaller within than between monozygous twin pairs, even if only one of them had IBD [31]. However, this was not seen in dizygous twins with one suffering from IBD and the other being healthy and having partly discordant genetics, supporting the role of genetic factors [31]. By contrast, similar ASCA levels were observed only in a subgroup of monozygous twins both having IBD [31,32]. Bearing this in mind, it is interesting that we found stronger associations of anti-I2 positivity between the relatives from the same family than between the families, whereas with ASCA this was seen only when autoantibody-positive relatives were included in the analysis. Taken together, it seems that both genetic and environmental factors have a role in the antibody production, with this varying depending on the microbial marker, but further studies are needed.

Environmental factors including gluten intake [33,34] and infections in early life [35–37] have also been associated with increased CD risk. Other suggested, although controversial [38,39], risk factors include bacterial infections and frequent use of antibiotics [40,41]. Interestingly, the incidence has been reported to vary depending on socioeconomic circumstances [10], leading to the hypothesis that slight microbial exposure increases CD risk by driving immune reactions toward autoantigens and dietary components [42]. Close relatives usually share the living milieu and may, thus, experience similar environmental modulatory effects on the microbiota and immune system that, in addition to genetics, could give rise to parallel responses to microbial antigens. It remains unclear, however, which external factors drive these responses and whether the microbial markers have a causal role [43]. It is likely that a complex interaction between multiple factors, such as dysregulation of the immune system, changes in the epithelial barrier, and dysbiosis causes the loss of tolerance to microbial antigens [13,44–46]. In addition, a very recent study showed that *Pseudomonas fluorescens* peptides mimic gluten epitopes and activate gliadin-reactive T cells, with this cross-reactivity possibly contributing to the onset of CD [47].

We previously found most of the potential CD patients to already exhibit the microbial markers before the development of villous damage or autoantibodies [21], reflecting the situation in the relatives in the present study. Interestingly, Torres and colleagues recently showed that ASCA also predicts forthcoming Crohn's disease up to five years before the diagnosis [48]. More studies are needed to determine the role of these markers in early development of CD and whether they could be utilized to predict the disease in at-risk groups.

The main strengths of our study include the large and well-defined cohort of relatives of CD patients who underwent systematic screening for CD-associated HLA and autoantibodies and the representative control groups. As a weakness, however, large differences between the group sizes could have influenced the results. Furthermore, only the groups with relatives contained pediatric subjects, although the results remained unchanged after excluding children from the analyses. Genetic data of the non-HLA alleles were also lacking, which could be an even more significant limitation among relatives with a less marked HLA predisposition to CD. Since we did not have detailed information on the health condition of the relatives, and the histological status of their intestines remains unknown, it is possible that some of them had unreported CD or another disease affecting the results. Furthermore, dietary data of the relatives was lacking, and it is possible that cross-reactions between food antigens influenced the microbial antibody levels. ASCA is known to cross-react with other yeast strains [49], and the lack of correlation between ASCA antibodies and *Saccharomyces cerevisiae* DNA on intestinal mucosa [50] indicates the possibility of some yet-unidentified cross-reactive antigens. In accord with our previous study [51], for currently unclear reasons, ASCA levels were generally higher in the IgG class than the IgA class. By contrast, IgA class ASCA seems to be more consistently elevated in IBD [48,52]. Which of these two antibody classes is the more useful marker in CD would be an interesting subject for further research. The median duration of GFD in the CD group was only one year, which may have biased the serological results, as histological and serological recovery often take longer despite a strict diet [53]. Finally, a few adults here had surprisingly high anti-OmpW values compared with our previous studies. Although we still believe that the used cutoff was valid, we recommend that it be confirmed in other populations.

In conclusion, we found increased seroreactivity to serum microbial markers, particularly ASCA and anti-I2, in relatives of CD patients even in the absence of the disease-specific autoantibodies or other signs of active CD. This observation was not explained by the presence or absence of predisposing HLA haplotypes, thereby suggesting the role of other genetic and environmental factors.

Author Contributions: L.V.: Conceptualization, writing the original draft; S.I.: Conceptualization, supervision, writing—review and editing; H.H.: Conceptualization, methodology, writing—review and editing; P.S.: Conceptualization, methodology, writing—review and editing; K.K. (Katri Kaukinen): Conceptualization, funding acquisition, writing—review and editing; K.L.: Conceptualization, methodology, funding acquisition, writing—review and editing; K.K. (Kalle Kurppa): Conceptualization, funding acquisition, supervision, writing—review and editing. All authors have read and agreed to the published version of the manuscript.

Funding: This study was supported by the Academy of Finland, the Finnish Medical Foundation, the Sohlberg Foundation, the Paulo Foundation, the Sigrid Juselius Foundation, the Foundation for Pediatric Research, the Competitive State Research Financing of the Expert Area of Tampere University Hospital, Helsinki University Hospital Research Fund, the Finnish Medical Foundation and the Finnish Coeliac Society.

Acknowledgments: We would like to acknowledge Jonathan Braun, Pathology and Laboratory Medicine, UCLA David Geffen School of Medicine, UCLA Health System for his support.

Conflicts of Interest: The authors declare no conflicts of interest.

References

1. Kapitany, A.; Toth, L.; Tumpek, J.; Csipo, I.; Sipos, E.; Woolley, N.; Partanen, J.; Szegedi, G.; Olah, E.; Sipka, S.; et al. Diagnostic significance of HLA-DQ typing in patients with previous coeliac disease diagnosis based on histology alone. *Aliment. Pharmacol. Ther.* **2006**, *24*, 1395–1402. [CrossRef] [PubMed]
2. Wijmenga, C.; Gutierrez-Achury, J. Celiac disease genetics: Past, present and future challenges. *J. Pediatr. Gastroenterol. Nutr.* **2014**, *59*, S4–S7. [CrossRef] [PubMed]
3. Trynka, G.; Hunt, K.A.; Bockett, N.A.; Romanos, J.; Mistry, V.; Szperl, A.; Bakker, S.F.; Bardella, M.T.; Bhaw-Rosun, L.; Castillejo, G.; et al. Dense genotyping identifies and localizes multiple common and rare variant association signals in celiac disease. *Nat. Genet.* **2011**, *43*, 1193–1201. [CrossRef] [PubMed]
4. Coleman, C.; Quinn, E.M.; Ryan, A.W.; Conroy, J.; Trimble, V.; Mahmud, N.; Kennedy, N.; Corvin, A.P.; Morris, D.W.; Donohoe, G.; et al. Common polygenic variation in coeliac disease and confirmation of ZNF335 and NIFA as disease susceptibility loci. *Eur. J. Hum. Genet.* **2016**, *24*, 291–297. [CrossRef]
5. Singh, P.; Arora, S.; Lal, S.; Strand, T.A.; Makharia, G.K. Risk of celiac Disease in the first- and second-degree relatives of patients with celiac disease: A systematic review and meta-analysis. *Am. J. Gastroenterol.* **2015**, *110*, 1539–1548. [CrossRef]
6. Lohi, S.; Mustalahti, K.; Kaukinen, K.; Laurila, K.; Collin, P.; Rissanen, H.; Lohi, O.; Bravi, E.; Gasparin, M.; Reunanen, A.; et al. Increasing prevalence of coeliac disease over time. *Aliment. Pharmacol. Ther.* **2007**, *26*, 1217–1225. [CrossRef] [PubMed]
7. Mäki, M.; Mustalahti, K.; Kokkonen, J.; Kulmala, P.; Haapalahti, M.; Karttunen, T.; Ilonen, J.; Laurila, K.; Dahlbom, I.; Hansson, T.; et al. Prevalence of celiac disease among children in Finland. *N. Engl. J. Med.* **2003**, *348*, 2517–2524. [CrossRef]
8. Hervonen, K.; Karell, K.; Holopainen, P.; Collin, P.; Partanen, J.; Reunala, T. Concordance of dermatitis herpetiformis and celiac disease in monozygous twins. *J. Invest. Dermatol.* **2000**, *115*, 990–993. [CrossRef]
9. Kuja-Halkola, R.; Lebwohl, B.; Halfvarson, J.; Wijmenga, C.; Magnusson, P.K.; Ludvigsson, J.F. Heritability of non-HLA genetics in coeliac disease: A population-based study in 107,000 twins. *Gut* **2016**, *65*, 1793–1798. [CrossRef]
10. Kondrashova, A.; Mustalahti, K.; Kaukinen, K.; Viskari, H.; Volodicheva, V.; Haapala, A.M.; Ilonen, J.; Knip, M.; Mäki, M.; Hyöty, H.; et al. Lower economic status and inferior hygienic environment may protect against celiac disease. *Ann. Med.* **2008**, *40*, 223–231. [CrossRef]
11. Rubio-Tapia, A.; Kyle, R.A.; Kaplan, E.L.; Johnson, D.R.; Page, W.; Erdtmann, F.; Brantner, T.L.; Kim, W.R.; Phelps, T.K.; Lahr, B.D.; et al. Increased prevalence and mortality in undiagnosed celiac disease. *Gastroenterology* **2009**, *137*, 88–93. [CrossRef] [PubMed]
12. Catassi, C.; Kryszak, D.; Bhatti, B.; Sturgeon, C.; Helzlsouer, K.; Clipp, S.L.; Gelfond, D.; Puppa, E.; Sferruzza, A.; Fasano, A. Natural history of celiac disease autoimmunity in a USA cohort followed since 1974. *Ann. Med.* **2010**, *42*, 530–538. [CrossRef] [PubMed]
13. Nadal, I.; Donat, E.; Ribes-Koninckx, C.; Calabuig, M.; Sanz, Y. Imbalance in the composition of the duodenal microbiota of children with coeliac disease. *J. Med. Microbiol.* **2007**, *56*, 1669–1674. [CrossRef] [PubMed]
14. De Palma, G.; Nadal, I.; Medina, M.; Donat, E.; Ribes-Koninckx, C.; Calabuig, M.; Sanz, Y. Intestinal dysbiosis and reduced immunoglobulin-coated bacteria associated with coeliac disease in children. *BMC Microbiol.* **2010**, *10*, 63. [CrossRef] [PubMed]
15. Collado, M.C.; Donat, E.; Ribes-Koninckx, C.; Calabuig, M.; Sanz, Y. Specific duodenal and faecal bacterial groups associated with paediatric coeliac disease. *J. Clin. Pathol.* **2009**, *62*, 264–269. [CrossRef] [PubMed]
16. Iltanen, S.; Tervo, L.; Halttunen, T.; Wei, B.; Braun, J.; Rantala, I.; Honkanen, T.; Kronenberg, M.; Cheroutre, H.; Turovskaya, O.; et al. Elevated serum anti-I2 and anti-OmpW antibody levels in children with IBD. *Inflamm. Bowel Dis.* **2006**, *12*, 389–394. [CrossRef] [PubMed]
17. Ashorn, S.; Honkanen, T.; Kolho, K.L.; Ashorn, M.; Välineva, T.; Wei, B.; Braun, J.; Rantala, I.; Luukkaala, T.; Iltanen, S. Fecal calprotectin levels and serological responses to microbial antigens among children and adolescents with inflammatory bowel disease. *Inflamm. Bowel Dis.* **2009**, *15*, 199–205. [CrossRef]
18. Landers, C.J.; Cohavy, O.; Misra, R.; Yang, H.; Lin, Y.C.; Braun, J.; Targan, S.R. Selected loss of tolerance evidenced by Crohn's disease-associated immune responses to auto- and microbial antigens. *Gastroenterology* **2002**, *123*, 689–699. [CrossRef]

19. Ashorn, S.; Raukola, H.; Välineva, T.; Ashorn, M.; Wei, B.; Braun, J.; Rantala, I.; Kaukinen, K.; Luukkaala, T.; Collin, P.; et al. Elevated serum anti-*Saccharomyces cerevisiae*, anti-I2 and anti-OmpW antibody levels in patients with suspicion of celiac disease. *J. Clin. Immunol.* **2008**, *28*, 486–494. [CrossRef]

20. Ashorn, S.; Välineva, T.; Kaukinen, K.; Ashorn, M.; Braun, J.; Raukola, H.; Rantala, I.; Collin, P.; Mäki, M.; Luukkaala, T.; et al. Serological responses to microbial antigens in celiac disease patients during a gluten-free diet. *J. Clin. Immunol.* **2009**, *29*, 190–195. [CrossRef]

21. Viitasalo, L.; Niemi, L.; Ashorn, M.; Ashorn, S.; Braun, J.; Huhtala, H.; Collin, P.; Mäki, M.; Kaukinen, K.; Kurppa, K.; et al. Early microbial markers of celiac disease. *J. Clin. Gastroenterol.* **2014**, *48*, 620–624. [CrossRef] [PubMed]

22. Ladinser, B.; Rossipal, E.; Pittschieler, K. Endomysium antibodies in coeliac disease: An improved method. *Gut* **1994**, *35*, 776–778. [CrossRef] [PubMed]

23. Hadithi, M.; von Blomberg, B.M.; Crusius, J.B.; Bloemena, E.; Kostense, P.J.; Meijer, J.W.; Mulder, C.J.; Stehouwer, C.D.; Pena, A.S. Accuracy of serologic tests and HLA-DQ typing for diagnosing celiac disease. *Ann. Intern. Med.* **2007**, *147*, 294–302. [CrossRef] [PubMed]

24. Monsuur, A.J.; de Bakker, P.I.; Zhernakova, A.; Pinto, D.; Verduijn, W.; Romanos, J.; Auricchio, R.; Lopez, A.; van Heel, D.A.; Crusius, J.B.; et al. Effective detection of human leukocyte antigen risk alleles in celiac disease using tag single nucleotide polymorphisms. *PLoS ONE* **2008**, *28*, e2270.

25. Koskinen, L.; Romanos, J.; Kaukinen, K.; Mustalahti, K.; Korponay-Szabo, I.; Barisani, D.; Bardella, M.T.; Ziberna, F.; Vatta, S.; Szeles, G.; et al. Cost-effective HLA typing with tagging SNPs predicts celiac disease risk haplotypes in the Finnish, Hungarian, and Italian populations. *Immunogenetics* **2009**, *61*, 247–256. [CrossRef]

26. Sutton, C.L.; Kim, J.; Yamane, A.; Dalwadi, H.; Wei, B.; Landers, C.; Targan, S.R.; Braun, J. Identification of a novel bacterial sequence associated with Crohn's disease. *Gastroenterology* **2000**, *119*, 23–31. [CrossRef]

27. Wei, B.; Dalwadi, H.; Gordon, L.K.; Landers, C.; Bruckner, D.; Targan, S.R.; Braun, J. Molecular cloning of a bacteroides caccae TonB-linked outer membrane protein identified by an inflammatory bowel disease marker antibody. *Infect. Immunol.* **2001**, *69*, 6044–6054. [CrossRef]

28. Da Silva Kotze, L.M.; Nisihara, R.M.; Nass, F.R.; Theiss, P.M.; Silva, I.G.; da Rosa Utiyama, S.R. Anti-*Saccharomyces cerevisiae* antibodies in first-degree relatives of celiac disease patients. *J. Clin. Gastroenterol.* **2010**, *44*, 308. [CrossRef]

29. Setty, M.; Discepolo, V.; Abadie, V.; Kamhawi, S.; Mayassi, T.; Kent, A.; Ciszewski, C.; Maglio, M.; Kistner, E.; Bhagat, G.; et al. Distinct and synergistic contributions of epithelial stress and adaptive immunity to functions of intraepithelial killer cells and active celiac disease. *Gastroenterology* **2015**, *149*, 681–691. [CrossRef]

30. Bosca-Watts, M.M.; Minguez, M.; Planelles, D.; Navarro, S.; Rodriguez, A.; Santiago, J.; Tosca, J.; Mora, F. HLA-DQ: Celiac disease vs inflammatory bowel disease. *World J. Gastroenterol.* **2018**, *24*, 96–103. [CrossRef]

31. Amcoff, K.; Joossens, M.; Pierik, M.J.; Jonkers, D.; Bohr, J.; Joossens, S.; Romberg-Camps, M.; Nyhlin, N.; Wickbom, A.; Rutgeerts, P.J.; et al. Concordance in anti-OmpC and anti-I2 indicate the influence of genetic predisposition: Results of a European study of twins with crohn's disease. *J. Crohns. Colitis.* **2016**, *10*, 695–702. [CrossRef] [PubMed]

32. Halfvarson, J.; Standaert-Vitse, A.; Jarnerot, G.; Sendid, B.; Jouault, T.; Bodin, L.; Duhamel, A.; Colombel, J.F.; Tysk, C.; Poulain, D. Anti-*Saccharomyces cerevisiae* antibodies in twins with inflammatory bowel disease. *Gut* **2005**, *54*, 1237–1243. [CrossRef] [PubMed]

33. Andren Aronsson, C.; Lee, H.S.; Koletzko, S.; Uusitalo, U.; Yang, J.; Virtanen, S.M.; Liu, E.; Lernmark, A.; Norris, J.M.; Agardh, D.; et al. Effects of gluten intake on risk of celiac disease: A case-control study on a Swedish birth cohort. *Clin. Gastroenterol. Hepatol.* **2016**, *14*, 403–409. [CrossRef] [PubMed]

34. Andren Aronsson, C.; Lee, H.S.; Hard Af Segerstad, E.M.; Uusitalo, U.; Yang, J.; Koletzko, S.; Liu, E.; Kurppa, K.; Bingley, P.J.; Toppari, J.; et al. Association of gluten intake during the first 5 years of life with incidence of celiac disease autoimmunity and celiac disease among children at increased risk. *JAMA* **2019**, *322*, 514–523. [CrossRef]

35. Kemppainen, K.M.; Lynch, K.F.; Liu, E.; Lönnrot, M.; Simell, V.; Briese, T.; Koletzko, S.; Hagopian, W.; Rewers, M.; She, J.X.; et al. Factors that increase risk of celiac disease autoimmunity after a gastrointestinal infection in early life. *Clin. Gastroenterol. Hepatol.* **2017**, *15*, 694–702. [CrossRef]

36. Kahrs, C.R.; Chuda, K.; Tapia, G.; Stene, L.C.; Marild, K.; Rasmussen, T.; Ronningen, K.S.; Lundin, K.E.A.; Kramna, L.; Cinek, O.; et al. Enterovirus as trigger of coeliac disease: Nested case-control study within prospective birth cohort. *BMJ* **2019**, *364*, 231. [CrossRef]

37. Stene, L.C.; Honeyman, M.C.; Hoffenberg, E.J.; Haas, J.E.; Sokol, R.J.; Emery, L.; Taki, I.; Norris, J.M.; Erlich, H.A.; Eisenbarth, G.S.; et al. Rotavirus infection frequency and risk of celiac disease autoimmunity in early childhood: A longitudinal study. *Am. J. Gastroenterol.* **2006**, *101*, 2333–2340. [CrossRef]
38. Marild, K.; Kahrs, C.R.; Tapia, G.; Stene, L.C.; Stordal, K. Infections and risk of celiac disease in childhood: A prospective nationwide cohort study. *Am. J. Gastroenterol.* **2015**, *110*, 1475–1484. [CrossRef]
39. Kemppainen, K.M.; Vehik, K.; Lynch, K.F.; Larsson, H.E.; Canepa, R.J.; Simell, V.; Koletzko, S.; Liu, E.; Simell, O.G.; Toppari, J.; et al. Association between early-life antibiotic use and the risk of islet or celiac disease autoimmunity. *JAMA Pediatr.* **2017**, *171*, 1217–1225. [CrossRef]
40. Sander, S.D.; Nybo Andersen, A.M.; Murray, J.A.; Karlstad, O.; Husby, S.; Stordal, K. Association between antibiotics in the first year of life and celiac disease. *Gastroenterology* **2019**, *156*, 2217–2229. [CrossRef]
41. Riddle, M.S.; Murray, J.A.; Cash, B.D.; Pimentel, M.; Porter, C.K. Pathogen-specific risk of celiac disease following bacterial causes of foodborne illness: A retrospective cohort study. *Dig. Dis. Sci.* **2013**, *58*, 3242–3245. [CrossRef] [PubMed]
42. Bach, J.F. The hygiene hypothesis in autoimmunity: The role of pathogens and commensals. *Nat. Rev. Immunol.* **2018**, *18*, 105–120. [CrossRef] [PubMed]
43. Verdu, E.F.; Galipeau, H.J.; Jabri, B. Novel players in coeliac disease pathogenesis: Role of the gut microbiota. *Nat. Rev. Gastroenterol. Hepatol.* **2015**, *12*, 497–506. [CrossRef] [PubMed]
44. Kupfer, S.S.; Jabri, B. Pathophysiology of celiac disease. *Gastrointest. Endosc. Clin. N. Am.* **2012**, *22*, 639–660. [CrossRef] [PubMed]
45. Schumann, M.; Siegmund, B.; Schulzke, J.D.; Fromm, M. Celiac disease: Role of the epithelial barrier. *Cell Mol. Gastroenterol. Hepatol.* **2017**, *3*, 150–162. [CrossRef]
46. Kalliomäki, M.; Satokari, R.; Lähteenoja, H.; Vahamiko, S.; Gronlund, J.; Routi, T.; Salminen, S. Expression of microbiota, Toll-like receptors, and their regulators in the small intestinal mucosa in celiac disease. *J. Pediatr. Gastroenterol. Nutr.* **2012**, *54*, 727–732. [CrossRef]
47. Petersen, J.; Ciacchi, L.; Tran, M.T.; Loh, K.L.; Kooy-Winkelaar, Y.; Croft, N.P.; Hardy, M.Y.; Chen, Z.; McCluskey, J.; Anderson, R.P.; et al. T cell receptor cross-reactivity between gliadin and bacterial peptides in celiac disease. *Nat. Struct. Mol. Biol.* **2020**, *27*, 49–61. [CrossRef]
48. Torres, J.; Petralia, F.; Sato, T.; Wang, P.; Telesco, S.E.; Choung, R.S.; Strauss, R.; Li, X.J.; Laird, R.M.; Gutierrez, R.L.; et al. Serum biomarkers identify patients who will develop inflammatory bowel diseases up to 5 y before diagnosis. *Gastroenterology* **2020**, (in press). [CrossRef]
49. Schaffer, T.; Muller, S.; Flogerzi, B.; Seibold-Schmid, B.; Schoepfer, A.M.; Seibold, F. Anti-*Saccharomyces cerevisiae* mannan antibodies (ASCA) of Crohn's patients crossreact with mannan from other yeast strains, and murine ASCA IgM can be experimentally induced with *Candida albicans. Inflamm. Bowel Dis.* **2007**, *13*, 1339–1346. [CrossRef]
50. Mallant-Hent, R.C.; Mooij, M.; von Blomberg, B.M.; Linskens, R.K.; van Bodegraven, A.A.; Savelkoul, P.H. Correlation between *Saccharomyces cerevisiae* DNA in intestinal mucosal samples and anti-*Saccharomyces cerevisiae* antibodies in serum of patients with IBD. *World J. Gastroenterol.* **2006**, *12*, 292–297. [CrossRef]
51. Viitasalo, L.; Kurppa, K.; Ashorn, M.; Saavalainen, P.; Huhtala, H.; Ashorn, S.; Mäki, M.; Ilus, T.; Kaukinen, K.; Iltanen, S. Microbial Biomarkers in Patients with Nonresponsive Celiac Disease. *Dig. Dis. Sci.* **2018**, *63*, 3434–3441. [CrossRef] [PubMed]
52. Pekki, H.; Kurppa, K.; Mäki, M.; Huhtala, H.; Sievänen, H.; Laurila, K.; Collin, P.; Kaukinen, K. Predictors and significance of incomplete mucosal recovery in celiac disease after 1 year on a gluten-free diet. *Am. J. Gastroenterol.* **2015**, *110*, 1078–1085. [CrossRef] [PubMed]
53. Yao, F.; Fan, Y.; Lv, B.; Ji, C.; Xu, L. Diagnostic utility of serological biomarkers in patients with Crohn's disease: A case-control study. *Medicine* **2018**, *97*, e11772. [CrossRef] [PubMed]

Article

Patient Perspectives of Living with Coeliac Disease and Accessing Dietetic Services in Rural Australia: A Qualitative Study

Rachelle Lee [1,2], Elesa T. Crowley [3,4], Surinder K. Baines [1], Susan Heaney [1,3] and Leanne J. Brown [3,*]

[1]   School of Health Sciences, College of Health, Medicine and Wellbeing, University of Newcastle, Callaghan, NSW 2308, Australia; rachelle.lee@health.nsw.gov.au (R.L.); surinder.baines@newcastle.edu.au (S.K.B.); susan.heaney@newcastle.edu.au (S.H.)
[2]   Dubbo Health Service, Western NSW Local Health District, Dubbo, NSW 2830, Australia
[3]   Department of Rural Health, College of Health Medicine and Wellbeing, University of Newcastle, Tamworth, NSW 2340, Australia; elesa.crowley@health.nsw.gov.au
[4]   Tamworth Rural Referral Hospital, Hunter New England Local Health District, Tamworth, NSW 2340, Australia
*   Correspondence: leanne.brown@newcastle.edu.au; Tel.: +61-2-67553540

Abstract: Adapting to living with coeliac disease requires individuals to learn about and follow a strict gluten-free diet. Utilising a qualitative inductive approach, this study aimed to explore the perspectives of adults diagnosed with coeliac disease who have accessed dietetic services in a rural outpatient setting. A purposive sample of adults with coeliac disease who had accessed dietetic services from two rural dietetic outpatient clinics were recruited. Semi-structured interviews were conducted by telephone. Data were thematically analysed. Six participants were recruited and interviewed. Three key themes emerged: (i) optimising individualised support and services, (ii) adapting to a gluten-free diet in a rural context, and (iii) managing a gluten-free diet within the context of interpersonal relationships. Key issues identified in the rural context were access to specialist services and the increased cost of gluten-free food in more remote areas. The findings of this study have highlighted the difficulties associated with coeliac disease management and how dietetic consultation has the potential to influence confidence in management and improve lifestyle outcomes. Further qualitative research is required to expand on the findings of this study and inform future dietetic practice that meets the expectations and individual needs of people with coeliac disease in rural settings.

Keywords: coeliac disease; dietitian; rural health services

Citation: Lee, R.; Crowley, E.T.; Baines, S.K.; Heaney, S.; Brown, L.J. Patient Perspectives of Living with Coeliac Disease and Accessing Dietetic Services in Rural Australia: A Qualitative Study. *Nutrients* **2021**, *13*, 2074. https://doi.org/10.3390/nu13062074

Academic Editors: Isabel Comino, Carolina Sousa and David S. Sanders

Received: 31 March 2021
Accepted: 12 June 2021
Published: 17 June 2021

Publisher's Note: MDPI stays neutral with regard to jurisdictional claims in published maps and institutional affiliations.

1. Introduction

Coeliac disease is often misdiagnosed or delayed in diagnosis, with dietary restrictions that may not be suitable or well understood. Managing and adapting to a chronic disease with strict dietary requirements is a challenge. Having to do so in a rural context may be further challenged due to a disparity of access to health care [1] and increased costs of food and availability issues [2]. The needs of the person requiring a specialised diet may compete with the food needs of other members of the household as well as expectations in social interactions [3].

Current treatment for coeliac disease includes strict life-long adherence to a gluten-free diet, [4] requiring complete avoidance of gluten-containing grains (wheat, barley, and rye) and their by-products [5]. As well as avoiding gluten, a balanced diet with adequate vitamins, fiber, and calcium is essential [6]. This permanent dietary restriction has a major impact on the nutritional adequacy of the diet and the quality of life of both the person with coeliac disease and those around them [7,8]. Factors that predict or influence long-term health outcomes include genetics, environmental factors, ongoing inflammation of the small intestine, and nutritional deficiencies [9]. Common problems associated with adjusting to a gluten-free diet include a lengthy education process and the identification of gluten-free foods that are affordable and enjoyable [8,10–12]. Eating out of the home and socialisation may also become more difficult under the constraints of the gluten-free diet, and feelings of social isolation, worry, and neglect are also commonly reported [11,12].

It has been suggested that those with coeliac disease require encouragement, motivation, and support from a collaborative medical and dietetic team to ensure adherence to the gluten-free diet and subsequent progressive treatment outcomes [13]. Improvements in practitioners' abilities to educate about coeliac disease has been linked to potential for improved adherence to a GF diet [14]. The British Society of Gastroenterology recommends that individuals with coeliac disease should attend a dietetic consultation and counselling session upon diagnosis, at three and six-months post-diagnosis, and then be reviewed annually by both a dietitian and the treating physician [13,15]. Despite these recommendations, some literature indicates that the availability and use of dietetic services in the management of coeliac disease is less than adequate [15,16].

Coeliac disease is a complex condition that can be difficult to manage [17], particularly in a rural or regional setting, where access to a wide variety of gluten-free foods and specialised health care services is limited. Current research has highlighted the importance of dietetic consultation post-coeliac disease diagnosis [13]; however, a number of studies indicate that individuals with coeliac disease are not being followed up adequately [10,16]. There has been limited exploration of patient expectations and satisfaction with dietitian consultations.

Due to the complexity of the gluten-free diet, specialist advice and dietary education are important in preventing inadvertent gluten consumption and persistence of symptoms [7,18]. Few dietitians specialise in coeliac disease management [5,15,16], and rural-based dietitians may be less experienced. Rural-based dietitians may have a generalist role with a broad case-load and limited opportunities for specialisation and consequently, less experience with the dietary management of coeliac disease [19,20].

Current evidence investigating coeliac disease from the qualitative perspective [8,11, 12,17,21,22] has focused on patients' experiences [11,12], perspectives of close relatives [8], diagnosis and realities of living with coeliac disease [17], and motives for adherence to a gluten-free diet [21]. Qualitative explorations specific to health professional care have explored experiences of dietetic consultations [22,23]. Further research is required to develop a greater awareness of the condition and its impact, especially in terms of access to dietetic services in different settings [8,12,17]. This study aimed to explore patient perspectives of use of dietetic services and their ongoing management of coeliac disease in a rural setting.

2. Materials and Methods

Study design: This qualitative study used a general inductive approach [24] to explore the perspectives of adults with coeliac disease. Patients who attended an outpatient dietetic clinic in a regional or rural setting of a Local Health District (LHD) regarding dietary advice for coeliac disease (between July 2015 and May 2019) were purposefully selected. The LHD sites were located in Modified Monash Model locations [25] classified as MM3 (large rural town) and MM4 (medium rural town). From two health care settings and a total of 19 eligible participants, six agreed to be interviewed. Subjects were eligible if they had been clinically diagnosed with coeliac disease, attended a dietetic consult within the past five years, and if they were able to participate in a telephone interview. Those under the age of 18 years at the time of consult were excluded.

Ethics: Ethics approval was obtained through the relevant NSW Health Local Health District Human Research Ethics Committee (15/05/20/5.01) and the University of Newcastle Human Research Ethics Committee (H-2015-0165). Anonymity of potential participants was maintained by having a member of hospital administration staff manage the mailed invitations. Consent was sought from all participants for both the data they provided to be utilised for research and their interview to be audio recorded. Analysis of results was undertaken by researchers who were not known to the participants (RL and LB).

Recruitment: An information statement explaining the inclusion and exclusion criteria, expectations of participants, rights of participation, and privacy were mailed out to all eligible participants along with a consent form. Those who consented were contacted by telephone for an interview after they returned the consent form via a pre-paid envelope.

Development of the data collection tool. A review of the literature informed the development of the semi-structured interview protocol. A search of databases Medline, CINAHL, and PubMed enabled the identification of common themes around issues in the management of coeliac disease and revealed gaps in current knowledge. These gaps were addressed through the development of interview questions about their use of dietetic services, health outcomes, symptom management, and quality of life in an attempt to improve the knowledge base and provide a more holistic understanding of the perspectives of individuals with coeliac disease.

Data collection: Semi-structured interviews recorded and ranged from 15 to 45 min in duration (Supplementary File). With verbal confirmation of consent, the interviews were audio recorded and transcribed verbatim. A series of probes and prompts were used during the interviews. Transcripts were checked for accuracy against the original recording and assigned a non-identifiable code. Interviewer field notes were completed for each interview by the interviewer. Transcripts were checked against the audio recording for accuracy. Participants were provided the opportunity for member checking, to ensure the information derived from the transcription was valid. Transcripts were read and coding developed with the assistance of NVivo 10 (QSR International Pty Ltd. 2015, Doncaster, Victoria, Australia). Categories and themes were developed and revised (RL) with another a second experienced researcher (LB). Themes were developed using a general inductive approach [26]. Excerpts from the interviews were edited for clarity and anonymity. The Standards for Reporting Qualitative Research checklist was utilised when reporting this research [24].

The size of the sample was largely determined by the availability of respondents who were willing to participate in an interview. A sample of six participants was deemed appropriate because of the exploratory nature of this research and the focus on identifying the experiences of those in a rural area. This study did not attempt to examine a representative sample to provide a comprehensive understanding of all possible experiences, but rather to explore issues within a rural context with a purposeful sample from a rural area with access to a specific dietetic service.

3. Results

Six people were interviewed, three male and three female, with an age range from 38 to 77 years; number of years since diagnosis ranged from <1 year to 10 years. Participants resided in MM3 (large rural towns) or MM5 locations (small rural towns). A summary of participant (P1–P6) demographics, their symptoms at diagnosis, and their dietetic management is provided in Table 1. Half of the participants attended more than one appointment with a dietitian, with one seeking follow-up appointments due to later tests showing ongoing inflammation of the small bowel.

Three key themes were identified through analysis of the data. These related to (i) optimising individualised support and services, (ii) adapting to a gluten-free diet in a rural context, and (iii) managing coeliac disease within the context of interpersonal relationships. Table 2 provides a summary of the themes and sub-themes with supportive quotes.

Table 1. Participant demographic information, symptoms at diagnosis, number of dietetic consultations, and other descriptive information.

Participant. MMM Location	Gender Age (Years)	Years since Diagnosis at Time of Interview	Length of Symptoms Prior to Diagnosis	Symptoms at Diagnosis	Number of Dietetic Consultations since Diagnosis	Household Description	Other Descriptive Information
P1 MM3	Female 38	$3\frac{1}{2}$	2 years	Diarrhoea, fatigue, 'cognitive fogginess', vitamin D and iron deficiencies	1	Living with husband and young children	Finds family life and socialising difficult within the constraints of the GF diet and often gives priority to the dietary needs of her family over her own.
P2 MM5	Male 69	3	20–30 years	Abdominal pain and distension	2	Living alone	Spends a considerable amount of time outside of the home due to travelling occupation, relying heavily on takeaway food.
P3 MM3	Male 77	2	N/A	No symptoms reported. Diagnosis as a result of routine gastroscopy	2	Living with wife	Very dependent on wife for support, dietary knowledge and meal provision,
P4 MM3	Male 66	≤1	1 year	Constipation, wind, abdominal pain	1	Living alone	Performs all cooking and shopping duties autonomously.
P5 MM5	Female 72	10	unclear	Abdominal symptoms and reflux	1	Living with husband	Lives in small rural town, distant from a major regional centre. Finds accessing GF food difficult locally.
P6 MM3	Female 58	8	Not specified	Fatigue, explosive loose bowel motions, anaemia, bloating and abdominal discomfort	5	Living with husband	Cooks at home, does not eat out often, finds food access reasonable. Husband chooses to eat GF at home.

GF—gluten-free, MM—Modified Monash classification.

Table 2. Themes, sub-themes and supportive quotes regarding perspectives of adults living with coeliac disease and accessing dietetic services in rural Australia.

Themes and Sub Themes	Supportive Quotes
3.1 Optimising individualised support and services	
	'They were able to point me in the right direction as to what foods are suitable for a gluten-free diet. So that was helpful … they carried out their service in a very informative and professional manner.' (P6)
	'I think it [the dietetic consult] was a two-way process. They're teaching me about things but they're also good in as much as that they ask questions and they take on board the information that you give them. So I think it's a repository for information backwards and forwards. That's been quite worthwhile.' (P3)
3.1.1 Provision of support and services relevant to patient needs	'I was pleasantly surprised. I got attention and good advice. I think it was reinforced that I was actually on the right track. I was pretty happy with my experience.' (P4)
	'We've gone into detail on my diet and how this could be happening and whether there's any cross-contamination and things like that. So I can't really say that I've been unhappy or unsatisfied with the service I've got from the dietitian.' (P6)
	'It was just really basic information that everyday people know about … but you could sort of see I guess by looking at me and by the way I spoke that I didn't need to be educated on basic healthy eating.' (P1)
	'She didn't give me all that much information. I mainly got it all out of the books.' (P5) and that access to a dietitian was difficult due to the lack of a local service 'a dietitian only comes here if you want to see one … at least once a month … if I wanted to see one I'd have to go to [regional centre] (P5).
3.1.2 Meeting the expectations of the patient	'There's some lipsticks you shouldn't use, some shampoos and things like that that can affect your coeliac … there's medication that has gluten in it … I would expect my dietitian to point [that] out to me' (P1).
	' … having the support and learning the skills and getting the tools to do that, which I feel the dietitians here where I live have done that … they gave me all the tools and skills to go out into the big wide world and hit the supermarket aisles.' (P6)
3.1.3 Consistency in communication and coordination of care	'I think there needs to be a lot more conversing, communication between the doctors and the dietitians … at the beginning there … there just seemed to be no communication or something… he said, "Your dietitian will tell you. There's lots of things that you don't realise that have gluten" … I don't think that he realised that I wasn't actually told all that sort of stuff'. (P1)
	'I guess in some ways you don't want someone to be sympathetic to you, but they are also understanding that it's challenging and helping you to move forward with this new diet that you need to follow, and pointing out … as the GP did as well, like we all have choices.' (P6)
3.1.4 Improving services and resources	'If there was someone that … says they're an expert in gluten-free cooking, we'd love to be aware of that' (P3).
	' … the Coeliac Society has a good website and there's another book … on gluten-free cooking … one that I use a lot.' (P6)
3.2 Adapting to a gluten-free diet in a rural context	
3.2.1 Confidence around managing ones' coeliac disease	'I'm getting better at it [managing the gluten-free diet], definitely, but I'm still learning the whole time.' (P2)
	'There's a lot of food that is actually gluten-free but they don't state it. I'm thinking to myself I should get that … Oh I better not because it doesn't have the 'gluten-free' [label] on it.' (P1)
	'You miss out on the pleasure of tasting something simply because you don't enquire enough or recognise that it is within the bounds of what you can and can't consume.' (P3)

Table 2. *Cont.*

Themes and Sub Themes	Supportive Quotes
3.2.2 Adapting to and maintenance of the gluten-free diet	'I'm certainly better and I feel better … ' (P4) and 'I feel a lot of those symptoms have gone … ' (P6) 'There may be the very odd occasion that I might have some gluten in my diet purely because I might be in a social situation … I just the pay the price a little bit later after eating gluten, if you know what I mean. So maybe in a social situation I might have a little bit of gluten very, very occasionally.' (P6) 'If it's something new I've got to read the label … I've never read a label before, now I've got to.' (P2) 'It shouldn't be that hard I don't think … It should only take six months really.' (P1) 'Trouble is, if I do get an episode where I've got something [gluten] slipping through … It seems to take a little while before the symptoms disappear.' (P2) 'The cost is usually double... Everything's a lot more expensive.' (P2) 'I cook myself, but I don't bake that much. So I buy the bread, and that's $6.99 [AUS$] a loaf … and you can only get about a dozen slices … I think it's too expensive. I really do and especially in country areas where we are.' (P5) 'I struggle with the price because everything that seems to be gluten-free, such as a loaf of bread for instance is three times the price of a normal loaf of bread.' (P6) 'I'm not a high income earner … I had to take [that] into consideration and to learn how to shop within my budget.' (P6) 'I miss a good meat pie … and bread. I get a second-rate meat pie and I get the worst bread.' (P2)
3.2.3 Food as a major aspect of life	'I find eating out is just awful, I dread it sometimes.' (P1) 'Eating is one of those things that you either do automatically or you do because you like it and I don't know that I've ever been one that ate automatically... So I don't think I'll ever forget the sensation of tasting something nice. So when you do eat something and you think to yourself, oh, that wasn't much of a thrill, was it?' (P3) 'Everything is gluten-free' [at home] but when out 'I cannot follow it when I'm out, if I have somewhere to eat. You know what I mean?' (P5) 'Well to be quite honest, I haven't really found it all that difficult because I guess I eat a pretty basic sort of diet and so there are some gluten-free products that I really don't like and I just don't eat but … I look at alternatives.' (P6) 'I found it difficult initially because a lot of the products had extra sugar and other components in there, so I was then struggling with the weight thing because of that. So instead of having some of these more processed foods … I look at more natural foods.' (P6) 'It's very hard when you're living in the country with only 1200 people here and you have to go out of town and … I can't afford to keep going out of town to get the food … in the supermarket it's good but you don't get an awful lot of gluten-free. It's mainly breakfast you can't get … I cannot get the stuff I want to eat and what they've got, I don't like.' (P5) 'My husband and I have travelled around Australia over the last few years so that was more challenging going into some pretty remote places where there was very little gluten-free products and if they were, you paid an enormous amount of money for them. So I guess again it was challenging because your fresh fruit and veggies that you would have normally were almost—even that was difficult to get hold of because of the availability and the price but you'd work around it.' (P6)
3.2.4 Resignation to the diet and lifestyle changes	'I'm still waiting for someone to ring me up and say, "Okay [refers to self], you're right to go back to eating whatever you like" and sometimes I get this smack in the face and I just go, "Oh my god, that's never going to happen. This is forever." I'm never actually going to be able to just go and have fish and chips freely when I go on holidays again.' (P1) 'It's one of these things. It's like getting your leg cut off, there's not much point wishing it hadn't happened.' (P2) 'Your life doesn't need to stop [when you have coeliac disease]; there are always ways around accommodating your likes and dislikes.' (P3) 'I just accept it. It's no different than being a diabetic. It's just what I need to do and I just do it.' (P6) 'I do enjoy baking, so that's probably been the biggest hurdle … is learning how to bake with the gluten-free flour and different flours on offer. But in time you manage it and you get accustomed to it and you can get some pretty good results.' (P6)

Table 2. *Cont.*

Themes and Sub Themes	Supportive Quotes
3.3 Managing coeliac disease within the context of interpersonal relationships	
3.3.1 The role of others	'A friend of mine invited all her friends and me over for dinner the other night and I just felt like the biggest pain in the neck. I always have to just say, "Is it gluten-free, what you're cooking" and they all go, "Oh, I forgot." I just feel like a burden all the time and that sort of gets you down.' (P1) 'Socially, it's a little bit embarrassing for the people mainly because if they put [food out] and [I] say, "Sorry I can't eat it because I'm gluten-free" … I tend not to go out much anyway.' (P2) 'Other family members that invite you over and they sit you down and you've got a gluten meal in front of you and you think why don't you get it? Why don't you understand I cannot have gluten? … If I don't eat that, I go hungry and I don't particularly want to say well I'll come as long as you cook me a gluten-free meal because that then—it just becomes too hard for them and I don't want to—I don't want that to be the case.' (P6) 'I'm lucky that I've been associated with these people for probably 30 years so they say "Well what can we do [to help]?" … .and they're quite prepared to do that.' (P3) 'If it [the gluten-free diet] affects other people, too bad, because I've got to look after number one. That's the way I look at it. And really, true friends will support that anyway.' (P4) ' … they certainly know that I'm coeliac and they came for morning tea the other day and had brought a cake which was from a supermarket which wasn't gluten-free. So I cut it up and I didn't have a piece … 'Why aren't you having a piece?' … And I don't like to draw attention to myself. I don't want to be a stick in the mud. And I'm going, 'I'm coeliac, I can't eat it.' … 'Oh, I'm so sorry.' (P6) 'I've had café [staff] actually just like roll their eyes at me when I've said, "I need gluten-free" and it's not until I've actually said I'm allergic to it.' (P1) 'It's a more recognised complaint now and the supermarkets and whatever cater for it … the range has just expanded so much.' (P3)
3.3.2 Management of the diet and disease within the family unit	'I find it's difficult when you've got little kids and I don't know whether to go fully gluten-free in this house. For example, the butter dish, it's always full of breadcrumbs, so I've gone down the track of doing two butter dishes. It's really difficult. Do I have two butter dishes, two jam dishes, two honey? You know, it just gets a little bit out of control.' (P1) 'There's no issues... if I was to go to my daughter's houses or them to come here or vice versa there's never any issue.' (P6) 'I know it's my husband's choice but at the time same time he pretty much has a gluten-free diet as well and doesn't eat those sorts of foods in front of me.' (P6)
3.3.3 Responsibility for the diet and disease	'Other people don't—which they don't need to … take responsibility for what you put in your mouth … but that's fine, I mean it's hard enough sometimes me getting my head around it, let alone other people thinking about what I'm eating.' (P1) 'If I'm in an environment where I can control it … I can order something that's gluten-free … whereas if you go into somebody's home, it's one thing to take some cheese and crackers. It's another thing to take your own total meal. I wouldn't feel comfortable doing that.' (P6) 'I understand and I appreciate their hospitality, but at the same token I don't want sit up to two slices of white normal bread and then come home and be on the toilet all afternoon.' (P6) 'He [husband] brought home … A treat and I said "I assume you asked whether that's gluten-free?" and he's like "Um … "' (P1). 'I took my wife with me [to the dietitian] … she's the one that looks at what's on the shelf, reads the instructions, reads the contents and questions it... I'm lazy and she's not, so she picks up the information and applies it.'(P3) 'I'm blessed with a lovely wife who's a thinking person and capable. I think if you had somebody that wasn't capable of getting around these difficulties [coeliac disease management], it'd be a burden.' (P3)

3.1. Optimising Individualised Support and Services

3.1.1. Provision of Support and Services Relevant to Patient Needs

Experiences with dietetic consultations were varied. Most participants found their consultation helpful and were positive about the role of the dietitian in coeliac disease management. One participant who described a 'two-way' process was positive about the experience and considered it worthwhile, expressing contentment with the dietitian in tailoring the consultation to his needs, and this was reflected in his willingness to return for necessary follow-up consultations. Another participant who described that he was 'on the right track' with his diet was 'pretty happy' with his experience.

Contrary to this, a participant was less satisfied with her experience, suggesting that the education she received did not align with her expectations, nor was it tailored to her needs, suggesting that the approach taken was not individualised and that her level of understanding was already high. Those with limited access to consult with a dietitian, due to limited outreach services, relied on other sources of information. One participant felt they mainly obtained information from books and that access to a dietitian was difficult due to the lack of a local service.

3.1.2. Meeting the Expectations of the Patient

Satisfaction with the dietetic consultation was linked to initial expectations of what the service was going to offer. P1 had specific expectations around the type of information that was to be provided and expressed that '*I would expect my dietitian to point [that] out to me*'. Other participants reported a more positive experience and indicated that their expectations had been met, indicating that the dietitian '*gave me all the tools and skills the skills*'. Expectations were varied, with some expectations going beyond advice about food.

3.1.3. Consistency in Communication and Coordination of Care

The unmet expectations of P1 were linked with inconsistent communication within the health-care team, with the implication of a lack of communication between the doctor and the dietitian. '*I think there needs to be a lot more conversing, communication between the doctors and the dietitians . . . ' (P1).* This contributed to confusion around the roles of these health professionals and lack of confidence in the health-care team. Other participants identified ways in which health professionals could support people with coeliac disease by understanding and supporting them to make their own choices. P6 stating that the dietitian ' *. . . helping you to move forward . . . as the GP did as well, like we all have choices . . . .'*

3.1.4. Improving Services and Resources

Two participants (P1 and P3) provided suggestions to improve support and services offered to coeliac disease patients. These included increasing access to educational resources and other services to support those managing a gluten-free diet, for example a social media page. It was also suggested that management would become easier and more satisfying with education and skills around gluten-free cooking. Other participants mentioned other resources they used and found helpful such as the Coeliac Society website and gluten-free cookbooks. Resources related to gluten-free cooking were highlighted ' *. . . another book on gluten-free cooking . . . that I use a lot . . . ' (P6)* as very useful.

3.2. Adapting to a Gluten-Free Diet in a Rural Context

3.2.1. Confidence Around Managing Ones' Coeliac Disease

All participants indicated that confidence in management of their coeliac disease developed over time and with increasing education. Despite a sense of self-managing their coeliac disease and expressing no desire to return for follow-up consultations with a dietitian, P1, P2, and P4 expressed a lack of confidence in their ability to determine if a food was gluten-free or not. These participants limited the variety of their diet and relied heavily on familiar and packaged foods labelled gluten-free: '*If it's got 'gluten-free' on it, well, I buy it'* (P4). Another participant indicated further awareness of this, suggesting the importance of education around gluten-free products for greater dietary enjoyment and flexibility. *'You miss out on the pleasure of tasting something simply because you don't . . . . recognise that it is within the bounds of what you can . . . consume'* (P3).

3.2.2. Adapting to and Maintenance of the Gluten-Free Diet

All participants noted improvements in their physical health after commencing the gluten-free diet; however, most reported situations where the inadvertent consumption of gluten caused ongoing symptoms. Despite positive physical outcomes, frustration was expressed at the length of time taken to adjust to and feel better on the gluten-free diet. *'It shouldn't be that hard I don't think'* (P1).

Adaptations to social situations were described as more difficult to control and could lead to inadvertent gluten consumption. *' . . . I might have some gluten in my diet purely because I might be in a social situation . . . '* (P6). A participant also reported on the importance of developing a new skill of reading food labels for better management of their condition. *I've never read a label before, now I've got to'* (P2).

Adaptation to a gluten-free diet and management was made more difficult for some participants due to the higher cost and lower quality of gluten-free foods, with the cost reportedly *'usually double'* or *'three times the price'* that of a gluten containing option. Staple foods such as bread and baked products were identified as particularly expensive and difficult to access in some rural areas. The higher cost of gluten-free foods necessitated the development of skills in managing a low income and *learning 'how to shop within [a] budget* (P6). Most participants expressed their dissatisfaction with the majority of gluten-free products, particularly staple foods such as bread, *'I get the worst bread'*, and pastries: *'I get a second-rate meat pie'* (P2).

3.2.3. Food as a Major Aspect of Life

Most participants communicated difficulties in managing their coeliac disease with food being a major aspect of everyday life. P1 and P2 spoke of the effect of the gluten-free diet on their ability to attend social events: *'I find eating out is just awful, I dread it sometimes'* (P1). Meanwhile, P3 reported a decreased enjoyment in life secondary to the elimination of gluten-free foods. Another participant found it easier to manage at home where *'everything is gluten-free'* (P5).

A participant (P6) did not find the change particularly difficult due to a *'basic sort of diet'* she followed at home prior to diagnosis. Despite this, she still found she needed to look for suitable alternatives. The same participant expressed that the challenge was with the impact of the dietary change on her weight *'because a lot of the products had extra sugar.'*(P6)

While access to food was not always an issue in the location of residence *'where I live, there's a good range of gluten-free products'* (P6), others had more difficulties. Living in a small rural town was linked to limited gluten-free options and required travel to larger centres to access varied options. Some found travel difficult particularly in more remote areas where the food was more expensive and there are fewer fresh food options that are available.

3.2.4. Resignation to the Diet and Lifestyle Changes

The participants responded differently to the chronic nature of coeliac disease. One was forthcoming with emotions around the long-term nature of the disease, acknowledging feelings of sadness. *'It's a bit depressing, yeah, definitely'* (P1).

Some participants (P2, P3, P6) expressed a greater acceptance of the condition; they implied that withdrawal of emotional involvement was how they coped with coeliac disease. Another participant talked about the challenges of adapting her shopping and cooking and how she became 'accustomed to' the dietary changes over time, *' . . . in time, you manage . . . you get accustomed to it and you can get some pretty good results'* (P6).

Being newly diagnosed, P4 had less experience in long-term coeliac disease management but had a positive outlook for the future: *'I'm confident that I'll be able to manage the thing . . . time will tell'* (P4). Another participant with a long-standing diagnosis said, *'I just accept it . . . It's just what I need to do and I just do it'* (P6).

3.3. Managing Coeliac Disease within the Context of Interpersonal Relationships

3.3.1. The Role of Others

The way 'others' reacted to the demands of the gluten-free diet was raised by all participants. It was implied that these reactions affected the self-esteem and confidence of participants, particularly within the social context. This contributed to a decline in social activity and a reluctance to attend social events where food is a major component.

The embarrassment of others when unable to provide for those with special dietary needs was also a key issue, reinforcing the onerous impact of the gluten-free diet and its contributing to their social withdrawal. With P1 expressing that *'I just felt like the biggest pain in the neck'.*

Difficulties in getting some family members to understand dietary needs led to participants opting to choose to not make it 'too hard for them'. Whereas a male participant had a more positive experience with socialising, which was fostered through long-term relationships with a stable social group who were aware of his special dietary requirements and willing to assist.

Another male participant approached the involvement of others from a different perspective, suggesting their acceptance or otherwise should not affect his management. *'I've got to look after number one.'* (P4) In contrast, another participant (female) was more apologetic and conflicted by the imposition of her diet on others. *'It just becomes too hard for them and I don't want that to be the case'* (P6).

The social emergence of the gluten-free diet was raised in both a positive and negative light. P1 spoke of the stigma around the gluten-free diet and the difficulties associated with communicating the seriousness of the condition. *'I've had café [staff] actually just like roll their eyes at me . . . '* (P1). Meanwhile, others suggested that the popularity of the gluten-free diet has facilitated the availability of new food products and greater dietary variety for people with coeliac disease.

3.3.2. Management of the Diet and Disease within the Family Unit

Ideas around management of the diet within the family unit was raised by two participants. P3 described living in a household that was 'completely gluten-free', and he expressed empathy for his wife in missing out on 'all the good foods' (P3). P1 spoke of the difficulties associated with raising and feeding children in a partially gluten-free household, and some debated the decision of whether or not to eliminate gluten from the household all together. *'I don't know whether to go fully gluten-free in this house'* (P1). Others commented that there were no issues when eating with family members in various locations with family members opting to choose to eat gluten-free at home and in their presence.

3.3.3. Responsibility for the Diet and Disease

Two participants (P1 and P2) indicated that they were autonomous in managing their gluten-free diet and had no expectations for those around them to take on this responsibility.

The challenge of juggling social situations that impact on taking responsibility for one's own food intake and the health implications was highlighted by P6. Others made the decision to not put themselves in a position to have gastrointestinal symptoms due to relying on others to understand the requirements of a gluten-free diet

The role of the partner or spouse in the management of coeliac disease was also raised with some partners being engaged and involved in supporting their partner with their gluten-free dietary requirements and others less engaged. P1 expressed the difficulties in educating her husband about gluten-free foods. In contrast to this, P3 described being completely reliant on his wife for his food and dietary management. This participant expressed his gratitude for his wife's involvement and indicated that he should learn to take more control of his coeliac disease and demonstrated an awareness of how difficult self-management would be.

4. Discussion

This study has been the first of its type to provide insights into the perspectives of individuals with coeliac disease who have accessed dietetic services in a rural outpatient setting. This research has described the difficulties associated with living with coeliac disease and the impact this has on an individual's social interactions and relationships. The findings have revealed some strengths and inconsistencies in the dietetic management of coeliac disease in a rural setting. The explorative findings from this study can inform future dietetic practice in coeliac disease management in rural settings.

In this study, some participants experienced lengthy delays to diagnosis, which is common for patients with coeliac disease [7,11,16]. Misdiagnoses are also common [17], which can be traumatising and frustrating for many people with coeliac disease [17]. Post-diagnosis, participants had variable experiences and levels of expressed satisfaction with the dietetic services provided. The dissatisfaction reported by one participant appeared to originate from a dietetic consultation that failed to meet her expectations and inadequate communication within the health-care team. This was also identified in another qualitative study based in Victoria, Australia [17], where 10 women (aged 31 to 60 years) with coeliac disease who were members of a state-based coeliac society were interviewed. Others have reported on the importance of considering patient expectations [22] in general dietetics consultations. These concepts have not previously been explored in the management of coeliac disease. Our study also suggests the need for greater interdisciplinary communication for improved patient outcomes. This also raises issues of understanding health care roles delineation (i.e., who covers what issues) and dietary versus non-dietary topics of discussion. This may be particularly important in rural areas where clinicians may have less opportunity to specialise or gain regular experience in particular fields of practice.

Research has found that people with coeliac disease expect to be provided with information specific to their lifestyle when attending dietitian consultations [23]. The three participants who returned for follow-up had positive reports about their dietetic consultations; they indicated that the interventions were well suited to their lifestyle and that the provision of skills and knowledge was adequate. Despite occasional difficulties in management, these participants felt they required no further follow-up. This was secondary to issues including rurality, which made access to dietetic services more difficult for one participant. These findings have not been previously explored through qualitative research.

All participants in this study reported improvements in physical health after commencement of the gluten-free diet. Other evidence demonstrates holistic improvements in the physical and mental health of those with coeliac disease after gluten-free diet commencement [17]. Confidence in coeliac disease management appears to develop over time as knowledge and insight into the requirements of the gluten-free diet increase. In the early years after coeliac disease diagnosis, one of the greatest difficulties experienced by patients is determining which foods are safe to eat. The accidental consumption of gluten containing foods is also common [12], which is a major source of dissatisfaction and self-reported poorer treatment outcomes among those with coeliac disease in this study. Decreased enjoyment in food is another key factor affecting the quality of life of patients with coeliac disease found in this and other studies [11,12,17]. In agreement with other research [17], the need to be compliant with the gluten-free diet and limited variety of gluten-free foods, particularly in rural areas, was a source of constant disappointment for most participants in this study. Interestingly, male participants appeared to be more accepting and less apologetic about their specialised dietary needs, while female participants tended to be apologetic or accommodating to fit in with their family members and social group.

The popularity of the gluten-free diet to treat ailments apart from coeliac disease is rising [17], which has also led to an increase in the availability and variety of gluten-free food products. King et al. [27] refer to this as the "double edged sword" of greater availability of gluten-free options but also the risk of contamination due to the undermining of the seriousness of the strict dietary needs of people with coeliac disease.

The sense of self-worth for patients with coeliac disease appears to be shaped by the reactions of those who play a meaningful role in their lives. Participants in this and other studies report feeling like a burden on those around them [8,12,17]. This study indicates that people with coeliac disease seem more comfortable with those who understand and are willing to cater for their dietary requirements, which another study suggests, increases their willingness to socialise [17].

The one female participant of this study reported an adverse emotional reaction to having coeliac disease. This issue has also been explored by other researchers [28], who suggest women find the disease more socially confining and tend to struggle with feelings of loss of control [29]. Furthermore, management of the gluten-free diet seems easier when responsibility is shared with a spouse or significant other, as reported by some participants in this study. Taylor et al. [17] have also reported that the burden of the diet and lifestyle were lessened when the load was shared within the relationship [17]. Access to quality resources such as those provided by the Coeliac Society were identified by participants as useful. Membership of a coeliac association has been linked to improvements in adherence to a gluten-free diet in a systematic review [14].

This purposeful sample provides insights into the experiences of rural-based people with coeliac disease accessing a free dietetic service. While a lack of information about education level and a limited age range of participants (38 to 77) in this study limits the generalisability of the findings, the purpose of the study was be exploratory and not to make the findings generalisable. The ages of the participants in this study may reflect the nature of those attending a free dietitian clinic at a rural LHD. Private practice dietetic services may be an alternative source of service provision for those who have the ability to pay a full consultation fee or who have private health cover. Although our sample size was considered sufficient for this exploratory study, further research with a range of diverse participants may provide additional insights. Further research may be required to explore the experiences of a diverse range of people with coeliac disease.

Due to the qualitative nature of this research, any associations made are open for interpretation and therefore subject to a degree of researcher bias. To minimise this, coding was cross-checked by at least two researchers. It could be considered that there may be a non-responder bias in that those less satisfied with the service did not respond to the invitation to participate. Despite this, there was a diverse range of experiences, and opinions were found in this pragmatic study with the sample of participants, which represent one-third of all who were eligible in the pragmatic timeframe of this study. Finally, this study did not explore the experiences and those living in a rural area who had not consulted a dietitian or those who may have sought consultation from a private practice dietitian. Individuals who did not access dietetic services or who sought private practice services may have other diverse experiences that are yet to be explored.

Findings from the current study suggest a need for a consistent but individualised approach to coeliac disease management, as the number of dietetic consultations was variable among the participants and participant needs were varied. Individualised counselling based on the expressed needs and existing knowledge of patients was also identified as important for patient engagement. Finally, greater interdisciplinary communication and a consistent and comprehensive nutrition assessment is needed to gain a deeper understanding of the priorities and expectations of patients; this could lead to optimal patient satisfaction.

5. Conclusions

The findings of this study highlight that adapting to a gluten-free diet to manage coeliac disease can be challenging for some people, and this can be exacerbated by living in a rural context. The experiences of a dietetic consultation have the potential to contribute to self-confidence with managing and transitioning to lifelong dietary changes. In order to achieve this, more research is needed to provide greater insight into the perspectives of individuals with coeliac disease and inform how dietitians can best assist in improving the lives of those living with coeliac disease in rural Australia.

Supplementary Materials: The following are available online at https://www.mdpi.com/article/10.3390/nu13062074/s1, Semi-structured interview questions.

Author Contributions: Conceptualisation, R.L., L.J.B. and S.K.B.; methodology, L.J.B. and R.L.; formal analysis, R.L. and L.J.B.; investigation, R.L. and L.J.B.; data curation, R.L. and L.J.B.; writing—original draft preparation, R.L.; writing—review and editing, R.L., L.J.B., S.H., S.K.B. and E.T.C.; supervision, L.J.B., E.T.C., S.K.B.; project administration, L.J.B. and E.T.C.; funding acquisition, L.J.B. All authors have read and agreed to the published version of the manuscript.

Funding: A component of this research was funded by the University of Newcastle Department of Rural Health through internal UON research funding.

Institutional Review Board Statement: The study was conducted according to the guidelines of the Declaration of Helsinki, and approved by NSW Health Local Health District Human Research Ethics Committee (15/05/20/5.01, 11 May 2015) and the University of Newcastle Human Research Ethics Committee (H-2015-0165, 3 June 2015).

Informed Consent Statement: Informed consent was obtained from all subjects involved in the study.

Data Availability Statement: The data presented in this study are available on request from the corresponding author. The data are not publicly available due to participant privacy concerns as included data has been appropriately de-identified.

Acknowledgments: The authors would like to thank others who have contributed to supporting this research study (Camille Kelly) and to those who have provided feedback on drafts of this manuscript (Karin Fisher).

Conflicts of Interest: E.T.C. provides services in rural sites of the LHD. and may have consulted with some of the potential participants. This was potential conflict of interest was addressed by having two external researchers (R.L. and L.J.B), with interviews and data analysis being undertaken by external researchers. Funders had no role in the design of the study; in the collection, analyses, or interpretation of data; in the writing of the manuscript, or in the decision to publish the results.

References

1. Australian Institute of Health and Welfare. Rural and Remote Health. 2019. Available online: https://www.aihw.gov.au/reports/rural-health/rural-remote-health/contents/rural-health (accessed on 11 March 2021).
2. Burns, C.; Gibbon, P.; Boak, R.; Baudinette, S.; Dunbar, J. Food cost and availability in a rural setting in Australia. *Rural Remote Health* **2004**, *4*, 311.
3. Crowley, E.; Williams, L.T.; Brown, L.J. How do mothers juggle the special dietary needs of one child while feeding the family? *Nutr. Diet.* **2012**, *69*, 272–277. [CrossRef]
4. DiSabationo, A.; Corazza, G.R. Coeliac disease. *Lancet* **2009**, *373*, 1480–1493. [CrossRef]
5. Green, P.H.R.; Jabri, B. Coeliac Disease. *Lancet* **2003**, *362*, 383–391. [CrossRef]
6. Ciacci, C.; Ciclitira, P.; Hadjivassiliou, M.; Kaukinen, K.; Ludvigsson, J.F.; McGough, N.; Sanders, D.S.; Woodward, J.; Leonard, J.N.; Swift, G.L. The gluten-free diet and its current application in coeliac disease and dermatitis herpetiformis. *United Eur. Gastroenterol. J.* **2014**, *3*, 121–135. [CrossRef] [PubMed]
7. Cranney, A.; Zarkadas, M.; Graham, I.D.; Butzner, J.; Rashid, M.; Warren, R.; Molloy, M.; Case, S.; Burrows, V.; Switzer, C. The Canadian celiac health survey. *Digest Dis. Sci.* **2007**, *52*, 1087–1095. [CrossRef]
8. Sverker, A.; Östlund, G.; Hallert, C.; Hensing, G. Sharing life with a gluten-intolerant person–the perspective of close relatives. *J. Hum. Nutr. Diet.* **2007**, *20*, 412–422. [CrossRef] [PubMed]
9. Haines, M.L.; Anderson, R.P.; Gibson, P.R. Systematic review: The evidence base for long-term managament of coeliac disease. *Aliment. Pharmacol. Ther.* **2008**, *28*, 1042–1066. [CrossRef] [PubMed]
10. Lee, A.; Newman, J.M. Celiac diet: Its impact on quality of life. *J. Am. Diet. Assoc.* **2003**, *103*, 1533–1535. [CrossRef] [PubMed]
11. Rose, C.; Howard, R. Living with coeliac disease: A grounded theory study. *J. Hum. Nutr. Diet.* **2014**, *27*, 30–40. [CrossRef]
12. Sverker, A.; Hensing, G.; Hallert, C. Controlled by food'- lived experiences of coeliac disease. *J. Hum. Nutr. Diet.* **2005**, *18*, 171–180. [CrossRef] [PubMed]
13. Ludvigsson, J.F.; Bai, J.C.; Biagi, F.; Card, T.R.; Ciacci, C.; Ciclitira, P.J.; Green, P.H.; Hadjivassiliou, M.; Holdoway, A.; Van Heel, D.A.; et al. Diagnosis and management of adult coeliac disease: Guidelines from the British Society of Gastroenterology. *Gut* **2014**, *63*, 1–20. [CrossRef]

14. Abu-Janb, N.; Jaana, M. Facilitators and barriers to adherence to gluten-free diet among adults with celiac disease: A systematic review. *J. Hum. Nutr. Diet.* **2020**, *33*, 786–810. [CrossRef] [PubMed]

15. Zarkadas, M.; Cranney, A.; Case, S.; Molloy, M.; Switzer, C.; Graham, I.D.; Butzner, J.D.; Rashid, M.; Warren, R.E.; Burrows, V. The impact of a gluten-free diet on adults with coeliac disease: Results of a national survey. *J. Hum. Nutr. Diet.* **2006**, *19*, 41–49. [CrossRef]

16. Mahadev, S.; Simpson, S.; Lebwohl, B.; Lewis, S.K.; Tennyson, C.A.; Green, P.H. Is dietitian use associated with celiac disease outcomes? *Nutrients* **2013**, *5*, 1585–1594. [CrossRef] [PubMed]

17. Taylor, E.; Dickson-Swift, V.; Anderson, K. Coeliac disease: The path to diagnosis and the reality of living with the disease. *J. Hum. Nutr. Diet.* **2013**, *26*, 340–348. [CrossRef]

18. Sharp, K.; Walker, H.; Coppell, K.J. Coeliac disease and the gluten-free diet in New Zealand: The New Zealand coeliac health survey. *Nutr. Diet.* **2014**, *71*, 223–228. [CrossRef]

19. National Rural Health Alliance Inc. Under Pressure and under-Valued: Allied Health Professionals in Rural and Remote Areas. 2004. Available online: ruralhealth.org.au/sites/default/files/position-papers/position-paper-04-11-05.pdf (accessed on 11 March 2021).

20. Brown, L.; Williams, L.; Capra, S. Dietetic workload in rural acute care settings. *Nutr. Diet.* **2010**, *67*, 10.

21. Dowd, A.J.; Tamminen, K.A.; Jung, M.E.; Case, S.; McEwan, D.; Beauchamp, M.R. Motives for adherence to a gluten-free diet: A qualitative investigation invovling adults with coeliac disease. *J. Hum. Nutr. Diet.* **2014**, *27*, 542–549. [CrossRef]

22. Hancock, R.E.; Bonner, G.; Hollingdale, R.; Madden, A.M. 'If you listen to me properly, I feel good': A qualitative examination of patient experiences of dietetic consultations. *J. Hum. Nutr. Diet.* **2012**, *25*, 275–284. [CrossRef]

23. Madden, A.M.; Riordan, A.M.; Knowles, L. Outcomes in coeliac disease: A qualitative exploration of patients' views on what they want to achieve when seeing a dietitian. *J. Hum. Nutr. Diet.* **2016**, *29*, 607–616. [CrossRef] [PubMed]

24. O'Brien, B.C.; Harris, I.B.; Beckman, T.J.; Reed, D.A.; Cook, D.A. Stanards for reporting qualitative research: A synthesis of recommendations. *Acad. Med.* **2014**, *89*, 1245–1251. [CrossRef]

25. Australian Government Department of Health. Modified Monash Model. 2019. Available online: https://www.health.gov.au/sites/default/files/documents/2019/12/modified-monash-model---fact-sheet.pdf (accessed on 11 March 2021).

26. Thomas, D.R. A General Inductive Approach for Analyzing Qualitative Evaluation Data. *Am. J. Eval.* **2006**, *27*, 237–246. [CrossRef]

27. King, J.A.; Kaplan, G.G.; Godley, J. Experiences of coeliac disease in a changing gluten-free landscape. *J. Hum. Nutr. Diet.* **2018**, *32*, 72–79. [CrossRef]

28. Ford, S.; Howard, R.; Oyebode, J. Psychosocial aspects of coeliac disease: A cross-sectional survey of a UK population. *Br. J. Health Psychnol.* **2012**, *17*, 743–757. [CrossRef] [PubMed]

29. Hallert, C.; Sandlund, O.; Broqvist, M. Perceptions of health-related quality of life of men and women living with coeliac disease. *Scand. J. Caring Sci.* **2003**, *17*, 301–307. [CrossRef] [PubMed]

Review

Challenges of Monitoring the Gluten-Free Diet Adherence in the Management and Follow-Up of Patients with Celiac Disease

Herbert Wieser †, Ángela Ruiz-Carnicer, Verónica Segura, Isabel Comino and Carolina Sousa *

Department of Microbiology and Parasitology, Faculty of Pharmacy, University of Seville, 41012 Seville, Spain; acarnicer@us.es (Á.R.-C.); vsegura@us.es (V.S.); icomino@us.es (I.C.)
* Correspondence: csoumar@us.es; Tel.: +34-954-55-64-52
† Retired; h.wieser2@gmx.de.

Abstract: Celiac disease (CD) is a chronic gluten-responsive immune mediated enteropathy and is treated with a gluten-free diet (GFD). However, a strict diet for life is not easy due to the ubiquitous nature of gluten. This review aims at examining available evidence on the degree of adherence to a GFD, the methods to assess it, and the barriers to its implementation. The methods for monitoring the adherence to a GFD are comprised of a dietary questionnaire, celiac serology, or clinical symptoms; however, none of these methods generate either a direct or an accurate measure of dietary adherence. A promising advancement is the development of tests that measure gluten immunogenic peptides in stools and urine. Causes of adherence/non-adherence to a GFD are numerous and multifactorial. Inadvertent dietary non-adherence is more frequent than intentional non-adherence. Cross-contamination of gluten-free products with gluten is a major cause of inadvertent non-adherence, while the limited availability, high costs, and poor quality of certified gluten-free products are responsible for intentionally breaking a GFD. Therefore, several studies in the last decade have indicated that many patients with CD who follow a GFD still have difficulty controlling their diet and, therefore, regularly consume enough gluten to trigger symptoms and damage the small intestine.

Keywords: celiac disease; patients with CD; dietary adherence; gluten-free diet; symptoms

Citation: Wieser, H.; Ruiz-Carnicer, Á.; Segura, V.; Comino, I.; Sousa, C. Challenges of Monitoring the Gluten-Free Diet Adherence in the Management and Follow-Up of Patients with Celiac Disease. *Nutrients* **2021**, *13*, 2274. https://doi.org/10.3390/nu13072274

Academic Editor: Giacomo Caio

Received: 31 May 2021
Accepted: 26 June 2021
Published: 30 June 2021

Publisher's Note: MDPI stays neutral with regard to jurisdictional claims in published maps and institutional affiliations.

1. Introduction

Celiac disease (CD) is a chronic T-cell-mediated enteropathy caused by dietary exposure to the storage proteins of wheat, rye, barley, and some varieties of oats (called gluten in the field of CD) in genetically predisposed individuals [1–4]. Epidemiological data suggest a prevalence of approximately 1% in the general population of Western countries, Australia and New Zealand, but CD is also present in North Africa and major parts of Asia. To date, CD occurs rarely in people from other parts of Sub-Saharan Africa [5]. The precipitating gluten comprises hundreds of different proteins, which are roughly divided into the alcohol-soluble prolamins and the alcohol-insoluble glutelins [6]. Gluten proteins have been given the following cereal-specific names: wheat gliadins (prolamins) and glutenins (glutelins), rye secalins, barley hordeins, and oat avenins. They are all structurally characterized by unique repetitive amino acid sequences, rich in glutamine and proline, which are commonly considered the triggering factor of CD [7]. In particular, the high proline content makes these proteins resistant to complete digestion to ensure that long-chain immunogenic peptides reach the intestinal mucosa.

The pathogenesis of CD consists of the CD-specific passage of immunogenic gluten peptides through the small intestinal epithelium, and the combined adaptive and innate immune responses to the peptides in the lamina propria [2,8]. CD predominantly affects the duodenal intestine and induces a general flattening of the mucosa characterized by villous atrophy, crypt hyperplasia, and increased lymphocyte infiltration of the epithelium [2,9–13]. Moreover, CD is marked by a disease-specific antibody response to gluten peptides and tissue transglutaminase (autoantibodies). In addition to the ingestion of gluten and genetic predisposition, environmental factors such as infections, imbalanced intestinal microbiota, and increased intestinal permeability have been associated with the development of CD [4,14,15]. The clinical presentation of CD is extremely variable and can be divided into intestinal symptoms such as chronic diarrhea, abdominal pain, and, among children, the failure to grow normally as well as extra-intestinal manifestations including conditions caused by deficiencies of essential nutrients, neurological disorders, psychiatric complaints, dental enamel defects, liver abnormalities, joint manifestations, dermatitis herpetiformis, bone disease, problems in reproductive and endocrine systems, etc. [1,16,17]. A considerable number of patients present with atypical symptoms or even no symptoms despite the presence of a flattened small intestinal mucosa and CD-specific serum antibodies (asymptomatic CD) or present only CD-specific serum antibodies (potential CD). The diagnostic scheme of CD is usually based on clinical history and the presence of symptoms typical of CD, testing of CD-specific serum antibodies, histological judgement of small intestinal biopsies, and response to a gluten-free diet (GFD) [1,2]. In young children with clear symptoms and positive serology ($10\times$ the upper limit for normal antibody levels), CD diagnosis may be established without histological examination according to the diagnostic criteria of the European Society of Pediatric Gastroenterology, Hepatology, and Nutrition (ESPGHAN) [18,19].

Following a GFD creates difficulties and limitations in the life of patients with CD. Therefore, non-adherence to a GFD is a daily occurrence, which delays or prevents patient's healing. Compliance with a GFD among patients with CD, examined in the last few decades, is in the range of 45 to 90% [20]. Inadvertent gluten intake occurs more frequently than intentional intake, and gluten contamination in naturally or certified gluten-free foods and meals is likely to be one of the most important factors of inadvertent non-adherence to a GFD [21,22]. However, a strict GFD usually results in prompt relief of clinical symptoms, while recovery of small bowel mucosal damage may even take years [23]. A strict GFD is currently indicated in all cases of symptomatic CD and has also been recommended for asymptomatic patients. Recently, Ruiz-Carnicer et al. [24] have demonstrated that the fact that patients remain asymptomatic does not imply that they have not consumed gluten and that they are not at risk of developing histological lesions and complications as a result of their condition. In contrast, whether patients with potential CD should be treated with a GFD remains unclear [25].

This review focuses insight into the problematic issues of adherence to a strict GFD in patients with CD. Apart from highlighting the celiac dietary adherence methods, the recent literature on monitoring and the rate of GFD adherence, as well as on barriers to adherence, are presented.

2. Adherence to a GFD

Permanent, lifelong adherence to a strict GFD is the only available treatment for CD. Traditional cereal-based gluten-containing foodstuffs such as bread, pasta, and beer must be replaced by corresponding surrogates made from raw materials that do not contain gluten. However, a lifelong strict GFD is not easy, due to gluten ubiquity, cross-contamination of foods, improper labeling, and social constraints [20,26] and, therefore, a considerable portion of patients with CD do not adhere to a GFD. Numerous studies have investigated the factors influencing the compliance to a GFD, showing that adherence rates in patients with CD are well below optimal. A systematic review, summarizing the literature between 1980 and 2007, on the adherence to a GFD, had the following important findings: Rates for strict adherence ranged from 42 to 91%, depending on definition and method of assessment [27]. Adherence was most strongly associated with cognitive, emotional, and socio-cultural influences, membership to an advocacy group, and regular dietetic follow-up. Recent developments, including methods for monitoring adherence, the recently determined degree of adherence, understanding reasons for non-adherence, and interventions to improve adherence are outlined in the subsequent sections.

2.1. Monitoring Adherence

After diagnosis, it is important to monitor the adherence to a GFD to prevent ongoing symptoms and small intestinal damage. Non-responsiveness to a strict GFD, i.e., the presence of ongoing symptoms, could be caused by refractory CD or other complaints such as irritable bowel syndrome, lactose intolerance, or gastroesophageal reflux disease apart from non-adherence to a GFD [28–31]. Therefore, monitoring adherence is essential for identifying the cause on the ongoing symptoms. The following several procedures involving various approaches have been employed [32]: (a) periodic visits by expert nutritionists, (b) structured questionnaires, (c) clinical follow-up, (d) CD-specific antibodies, (e) gluten detection technologies that measure gluten prior to consumption in food samples, (f) gluten immunogenic peptides (GIP) in stools and/or urine, (g) serial endoscopies with collection of duodenal biopsies, and (h) and other endogenous markers such as fecal calprotectin (FC) measurements.

An endoscopy to collect intestinal biopsies is an invasive, expensive, and impractical procedure for frequent monitoring of GFD compliance. Additionally, since it may take up to two years for complete histological resolution of CD-related intestinal lesions, there is only a modest correlation between intestinal histology and the assessed dietary adherence that can be observed [33]. Similarly, symptomatic improvement during clinical follow-up may not accurately indicate adherence to a GFD because patients with proven strict adherence show more symptoms than healthy subjects [34]. Moreover, persistent symptoms may be induced, for example, by small-intestinal bacterial overgrowth, irritable bowel disease, microscopic colitis, and refractory CD, and therefore, are not markers for non-adherence. Additionally, a number of patients are asymptomatic, despite CD-specific small intestinal atrophy [23,35]. FC is used to diagnose and monitor inflammatory bowel diseases, such as ulcerative colitis and Crohn's disease, as well as in the differentiation of functional and organic intestinal pathologies [36]. The evidence currently shows a correlation between FC and CD activity in the pediatric population; however, there is a lack of studies in adult patients with CD [37,38].

2.1.1. Follow-Up by a Dietician

Studies have shown that patients who receive individual instructions on gluten-free foods and a GFD from healthcare providers are more likely to adhere to a GFD. Adherence is assessed by dietetic interviews supplemented by dietary questionnaires, such as the Standardized Dietician Evaluation (SDE). Measuring GFD adherence through patients´ self-reporting appears to be subjective and less accurate because it relies on the patient´s possibly limited knowledge of a GFD and gluten-free foods. In 2009, a simple validated Celiac Dietary Adherence Test (CDAT) for adults with CD was developed, which is one of the few validated measures available [39]. Items and domains, believed to be essential for successful GFD adherence, were used to develop an 85-item survey, which was administered to 200 individuals with biopsy-proven CD, who underwent standardized dietician evaluation and serological testing. A compacted 7-item questionnaire proved to be clinically relevant, easily administered, correlated highly with the SDE, and performed significantly better than serological testing [40].

A fast questionnaire, based on four simple questions with a five-level score (0–4; the Biagi score) was also shown to be a reliable and simple method of verifying adult patient compliance with a GFD [41]. The questionnaire was administered to 141 adult patients with CD on a GFD who were undergoing re-evaluation [42]. The score obtained was compared with the persistence of both villous atrophy and endomysial antibodies (EMAs). The rate of lower scores was higher among patients with the persistence of either villous atrophy or positive EMAs. For pediatric patients, a study of 151 children with CD demonstrated that short dietary questionnaires detected dietary transgressions only in 14% of patients, while a standardized dietary interview substantiated non-adherence in 52% of patients (Table 1) [43].

Table 1. Studies on the rate of adherence to GFD in children, adolescents, and adults in different countries. Anti-TGA, anti-tissue transglutaminase antibody; CD, celiac disease; CDAT, Celiac Dietary Adherence Test; GFD, gluten-free diet; GIP, gluten immunogenic peptides; NCGS, non-celiac gluten sensitivity.

	Country	Characteristics of Patients	*n*	Method	% Adherence to a GFD	References
Children	India	CD, >6 months on GFD	134	Questionnaires	66	Garg and Gupta, 2014 [44]
	Spain	CD	114	GIP	79	Comino et al., 2016 [45]
	Spain	CD and healthy subjects	65	GIP	55	Moreno et al., 2017 [46]
	Poland	CD, >2 years on GFD	102	Questionnaires Serology (anti-TGA)	67	Czaja and Bulsa, 2018 [47]
	Slovak Republic	CD	325	Questionnaires	69	Rimárová et al., 2018 [48]
	Australia	CD	151	BIAGI	86	Wessels et al., 2018 [43]
	Australia	CD	151	Standardized dietary interview	48	Wessels et al., 2018 [43]
	Spain	CD, >2 years on GFD	64	GIP	75	Comino et al., 2019 [49]
	Spain	CD	80	GIP	92	Fernández-Miaja et al., 2020 [50]
	Spain	CD, >6 months on GFD	43	GIP	65	Roca et al., 2020 [51]
	Italy	CD	200	BIAGI and Serology (anti-TGA)	84–100	Sbravati et al., 2020 [52]
Teenagers	Brazil	CD, >1 years on GFD	35	Questionnaires Serology	80	Rodrigues et al., 2018 [53]
	Italy	CD, >2 years on GFD	58	Questionnaires	36	Zingone et al., 2018 [54]
	Sweden	CD, >5 years on GFD	70	CDAT	86	Johansson et al., 2019 [55]
Teenagers and adults	Spain	CD	74	GIP	61	Comino et al., 2016 [45]
Adults	Italy	CD, >1 years on GFD	65	CDAT	82	Galli et al., 2014 [56]
	United States, US	CD, 10 years on GFD	355	CDAT	76	Villafuerte-Galvez et al., 2015 [57]
	Mexico	CD and NCGS, >3 years on GFD	80	CDAT	58	Ramírez-Cervantes et al., 2016 [58]
	Canada	CD, >4 years on GFD	222	CDAT	56	Silvester et al., 2016a [59]
	Canada	CD, 6 months on GFD	105	CDAT	91	Silvester et al., 2016b [60]

Table 1. *Cont.*

Country	Characteristics of Patients	*n*	Method	% Adherence to a GFD	References
Spain	CD and healthy subjects	69	GIP	52	Moreno et al., 2017 [46]
United Kingdom, UK	CD, >3 years on GFD	375	CDAT	53–81	Muhammad et al., 2017 [61]
Australia and New Zealand	CD, >6 months on GFD	5310	Online surveys	61	Halmos et al., 2018 [62]
Italy	CD, >5 years on GFD	750	Questionnaires Symptoms / Serology	90–91	Tovoli et al., 2018 [63]
Argentina	CD, >2 years on GFD	44	GIP	75	Costa et al., 2019 [64]
Italy	CD, >1 years on GFD	104	CDAT	65	Paganizza et al., 2019 [65]
Israel	CD, >4 years on GFD	301	BIAGI	82	Dana et al., 2020 [66]
Spain	CD, 7 years on GFD	271	CDAT	72	Fueyo-Diaz et al., 2020 [67]
Canada	CD	18	GIP	23	Silvester et al., 2020a; 2020b [68,69]
Spain	CD, >2 years on GFD	77	GIP	42	Ruiz-Carnicer et al., 2020 [24]
Argentina	CD, >2 years on GFD	53	GIP	11–62	Stefanolo et al., 2020 [70]
Spain	CD	76	GIP	21	Fernández-Bañares et al., 2021 [23]

However, there is considerable controversy regarding the validity of dietary questionnaires in assessing a GFD because patients do not intentionally record actual gluten consumption in the questionnaire. At the same time, there is evidence suggesting that the intervention of expert nutritionists cannot aid the detection of exposures in ~30% of the patients who present with mucosal damage until up to a third biopsy [71].

2.1.2. Serological Testing

The analysis of CD-specific serum antibody levels is a useful diagnostic tool in clinical practice and plays a supporting role in monitoring dietary compliance [72]. Indeed, there is evidence that persistently elevated levels of serum antibodies against gliadin (AGAs), transglutaminase 2 (TGAs), or deamidated gliadin peptide (DGPAs) can indicate non-adherence to a GFD. The performance of four different antibody collections (IgA DGPA, Ig A+G DGPA, IgA TGA, and IgA AGA) in detecting compliance with a GFD was tested in 95 Italian CD children with CD on a GFD >1 year [73]. Adherence interviews and serum collections were performed every three months. The Ig A+G DGPA level seemed to be the best for monitoring compliance with a GFD. The sensitivity to (i.e., ability to detect) transgressions from a GFD was 100% at 9 to 12 months and decreased to 76% after more than 1 year on a GFD. The IgA TGA and IgA AGA sensitivities were much lower (24 and 4%, respectively). To evaluate compliance with a GFD in a clinical ambulatory setting, a rapid IgA TGA assay, based on a whole-blood drop, was tested and compared with a conventional Enzyme-Linked ImmunoSorbent Assay (ELISA) and the patients´ interviews [74]. The results showed that the rapid test was just as reliable as a conventional ELISA and easy to perform in the ambulatory setting. However, patient interviews were shown to be more sensitive than serology in identifying patients who do not adhere to a GFD.

However, the normalization of antibody titers takes a long time, and these tests cannot identify incidents of occasional gluten exposure. Therefore, their use is limited to indicating a lack of adherence but is of no value for evaluating whether there is strict adherence to a GFD. Moreover, up to 10% of patients with CD are seronegative, despite the positive histology of duodenal biopsy samples. In 2007, a prospective comparative study, including 154 adult patients with CD on a GFD, demonstrated that serological tests cannot replace evaluations by trained nutritionist evaluation in the assessment of GFD adherence [75]. More recently, a comparison with a standardized evaluation by a registered dietician revealed that negative TGA levels are not necessarily indicative of good adherence to a GFD in pediatric patients with CD [76].

Altogether, the data clearly show that serology and dietary questionnaires at follow-up have a poor correlation with mucosal healing and, therefore, relying solely on these may underestimate the activity of CD [24,45,46,60,68,69,77–79].

2.1.3. Stools and Urine Testing

GIP, analyzed in stools and urine by using commercial ELISAs (monoclonal antibodies G12 and/or A1), have been proposed as new non-invasive biomarkers to detect gluten intake and to verify GFD compliance in patients with CD [45,46,49,80–85]. GIP are resistant to gastrointestinal digestion and responsible for immunogenic reactions in the T cells of patients with CD [83]. Unlike traditional methods for monitoring GFD adherence, which only evaluate the consequences of GFD transgressions, this non-invasive method enables a direct and quantitative assessment of gluten exposure.

These simple immunoassays could overcome some key unresolved scientific and clinical problems in CD management. Several prospective studies have been carried out to investigate the efficacy of GIP determination in stools. To assess the capacity to determine gluten ingestion and to monitor GFD compliance in patients with CD by the detection of GIP equivalents in stools, 53 children with CD (age range: 1–12 years) were enrolled [82]. Seven subjects had active disease and 46 subjects maintained a GFD for >2 years. After the controlled ingestion of a fixed amount of gluten (9–30 g), stools samples were analyzed using a G12 competitive ELISA. The results demonstrated that the method was a reliable tool for the detection of GFD transgressions in patients with active CD and CD in remission. A prospective multicenter study, including 188 patients with CD on a GFD and 73 healthy controls on a gluten-containing diet, revealed that 56 patients with CD (30%) had high levels of GIP in their stools [45]. All the controls except one (98.5%) had quantifiable amounts of GIP in stools. The results for patients with CD showed increasing dietary transgressions with advancing age (39% over 13 years) and gender (a predominance of males in this evaluation). Simultaneously, the study indicated limitations of traditional methods, such as food questionnaires or serological tests, for monitoring a GFD in patients with CD (Table 1).

To investigate the course of gluten intake after a diagnosis of CD and subsequent GFD, the stools of 64 pediatric patients with CD were analyzed for GIP at diagnosis and 6, 12, and 24 months thereafter [49]. Most of the children (97%) had detectable GIP at diagnosis. After GFD initiation, the rate of children with detectable GIP decreased to 13% at 6 months and increased to 25% at 24 months (Table 1). A recent examination of 25 patients with CD on a GFD for at least one year revealed that four patients (16%) tested positive for stool GIP [86]. Two of them complied strictly with a GFD according to the Biagi questionnaire and none of them manifested symptoms. The results demonstrated that stool GIP analysis identified those patients who did not comply with a GFD more accurately than a validated questionnaire. Therefore, monitoring the GIP in stools offers a direct objective quantitative assessment of exposure to gluten after CD diagnosis.

To compare the sensitivity and specificity of a rapid lateral flow technique (LFT) based on G12 and A1 monoclonal antibodies with the G12 ELISA method, stool samples from 54 healthy infants divided into two groups were analyzed [85]. Group 1 included infants aged 6 to 24 months, with an unrestricted consumption of gluten-containing cereals. Group 2 (negative controls) comprised of infants aged 0 to 6 months, either breastfed or formula fed, who had never ingested gluten. In group 1, all the infants had positive values using a conventional ELISA, while the LFT was negative in 5/20 cases. In group 2, all the samples were negative using both methods. Therefore, both methods were highly specific, while an ELISA had a higher sensitivity.

Urine can also be used to monitor GFD adherence, as shown in the following studies. A total of 76 healthy individuals (aged 3–57 years) (group 1) and 58 patients with CD (aged 3–64 years) (group 2) were subjected to different dietary conditions [46]. Urine samples were collected, concentrated, and analyzed for the presence and quantities of GIP by means of an LFT based on G12 antibodies. GIP were detectable in concentrated urines from group 1 individuals (previously subjected to a GFD) as early as 4–6 h after gluten intake (ingestion of at least a portion of pasta, bread, or whole grain from wheat, barley, and rye) and remained detectable for 1–2 days. The experiments indicated that the ingestion of >25 mg gluten could be detected in urine. The presence of GIP in the urine of group 2 (patients with CD on a GFD) revealed a high percentage of non-compliance with a GFD. GIP in urine were detectable in 48% of adults and 45% of children. An examination of duodenal biopsies revealed that most of the treated patients with CD without villous atrophy (89%) had no detectable GIP in their urine (Table 1), while all the patients with quantifiable GIP in their urine showed an incomplete intestinal mucosa recovery [46].

A simple and highly sensitive point-of-care (POC) device, based on a surface plasmon resonance biosensor and G12 and A1 monoclonal antibodies, enabled the rapid and efficient monitoring of a GFD directly in urine [87,88]. The excellent limits of detection of GIP (1.6–4.0 ng/mL) ensured the detection of gluten intake around the maximum amount tolerable for patients with CD (<50 mg). No sample pre-treatment, extraction, or dilution was required, and the analysis took less than 15 min. The assays had an excellent reproducibility (coefficient of variation: 3.6 and 11.3% for G12- and A1-based assays, respectively) and were validated with real samples.

Commercial ELISA (stool) and LFT (stool and urine) kits were used in parallel to assess the excretion of GIP in stools and urine [64]. Forty-four patients with CD, following a GFD >2 years, were asked to deliver stools and urine samples and were examined twice, 10 days apart. Considering the results for both assays, 11 patients (25%) had at least one positive GIP test. The ELISA and LFT were concordant (concomitantly positive or negative) in 67 out of 74 stools samples. To examine how often subjects with CD are still exposed to gluten, 53 Argentinian patients, who had been on a GFD for >2 years, collected stool each Friday and Saturday, and urine each Sunday for 4 weeks [70]. GIP were measured using a conventional ELISA (stool) and an LFT (urine). Overall, 159 of 420 samples (38%) were positive for GIP; 89% of patients had at least one positive stool or urine sample. On weekends, 70% of patients excreted GIP at least once compared with 62% during weekdays (Table 1).

Recently, an article by Ruiz-Carnicer et al. [24] provided additional data supporting the utility of GIP testing in the management of CD. The authors found that 58% of the patients with CD consuming a GFD had detectable GIP in their urine at least once, with higher rates of positivity on the weekend. The results demonstrated the high sensitivity (94%) and negative predictive value (97%) of GIP measurements in relation to duodenal biopsy findings. Additionally, they demonstrated that 25% of the patients on a GFD presented with Marsh type II–III duodenal lesions. If the authors had only considered serology, symptoms, or questionnaire scores, and had not performed a duodenal biopsy, 60–80% of these patients would have been overlooked. It was demonstrated that taking a GIP measurement on three days of the week, including the weekend, could be the best option to confirm GFD adherence in the short term and appears to accurately predict the absence of histological lesions. The introduction of GIP testing as an assessment technique for GFD adherence may help in ascertaining dietary compliance and to target the most suitable intervention during follow-up. Additionally, this would eliminate the uncertainty in predicting the reappearance of villous atrophy and its possible complications and the association between any nonspecific symptomatology and the underlying disease.

Moreover, a key future use of GIP may be the evaluation of nonresponsive CD. Since refractory CD type 1 is a diagnosis of exclusion based on a traditional dietary adherence assessment, the use of GIP could reveal unsuspected gluten exposures in this population as well and guide an intervention. It may also help distinguish between gluten exposures and irritable bowel syndrome or fermentable oligosaccharides, disaccharides, monosaccharides, and polyols (FODMAPs) intolerance among symptomatic individuals. In addition, these new technologies help to improve not only healing, but also quality-of-life outcomes such anxiety and depression in patients with CD.

2.2. Rate of Adherence

Numerous international studies on GFD adherence rates in patients with CD have been published in the last few decades. Italian research groups have been leading in the investigation of GFD adherence. The rates of gluten-free adherence presented in the literature are characterized by extreme variability among the populations studied. For example, 38 studies, published up to 2007, indicated an adherence rate ranging between 42 and 91% [89]. More recently, compliance with a GFD has been reported to be in the range of approximately 45 to 90% [20]. In the case of children, a systematic review of 49 studies, published up to 2018, revealed adherence rates ranging from 23 to 98% (Table 1) [90]. The broad variability in adherence rates reported in the literature may be explained by the different populations examined (e.g., adults, adolescents, children, ethnic minorities), but also by the different methods used for determining adherence, the quality of investigations, and the definition of adherence (e.g., strict, high, partial, fair, poor adherent, or non-adherent). Despite the importance of monitoring the adherence to a GFD, there are no clear guidelines for how to explore this. In the following, selected studies from the last decade present data on adherence rates grouped into examinations of adults, adolescents, children, and ethnic minorities. The adherence rate was predominantly evaluated using a CDAT (Table 1).

Adults. A total of 65 Italian patients newly diagnosed with CD were re-evaluated after one year on a GFD [56]. According to dietary interviews, based on a CDAT, 82% had adequate adherence to a GFD. To evaluate differences in the GFD adherence between the clinically diagnosed and screening-detected Italian patients with CD, the medical records of 750 subjects, diagnosed during 2004 and 2013, were evaluated [63]. The patients were considered to have complied with a GFD, if all the following criteria were satisfied: (1) no reported intentional or accidental gluten ingestion; (2) absence of CD-related symptoms; and (3) negative IgA-TGA. The results clearly demonstrated that both groups of patients shared similar GFD adherence (91% versus 90%) years after the diagnosis. In another Italian investigation, 104 patients with CD took part in a study focused on the relationship between adherence to a GFD and knowledge of the disease and its treatment in general [65]. By means of a CDAT, 65% of patients reported strictly adhering to a GFD (Table 1). Adherence was strongly and significantly associated with knowledge of CD and a GFD.

A total of 271 Spanish patients with CD completed a series of questionnaires regarding adherence to a GFD among other items; 72% of subjects indicated an excellent or good adherence (Table 1) [67]. Higher levels of adherence were particularly associated with CD-specific self-efficacy. Three studies from the UK, which compared the GFD adherence of white and South Asian patients with CD, indicated that white patients were adherent to a GFD in a range from 53 to 81%. To examine the GFD adherence of patients with CD in Israel, 301 subjects completed an anonymous online questionnaire sent via the Israeli Celiac Association and social networks [66]. According to the Biagi score, 82% of patients were found to be highly adherent to a GFD (score 3–4) and 18% were poorly adherent (score 0–2) (Table 1). Young age at diagnosis and smoking were significantly associated with non-adherence to a GFD.

To assess GFD adherence among a Canadian community, 222 patients with CD on a GFD completed a CDAT [59]. The results revealed that the degree of strict adherence was only 56%. Another group of Canadian adults with CD was enrolled to examine GFD adherence six months after diagnosis by means of a CDAT [60]. Of the 105 participants, 91% reported gluten exposure less than once a month and thus were consistent with adequate adherence. To determine long-term GFD adherence, 355 US patients with CD were re-evaluated after a mean time of 9.9 years on a GFD [57]. Adequate adherence, determined using a CDAT, was found in 76% of the patients (Table 1).

The evaluation of GFD adherence among Mexican adults with CD (*n* = 56) and non-celiac gluten sensitivity (NCGS) (*n* = 24) using a CDAT revealed that 58% of subjects perceived themselves as strictly adherent [58]. However, inadvertent gluten intake was frequent in both CD (39%) and NCGS (33%). The result of a CDAT provided to 5310 adult and adolescent Australians and New Zealanders with CD, showed that 61% were adherent to a GFD (Table 1) [62]. Older age, being male, symptoms after gluten ingestion, better food knowledge, and lower risk of psychological distress were independent predictors of adherence. In summary, thirteen studies from nine countries indicated adherence rates among adult patients with CD in a range from 53 to 91%.

Adolescents. Concerning dietary compliance, 58 young Italian patients with CD around the transition age were asked to answer the question: "Do you voluntarily eat gluten-containing food?" Nobody answered, "often or at times"; 16 subjects answered, "on special occasions"; 21, "rarely"; and 21, "never" [54]. Out of the 21 patients who declared no dietary lapses, five showed positive serology, which indicated that they were underestimating or not aware of gluten contamination in food (Table 1).

To investigate the GFD adherence of 70 Swedish adolescents with CD detected by screening, they filled in a CDAT and came to a five-year follow-up [55]. The evaluation showed that 86% of the adolescents were adherent to a GFD five years after screening (Table 1).

The rate of non-adherence to a GFD among 35 patients with CD under 20 years of age was assessed in a tertiary Brazilian referral center by means of a questionnaire and a serological test [53]. Despite dietary guidance, 20% of the patients reported non-adherence to the diet. Altogether, three studies on adolescents from three countries revealed adherence rates from 36 to 86% (Table 1).

Children. A Polish study compared frequency and cause of diet failure in 102 children with CD treated with a GFD for >2 years [47]. Dietary adherence was evaluated serologically (TGA test) and using a questionnaire. The results showed that one-third of the patients, mainly children aged 13–18 years, failed to follow a GFD. Younger children (up to 12 years) were less likely to abandon the diet. In this age group, inadvertent diet failure prevailed, while teenagers predominantly interrupted their diet intentionally. Personal questionnaires, completed by 325 parents or caregivers of pediatric patients with CD from the Slovak Republic, revealed that strict GFD adherence was maintained by 69% of children [48]. Adherence was significantly higher among girls compared to boys, younger children, children with a family history of CD, and children of parents with higher education.

The GFD adherence of 200 Italian children with CD was assessed to evaluate differences as a consequence of transition from a referral center (V1) to a general pediatrician (V2) [52]. Adherence was measured using the Biagi score and the IgA TGA test at the last follow-up at V1 and at an annual follow-up at V2. Adherence at V2 was significantly worse compared with V1: 84% vs. 95% (Biagi score) and 97% vs. 100% (TGA test), respectively. A study of 134 Indian children with CD using a questionnaire-based interview showed that 88 patients (66%) were adherent to a strict GFD [44]. Compliance was higher in children up to 9 years of age than in children aged >9 years. In summary, four studies from four countries showed adherence rates of children in a range from 66 to 84% (Table 1).

Ethnic minorities. Differences in GFD adherence between ethnically different patients were reported in three studies from the UK. After CD diagnosis, 71 South Asian and 67 white adult patients with CD from a single center in Southern Derbyshire were advised to maintain a GFD [91]. After six weeks on a GFD, the patients were classified by an experienced dietician as adhering strictly to a GFD (not ingesting any gluten) or taking gluten. Fifty-four white patients (81%) had strict diet compared with only 37 Asians patients (52%). A combined cross-sectional survey, based on a CDAT and CD adherence score, was used to determine the GFD adherence of 375 white and 38 South Asian patients with CD residing in the UK [92]. The results demonstrated an almost identical rate of adhering to a GFD (53 and 52%, respectively). The examination of ethnically diverse populations with CD in the North West of England was performed using the assessment of dietetic notes from follow-up visits with dieticians [93]. The results revealed that the rate of strict adherence to a GFD in the 33 South Asian patients was significantly lower (12%) than that of the 113 tested Caucasian patients (65%). Altogether, the adherence rates of ethnic minorities in the UK, assessed in three studies, ranged from 12 to 52% (Table 1).

Few studies have tried to separate intentional and inadvertent gluten ingestion. Identification of inadvertent gluten consumption, for example, using questionnaires or interviews, is more difficult or even impossible compared to intentional consumption. Therefore, the rate of inadvertent non-adherence, caused by, for example, contaminated naturally gluten-free products or hidden vital gluten, is likely to be highly underestimated and this fact should be particularly considered in the judgement of reported GFD adherence rates. A cross-sectional survey on intentional and inadvertent non-adherence was conducted using a self-completion questionnaire received from 287 adult patients with CD from the North East of England [90]. Intentional gluten consumption was common (40%), but not as frequent as inadvertent lapses. A multicenter study, including seven Spanish hospitals, investigated the adherence of 366 adult patients with CD using the Morisky questionnaire scale [94]. Results showed that 71.5% of patients reported perfect treatment adherence, 23.5% inadvertent poor adherence, and 5% intentional poor adherence. A total of 82 Canadian patients with CD, having a medium of 6 years GFD experience, completed a questionnaire with items related to GFD information including GFD adherence [95]. They reported strict adherence (55%), inadvertent gluten consumption (21%), and intentional gluten consumption (18%). In summary, three studies from three countries clearly demonstrated that intentional non-adherence to a GFD was less frequent than inadvertent non-adherence (Table 1).

In conclusion, studies from the last decade indicated that many patients with CD following a GFD still have difficulties in controlling their diet and hence regularly consume sufficient gluten to trigger symptoms and small intestinal damage. The rates of GFD adherence did not significantly change compared to previous decades. Ethnic minorities appear to be at the highest risk for non-adherence. Inadvertent lapses are distinctly more frequent than intentional lapses.

2.3. Factors Influencing Dietary Adherence

A comprehensive understanding of the facilitators and barriers associated with a strict GFD adherence is needed to develop strategies and resources to assist patients with CD following a GFD. Causes of adherence/non-adherence to a GFD are numerous and multifactorial, but limited evidence is available on their nature and magnitude (recently reviewed by Muhammad et al. [20] and Abu-Janb and Jaana [26]). A number of factors governing long-term GFD adherence have been identified by Leffler et al. [33] and Villafuerte-Galvez et al. [57]. An overview of the recent literature on factors that may limit or enhance the GFD adherence of patients with CD is presented in the following paragraphs, including a consideration of gluten cross-contamination in foods, problems in gluten analysis, knowledge of a GFD and gluten-free foods as well as availability, costs, and quality of dietetic gluten-free foods and a broad spectrum of individual factors.

2.3.1. Gluten Cross-Contamination

Gluten cross-contamination in gluten-free foods is omnipresent; therefore, it is complicated for patients with CD to maintain a zero-gluten diet. Cross-contamination contributes majorly to inadvertent non-adherence and affects both naturally and certified gluten-free foods and composite foods containing hidden gluten as an additive. Naturally gluten-free material can be contaminated with gluten-containing cereals in the field, during storage, transport, and food production. Usually, these foods are not analyzed for gluten and not labeled gluten-free, and thus, are virtually considered safe by patients with CD. However, a number of studies have demonstrated that naturally gluten-free foods can be heavily contaminated with gluten. For example, a pilot study revealed that some naturally gluten-free grains, seeds, and flours, used for gluten-free food production, contained gluten levels up to 2925 mg/kg [96]. Consequently, naturally gluten-free foods are major contributors to inadvertent GFD lapses, which can hardly be avoided.

Labeled dietetic gluten-free foods seem to be safe for patients with CD. Numerous investigations, however, have shown that gluten levels above the threshold of 20 mg/kg are a daily occurrence. For example, a systematic review including 24 cross-sectional studies revealed that, on average, 13% of industrial products labeled gluten- free and 42% of gluten-free products offered by food services exceeded the threshold of 20 mg/kg [97,98]. Fortunately, certified products rarely exceed gluten contents above 100 mg/kg but may contribute to an increased inadvertent intake of gluten.

Patients with CD following a GFD should be aware of numerous composite foods and medicines that contain hidden sources of vital gluten, which is frequently used as an additive to improve product quality. Composite foods, which increase the risk of gluten contamination, are ubiquitous, for example, soups and sauces at restaurants, coffee creamers at cafeterias, ice cream at ice cream parlors, and sausages at butcher shops. Eating at workplaces, schools, hospitals, assistive living facilities, and while visiting other people is an additional risk factor for hidden gluten intake. Inadvertent gluten intake via hidden channels can exceed the allowed amount by ten-fold or more. To prevent gluten intake, patients should study the label in case of prepacked products, where gluten has to be indicated as an allergen according to the Codex Standard 1–1985. In the case of unpacked products, patients should ask for information about gluten content.

Surprisingly, little is known about the quantity of gluten that is accidentally ingested by patients despite a GFD. Previous results of GIP analyses in stools and urine were used to estimate the amount of gluten consumed by patients with CD following a GFD for at least one year [99]. A total of 74 adults and adolescents ($\geq$13 years old) were invited for a follow-up visit, in which stool and blood samples were collected. The computed daily mean gluten consumption for adults and adolescents measured in stool was 244 mg, for older children (4–12 years old) it was 387 mg, and for younger children (0–3 years old) it was 155 mg. The computed daily mean gluten consumption measured in urine was 363 mg for adults and adolescents and 316 mg for children. Individual data showed that a small portion of patients consumed more than 600 mg gluten on a daily basis. The analysis of GIP in stools from 64 pediatric patients with CD was used to estimate gluten intake at diagnosis and after 6, 12, and 24 months on a GFD [49]. The mean gluten exposure dropped from 5543 mg/day at diagnosis to 144 mg/day at 6 months on a GFD and then increased to 606 mg/day by 24 months. Recently, Silvester et al. [68,69] confirmed, in a period of 10 days, that 67% of patients with CD showed gluten exposure, and that the excretion kinetics were highly variable among individuals. The results demonstrated that most patients with CD can, in actuality, only attain a gluten-reduced diet, and gluten exposure is common, intermittent, and usually silent.

In addition, one of the promising advances to improve adherence to diet in patients with CD are the LFT Nima ™ (Nima Labs, Inc, San Francisco, CA, USA) and EZ Glutent ™ gluten sensors (ELISA Technologies, Inc, Gainesville, FL, USA), devices that have been developed to integrate food processing, gluten detection, interpretation results, and data transmission into a portable consumer device.

In conclusion, cross-contamination of gluten-free products with gluten is likely to be the main cause of inadvertent non-adherence to a GFD. The amount of gluten in certified gluten-free-labeled foods is normally low (rarely above 100 mg/kg). In contrast, naturally gluten-free foods as well as composite foods and medicines with hidden gluten may contain gluten amounts far above 100 mg/kg and are, therefore, serious risk factors for patients' health. Stools and urine analyses have shown that patients with CD on a GFD are at risk for gluten consumption exceeding the allowed amounts by ten-fold or more. Non-adherence, caused by contaminated naturally gluten-free and composite foods, is rarely identified by dieticians and questionnaires; therefore, a promising advance could be the determination of gluten by consumers.

2.3.2. Knowledge of GFD

Knowledge of a GFD and gluten-free foods is one of the most significant facilitators of GFD adherence identified in the literature. For example, knowledge that a strict GFD avoids immediate reactions and prevents long-term complications is a primary reason for dietary adherence. However, many individuals with CD exhibit significant deficits in their knowledge of a GFD and the gluten contents of foods, as exemplified by the following studies. A total of 5912 members of the two Canadian Celiac Associations filled in a questionnaire regarding knowledge of a GFD among other items [100]. When asked to review a list of 15 foods and ingredients and to identify those that were not allowed on a GFD, only 49% correctly identified all seven non-allowed items, and only an additional 33% correctly identified six of the seven non-allowed items. In another study, 82 adult Canadian patients with CD, having a medium of 6 years GFD experience, were asked to find gluten-containing foods among 17 different common foods [95]. None of the participants identified the potential gluten content of all the foods, and only 25 participants identified at least 14 foods correctly. A total of 104 adult Italian patients with CD completed a questionnaire comprising 31 statements on CD in general and foods appropriate in a GFD [65]. The patients' knowledge was generally poor (only one patient answered all the questions correctly) and was significantly associated with poor GFD adherence.

Reliable, up-to-date, and comprehensive education could play an important role in improving knowledge and GFD adherence. Dietary and psychological counselling, for example through health professionals and dieticians, can essentially increase compliance with a GFD [20]. Studies of patients with CD, recruited from a CD clinic in New Delhi, showed that repeated counselling of patients with CD following a GFD remarkably increased the level of adherence [101]. At the beginning of the study, only 53% of subjects maintained excellent or good adherence to a GFD. After 6 months of repeated counselling, the level of adherence increased to 92%. The evaluation of 1832 US adults with CD (19–65 years old) revealed a highly significant association between dietary adherence to a GFD (indicated by a higher CDAT score) and having visited a healthcare provider [102].

Visiting healthcare providers may cause improved adherence, because they provide patients with a better understanding of how to implement the diet and appreciation of the diet's importance. However, striking deficiencies in the quality of information and in the level of support that patients receive from their healthcare providers have been reported [33]. A total of 154 adult patients with CD judged the rate of adequate information and support provided by their healthcare providers as follows: dieticians, 63%; gastroenterologists, 57%; primary care physicians, 36%; and pharmacists, 23%. A large cross-sectional Canadian study, including 5912 adult patients with CD, resulted in low ratings of the usefulness of information provided by dieticians (52%), gastroenterologists (43%), and family physicians (25%) [100]. As confidence in treatment advice impacts GFD adherence, education of healthcare professionals should be improved substantially.

Apart from communication with healthcare providers, patient advocacy groups, and other persons with CD, the Internet, print media, and cookbooks are the most commonly used sources of information about a GFD [95]. Several studies have demonstrated that membership of celiac societies is associated with a greater GFD adherence; they particularly offer practical advice and support regarding gluten-free foods and a GFD. Members are often a self-selected group of patients who may exhibit greater motivation to adhere to a GFD and have significantly better knowledge of gluten-free foods than non-members.

Moreover, psychological support counselling seems to be able to increase GFD compliance (Addolorato et al. [103]. A cohort of 66 Italian patients with CD with state anxiety and current depression were randomized in two groups. In group A, psychological support was started at the beginning of a GFD, while group B was free of psychological support. A follow-up after six months revealed that the subjects of group A had a significantly higher rate of GFD compliance (39.4%) compared to group B (9.1%).

Different online programs have been shown to be effective in significantly improving adherence and represent a promising resource for individuals with CD who are struggling to achieve adequate GFD adherence. To test the effectiveness of an interactive online intervention, 189 Australian adults with CD were recruited and divided into a group receiving the intervention ($n = 101$) and a control group without intervention ($n = 88$) [104]. The primary outcome measure was GFD adherence during a three-month follow-up. The online intervention showed significantly improved GFD adherence in the intervention group relative to the control group. The effectiveness of a smartphone app (MyHealthyGut, Vancouver, BC, Canada) in helping adults self-manage CD was assessed by Dowd et al. [105]. The participants of the study reported high levels of app usability and were satisfied with the features of the app. The vast majority of participants reported improvements in GFD adherence, gastrointestinal symptoms, quality of life, self-regulatory efficacy, anxiety, and depression.

As home cooking is among the top challenges for patients with CD and a frequent cause of non-adherence, the feasibility and acceptability of a cooking-based education intervention was assessed [106]. A total of 12 adult US patients with CD participated in two intervention sessions (2 consecutive days, 4.5 h each), co-led by the center dietician and a trained chef, and completed a follow-up interview. At the 1-month follow-up, participants had significantly improved GFD adherence and quality of life scores. All the participants agreed that the intervention was helpful, promoted eating foods they otherwise would not have tried, and made them more informed about gluten sources when eating out.

In summary, lack of knowledge about a GFD and gluten-free foods is an important cause of inadvertent non-adherence. Future interventions to improve adherence to a GFD should include methods with very high specificity and sensitivity to monitor gluten exposure in patients with CD, enhanced dietary and psychological counselling by healthcare providers with expertise in CD, as well as the promotion of education by online training programs.

2.3.3. Difficulties with Certified Gluten-Free Foods

Apart from the dilemmas experienced when eating out, travelling, and socializing with friends, limited availability, high costs, poor quality, and not-understanding the labeling of dietetic gluten-free foods are frequent reasons for breaking a strict GFD intentionally [107]. Previously, dietetic gluten-free foods were niche market products, available almost exclusively in health food shops, pharmacies, and through mail order companies. Over the past few decades, the market for certified gluten-free foods has grown enormously, and nowadays they are also offered in many supermarkets and online providers. Nevertheless, dietetic gluten-free foods are not available everywhere, and patients with CD still have difficulty finding gluten-free foods when shopping. Limited access to gluten-free meals in canteens, schools, and kindergartens or whilst travelling is an additional barrier to GFD adherence.

A survey on the availability of 10 wheat-based everyday foods and 10 corresponding gluten-free counterparts at 30 different stores in London revealed an average availability of 82% of gluten-free foods [108]. Regular supermarkets had a greater availability (90%), whereas budget supermarkets (9%) and corner shops (9%) had almost no gluten-free versions. The inspection across four venues and five US geographic regions demonstrated that the availability of certified gluten-free products varied by region and venue but remained limited compared to their wheat-based counterparts [109]. Availability was greatest (66%) in health food and upscale venues. Among 38 South Asian patients with CD living in the UK, 85% reported no gluten-free foods in their local Asian stores [92]. Regarding eating out and travelling, dining establishments are frequently unable to provide safe meals; thus, patients make mistakes on their GFD [110].

The higher costs of dietetic gluten-free foods are a further barrier to adherence. For example, the cost of gluten-free foods (*n* = 63) and their gluten-containing counterparts (*n* = 126) were compared in 12 different Austrian supermarkets [111]. On average, gluten-free products were substantially higher in cost, ranging from +205% (breakfast cereals) to +267% (bread and bakery products) compared to similar gluten-containing products. Original data on retail prices in four major UK supermarkets within the UK "Bread and Cereal" category showed that the average price of gluten-free products was increased by a factor of 1.9 compared to corresponding gluten-containing products [112]. In Italy, the gluten-free version of pasta, the traditional staple food, was sold in supermarkets with an average price equal to more than twice that of conventional pasta [113]. A premium price was particularly found for the following attributes: small packages, brands that specialized in gluten-free products, content in fiber, and the presence of quinoa as an ingredient. A Greek study compared the cost of gluten-free products from supermarkets and pharmacies with the cost of their conventional counterparts [114]. All the supermarket gluten-free products, except for one, were more expensive by 22 to 334%, and all the pharmacy gluten-free products were more expensive by 88 to 476%. The weekly economic burden of a GFD, calculated for one person, ranged from EUR 12 to 28 per week. Gluten-free products, purchased across five geographic US regions and four venues in 2018, were 183% more expensive than their wheat-based counterparts [109]. In comparison to a study in 2006, the cost of gluten-free products has declined from 240 to 183%.

Adherence to a GFD is often associated with receiving gluten-free foods on prescription. In an English study, 375 adult patients with CD who were, in part, supported by the prescription of gluten-free foods, completed a CDAT to measure their dietary adherence [61]. Of the patients not receiving gluten-free foods on prescription, 62% were classified as non-adhering to a GFD compared with 42% of those receiving gluten-free foods on prescription. Paul et al. [115] suggested that in resource-limited settings, medical professionals should be creative in formulating cheaper and locally sourced gluten-free options in close co-operation with the dieticians, thereby ensuring the availability of gluten-free food items at affordable prices and the improvement of GFD adherence.

Patients are frequently unsatisfied with the quality of dietetic gluten-free foods, for example, due to poor flavor, taste, texture, and mouthfeel. In particular, the replacement of wheat bread and other baked goods, pasta, and beer by gluten-free substitutes is one of the most critical aspects of a GFD. Despite the improved quality of gluten-free breads in the last number of years, most products on the market are still described as low-quality products: gluten-free breads often have a low volume, pale crust, crumbly texture, bland flavor, and high rate of stalling, and gluten-free pasta products have an inferior texture [116]. The taste and flavor of gluten-free beer, made with alternative ingredients such as sorghum, millet, or buckwheat are not acceptable to many patients with CD. Novel strategies for the production of high-quality gluten-free beers, made from ultra-low gluten barley lines [117] or enzymatically detoxified barley malts [118], are currently in development and may contribute to improved compliance with a GFD.

Not understanding food labels is frequently associated with poorer dietary adherence. To investigate whether patients with CD can identify gluten-free foods based on product labeling, 25 different food items were presented to 144 adult US patients at 6, 12, and 24 months after CD diagnosis [119]. The median overall accuracy scores were 84, 96, and 84% at 6, 12, and 24 months, respectively. An examination of 375 adult patients with CD from the UK revealed that 73% of those who reported not understanding food labels were classified as non-adhering to a GFD compared with 45% who understood food labels [61]. A combined cross-sectional survey, including 972 ethnically different patients with CD residing in the UK, confirmed that there were substantial issues with the understanding of food labels that impacted adherence to a GFD [92].

In summary, numerous reasons are responsible for intentionally breaking a GFD. In particular, the limited availability, high costs, and poor quality of certified gluten-free products mislead patients to consume corresponding gluten-containing products. The non-ability to read and interpret food labels is another cause of concern, as it leads to inadvertent GFD non-adherence.

2.3.4. Individual Factors

A number of studies have reported a broad spectrum of individual factors that impact adherence to a GFD in patients with CD, including gender, age, ethnicity, education, and mental health conditions [20,26]. The influence of gender on adherence rate has been judged differently. A systematic review, presented in 2013, reported no difference between men and women [89], whereas a large study of 2018 demonstrated that males are more adherent than females [62]. In contrast, a multicenter clinical trial studied the adherence of children and adults with CD to a GFD ($n = 188$) by measuring stool GIP [45]. When further stratified by gender, GFD adherence was found to be closely related to the patient's gender in certain age groups. More males $\geq$13-years old had positive GIP stools compared with females in the same age group (60% vs. 31.5%, respectively, $p = 0.034$). The higher proportion of non-adherent male patients compared with females could be attributed to milder symptoms found in men or to stricter self-control over the diet in women. Regarding the age of the patients, the majority (85.7%) of celiac children between zero and three years of age had stools negative for GIP. Among those $\geq$13 years old, the proportion rose up to 39.2% with positive GIP. Altogether, these data show how increasing control over the diet could yield an increase in dietary adherence, as demonstrated by the four-fold greater adherence seen in children $\leq$3 years old. They have strong parental control over their diet, but no social pressure as compared with the adherence of adolescent males who are under little parental control but are subject to strong social influences. An Indian study of 134 children with CD found that the percentage of compliant children dropped from 76% in children aged 2–5 years to 41% in children above 9 years of age [44]. These results were in accordance with a previous Swedish study, which reported adherence rates of 93% at 12 years of age, decreasing to 76% in the 15–17 years age group [120]. In comparison, patients diagnosed later in life had relatively good adherence. The examination of GFD adherence among 35 biopsy-proven Finnish patients with CD aged over 50 years revealed that 27 patients (77%) maintained a strict diet, 5 patients (14%) had occasional transgressions less often than once a month, and 3 patients (9%) did not start a GFD [121]. In a cohort of Italian patients with CD aged over 65 years ($n = 59$), adherence to a GFD was 90% [122].

For adolescents with CD, adherence to a GFD is linked to many difficulties, and non-adherence is common even among those aware of the risks. The majority of transgressions occur intentionally at home or at parties. The reasons for non-adherence are manifold. Adolescents are aware of being different from others when maintaining a GFD, which often requires discussions and special requests [123]. Public eating can produce stigmatizing experiences in adolescence, and thus, dietary non-adherence can be understood in terms of dealing with GFD concealment. The absence of symptoms after consuming a small amount of gluten, the absence of peer acceptance, and even more often troublesome diet administration are further common reasons for non-adherence [45,47].

The "Prague consensus" focused on the GFD difficulties during the transition of CD-affected adolescents to adulthood, which presents a fragile and high-risk period for intentional and inadvertent gluten intake [124]. Although young children with CD may adhere to a GFD because of parental influence, the situation remains complex in adolescents. Several mechanisms for improving GFD adherence among youth have been identified, including regular CD engagement with an experienced multidisciplinary team, electronic tool utilization, and awareness of accurate resources for self-guided education [125].

Asian patients with CD living in Western countries may find it more difficult to adhere strictly to a GFD for a number of reasons [91]. If their command of the local language is poor, their understanding of food labeling will be compromised. They often live within an extended family setting, which puts increased pressure on them to comply with their cultural norms and, therefore, neglect a GFD. In addition, making Asian foods with naturally gluten-free materials can be very difficult.

Individuals with a high level of education have been shown to have higher GFD adherence. A study on long-term GFD adherence among 355 adult US patients with CD demonstrated that the level of education differed significantly between the subjects with adequate and inadequate adherence [57]. A higher level of education was associated with adequate adherence ($p = 0.002$), even after controlling for household income. Furthermore, a significant inverse correlation was found between the CDAT score and education level.

The circumstances in which the temptation to break the diet is most likely are practical in nature: being busy, having limited time or a break from usual routine, and difficulty in finding gluten-free foods when eating away from home [126]. Physical and emotional factors may be being physically unwell, tired, lacking energy, and bored as well as being stressed, upset or down, and emotionally exhausted. The overall health of patients and their adherence levels were shown to be highly correlated [94]. The health-related quality of life score obtained by patients with CD who reported perfect GFD adherence was found to be significantly higher than that obtained by patients with CD who reported unintended lapses and patients with CD who reported intentional lapses. A systematic review with meta-analysis, including eight cross-sectional studies, demonstrated a moderate association between poor GFD adherence and self-reported depressive symptoms, but further studies are needed to confirm this association [127].

The relationship between the strength of motivation and GFD adherence has been shown in several studies. For example, a Treatment Self-Regulation Questionnaire and a CDAT, administered to 433 South Italian people with CD, aged between 18 and 79 years, demonstrated that motivation strongly correlated with GFD adherence [128]. Poor adherence can be associated with low self-regulation, self-efficacy, facilitation, support, and psychological distress, social fear, depression, or frequent self-control lapses. Therefore, it is necessary to consider these factors in the treatment of individuals with CD. Studies of 200 North American adults with CD revealed that self-compassion predicted stricter adherence to a GFD both directly and indirectly through self-regulatory efficacy [129]. These findings indicate that self-compassion and concurrent self-regulatory efficacy are important cognitions in understanding adherence to a GFD.

In conclusion, individual facilitators and barriers concerning GFD adherence are manifold and include various socio-demographic factors such as education, age, and ethnicity as well as mental health conditions such as motivation, self-efficacy, and depression. It is necessary to consider these factors in the counselling of individuals with CD to improve GFD adherence. However, the number of studies that have investigated this aspect is still low, and future research is urgently needed.

3. Conclusions

The rates of GFD adherence among patients with CD reported in the literature are highly variable and are determined by the degree of adherence, the methods to assess it, and the barriers to its implementation. Inadvertent dietary lapses are distinctly more frequent than intentional lapses. For these reasons, adherence to a GFD by patients with CD have been reported to be far away from the optimal. The methods for evaluating the adherence to a GFD are comprised of a dietary questionnaire, serological test, or clinical symptoms; however, none of these methods generate either a direct or an accurate measure of dietary adherence. A small-bowel biopsy is the "gold-standard" method for CD diagnosis. However, according to most clinical guidelines, its role in the follow-up of patients is limited to cases involving a lack of clinical response or the recurrence of symptoms. A promising advancement is the development of tests that measure GIP in stools and urine. The cross-contamination of gluten-free products with gluten is one of the main causes of inadvertent non-adherence. Therefore, adequate nutritional counselling as well as an assessment technique for a GFD are necessary for patients diagnosed with CD in order to help in ascertaining dietary compliance and to target the most suitable intervention during follow-up and prevent the risk of possible complications long term.

Author Contributions: Conceptualization, H.W. and C.S.; data curation, H.W. and C.S.; investigation, H.W. and C.S.; methodology, H.W. and C.S.; resources, H.W. and C.S.; writing—original draft, H.W., Á.R.-C., V.S., I.C. and C.S.; writing—review and editing, H.W., Á.R.-C., V.S., I.C. and C.S. All authors have read and agreed to the published version of the manuscript.

Funding: This work was supported by grants from Ministry of Economy, Knowledge, Business and University (project AT17_5489_USE) PAIDI 2020: Knowledge Transfer Activities and Fundación Progreso y Salud, Consejería de Salud, Junta de Andalucía (project PI-0053-2018).

Institutional Review Board Statement: Not applicable.

Informed Consent Statement: Not applicable.

Data Availability Statement: Not Applicable.

Conflicts of Interest: The authors declare no conflict of interest.

References

1. Caio, G.; Volta, U.; Sapone, A.; Leffler, D.A.; De Giorgio, R.; Catassi, C.; Fasano, A. Celiac disease: A comprehensive current review. *BMC Med.* **2019**, *17*, 142. [CrossRef]
2. Lindfors, K.; Ciacci, C.; Kurppa, K.; Lundin, K.E.; Makharia, G.K.; Mearin, M.L.; Murray, J.A.; Verdu, E.F.; Kaukinen, K. Coeliac disease. *Nat. Rev. Dis. Primers* **2019**, *5*, 3. [CrossRef]
3. Rej, A.; Aziz, I.; Sanders, D.S. Coeliac disease and noncoeliac wheat or gluten sensitivity. *J. Intern. Med.* **2020**, *288*, 537–549. [CrossRef]
4. Lebwohl, B.; Rubio-Tapia, A. Epidemiology, presentation, and diagnosis of celiac disease. *Gastroenterology* **2021**, *160*, 63–75. [CrossRef]
5. Kelly, C.P.; Dennis, M. Patient Education: Celiac Disease in Adults (Beyond the Basics). Available online: https://www.uptodate.com/contents/celiac-disease-in-adults-beyond-the-basics (accessed on 23 May 2021).
6. García-Molina, M.D.; Giménez, M.J.; Sánchez-León, S.; Barro, F. Gluten free wheat: Are we there? *Nutrients* **2019**, *11*, 487. [CrossRef]
7. Wieser, H.; Koehler, P.; Scherf, K.A. The Two Faces of Wheat. *Front Nutr.* **2020**, *7*, 517313. [CrossRef]
8. Sharma, N.; Bhatia, S.; Chunduri, V.; Kaur, S.; Sharma, S.; Kapoor, P.; Kumari, A.; Garg, M. Pathogenesis of celiac disease and other gluten related disorders in wheat and strategies for mitigating. *Them. Front. Nutr.* **2020**, *7*, 6. [CrossRef]
9. Qiao, S.W.; Iversen, R.; Raki, M.; Sollid, L.M. The adaptive immune response in celiac disease. *Semin. Immunopathol.* **2012**, *34*, 523–540. [CrossRef]
10. Sollid, L.M.; Jabri, B. Triggers and drivers of autoimmunity: Lessons from coeliac disease. *Nat. Rev. Immunol.* **2013**, *13*, 294–302. [CrossRef] [PubMed]
11. Escudero-Hernández, C.; Peña, A.; Bernardo, D. Immunogenetic pathogenesis of celiac disease and non-celiac gluten sensitivity. *Curr. Gastroenterol. Rep.* **2016**, *18*, 36. [CrossRef]
12. López-Casado, M.Á.; Lorite, P.; Ponce de León, C.; Palomeque, T.; Torres, M.I. Celiac disease autoimmunity. *Arch. Immunol. Ther. Exp. (Warsz)* **2018**, *66*, 423–430. [CrossRef]
13. Christophersen, A.; Risnes, L.F.; Dahal-Koirala, S.; Sollid, L.M. Therapeutic and diagnostic implications of T Cell scarring in celiac disease and beyond. *Trends Mol. Med.* **2019**, *25*, 836–852. [CrossRef]
14. Lebwohl, B.; Sanders, D.S.; Green, P.H.R. Coeliac disease. *Lancet* **2018**, *391*, 70–81. [CrossRef]
15. Murray, J.A.; Frey, M.R.; Oliva-Hemker, M. Celiac disease. *Gastroenterology* **2018**, *154*, 2005–2008. [CrossRef]
16. Leffler, D.A.; Green, P.H.R.; Fasano, A. Extraintestinal manifestations of coeliac disease. *Nat. Rev. Gastroenterol. Hepatol.* **2015**, *12*, 561–571. [CrossRef]
17. Therrien, A.; Kelly, C.P.; Silvester, J.A. Celiac disease: Extraintestinal manifestations and associated conditions. *J. Clin. Gastroenterol.* **2020**, *54*, 8–21. [CrossRef] [PubMed]
18. Husby, S.; Koletzko, S.; Korponay-Szabó, I.R.; Mearin, M.L.; Phillips, A.; Shamir, R.; Troncone, R.; Giersiepen, K.; Koninckx, C.; Ventura, A.; et al. European society for pediatric gastroenterology, hepatology, and nutrition guidelines for the diagnosis of coeliac disease. *J. Pediatr. Gastroenterol. Nutr.* **2012**, *54*, 136–160. [CrossRef]
19. Husby, S.; Koletzko, S.; Korponay-Szabó, I.; Kurppa, K.; Mearin, M.L.; Ribes-Koninckx, C.; Shamir, R.; Troncone, R.; Auricchio, R.; Castillejo, G.; et al. European society paediatric gastroenterology, hepatology and nutrition guidelines for diagnosing coeliac disease. *J. Pediatr. Gastroenterol. Nutr.* **2020**, *70*, 141–156. [CrossRef]
20. Muhammad, H.; Reeves, S.; Jeanes, Y.M. Identifying and improving adherence to the gluten-free diet in people with coeliac disease. *Proc. Nutr. Soc.* **2019**, *78*, 418–425. [CrossRef]
21. Mahadev, S.; Murray, J.A.; Wu, T.T.; Chandan, V.S.; Torbenson, M.S.; Kelly, C.P.; Maki, M.; Green, P.H.; Adelman, D.; Lebwohl, B. Factors associated with villus atrophy in symptomatic coeliac disease patients on a gluten-free diet. *Aliment. Pharmacol. Ther.* **2017**, *45*, 1084–1093. [CrossRef] [PubMed]

22. Itzlinger, A.; Branchi, F.; Elli, L.; Schumann, M. Gluten-free diet in celiac disease–forever and for all? *Nutrients* **2018**, *10*, 1796. [CrossRef]

23. Fernández-Bañares, F.; Beltrán, B.; Salas, A.; Comino, I.; Ballester-Clau, R.; Ferrer, C.; Molina-Infante, J.; Rosinach, M.; Modolell, I.; Rodríguez-Moranta, F.; et al. Persistent villous atrophy in de novo adult patients with celiac disease and strict control of gluten-free diet adherence: A multicenter prospective study (CADER Study). *Am. J. Gastroenterol.* **2021**, *116*, 1036–1043. [CrossRef] [PubMed]

24. Ruiz-Carnicer, Á.; Garzón-Benavides, M.; Fombuena, B.; Segura, V.; García-Fernández, F.; Sobrino-Rodríguez, S.; Gómez-Izquierdo, L.; Montes-Cano, M.A.; Rodríguez-Herrera, A.; Millán, R.; et al. Negative predictive value of the repeated absence of gluten immunogenic peptides in the urine of treated celiac patients in predicting mucosal healing: New proposals for follow-up in celiac disease. *Am. J. Clin. Nutr.* **2020**, *112*, 1240–1251. [CrossRef]

25. Trovato, C.M.; Montuori, M.; Valitutti, F.; Leter, B.; Cucchiara, S.; Oliva, S. The challenge of treatment in potential celiac disease. *Gastroenterol. Res. Pract.* **2019**, *2019*, 8974751. [CrossRef] [PubMed]

26. Abu-Janb, N.; Jaana, M. Facilitators and barriers to adherence to gluten-free diet among adults with celiac disease: A systematic review. *J. Hum. Nutr. Diet* **2020**, *33*, 786–810. [CrossRef]

27. Hall, N.J.; Rubin, G.; Charnock, A. Systematic review: Adherence to a gluten-free diet in adult patients with coeliac disease. *Aliment. Pharmacol. Ther.* **2009**, *30*, 315–330. [CrossRef] [PubMed]

28. Rubio-Tapia, A.; Murray, J.A. Classification and management of refractory coeliac disease. *Gut* **2010**, *59*, 547–557. [CrossRef]

29. Rishi, A.R.; Rubio-Tapia, A.; Murray, J.A. Refractory celiac disease. *Expert Rev. Gastroenterol. Hepatol.* **2016**, *10*, 537–546. [CrossRef] [PubMed]

30. Malamut, G.; Cellier, C. Refractory celiac disease. *Gastroenterol. Clin. N. Am.* **2019**, *48*, 137–144. [CrossRef]

31. Penny, H.A.; Baggus, E.M.R.; Rej, A.; Snowden, J.A.; Sanders, D.S. Non-responsive coeliac disease: A comprehensive review from the NHS England National Centre for refractory coeliac disease. *Nutrients* **2020**, *12*, 216. [CrossRef] [PubMed]

32. Rodrigo, L.; Pérez-Martinez, I.; Lauret-Braña, E.; Suárez-González, A. Descriptive Study of the Different Tools Used to Evaluate the Adherence to a Gluten-Free Diet in Celiac Disease Patients. *Nutrients* **2018**, *10*, 1777. [CrossRef]

33. Leffler, D.A.; Edwards-George, J.; Dennis, M.; Schuppan, D.; Cook, F.; Franko, D.L.; Blom-Hoffman, J.; Kelly, C.P. Factors that influence adherence to a gluten-free diet in adults with celiac disease. *Dig. Dis. Sci.* **2008**, *53*, 1573–1581. [CrossRef] [PubMed]

34. Laurikka, P.; Salmi, T.; Collin, P.; Huhtala, H.; Mäki, M.; Kaukinen, K.; Kurppa, K. Gastrointestinal symptoms in celiac disease patients on a long-term gluten-free diet. *Nutrients* **2016**, *8*, 429. [CrossRef] [PubMed]

35. Kaukinen, K.; Peräaho, M.; Lindfors, K.; Partanen, J.; Woolley, N.; Pikkarainen, P.; Karvonen, A.L.; Laasanen, T.; Sievänen, H.; Mäki, M.; et al. Persistent small bowel mucosal villous atrophy without symptoms in coeliac disease. *Aliment. Pharmacol. Ther.* **2007**, *25*, 1237–1245. [CrossRef]

36. Motaganahalli, S.; Beswick, L.; Con, D.; van Langenberg, D.R. Faecal calprotectin delivers on convenience, cost reduction and clinical decision making in inflammatory bowel disease: A real world cohort study. *Intern. Med. J.* **2018**, *49*, 94–100. [CrossRef] [PubMed]

37. Balamtekın, N.; Baysoy, G.; Uslu, N.; Orhan, D.; Akçören, Z.; Özen, H.; Gürakan, F.; Saltik-Temızel, İ.N.; Yüce, A. Fecal calprotectin concentration is increased in children with celiac disease: Relation with histopathological findings. *Turk J. Gastroenterol.* **2012**, *23*, 503–508. [CrossRef]

38. Biskou, O.; Gardner-Medwin, J.; Mackinder, M.; Bertz, M.; Clark, C.; Svolos, V.; Russell, R.K.; Edwards, C.A.; McGrogan, P.; Gerasimidis, K. Faecal calprotectin in treated and untreated children with coeliac disease and juvenile idiopathic arthritis. *J. Pediatr. Gastroenterol. Nutr.* **2016**, *63*, e112–e115. [CrossRef] [PubMed]

39. Leffler, D.A.; Dennis, M.; Edwards, J.B.; Jamma, S.; Magge, S.; Cook, E.F.; Schuppan, D.; Kelly, C.P. A simple validated gluten-free diet adherence survey for adults with celiac disease. *Clin. Gastroenterol. Hepatol.* **2009**, *7*, 530–536. [CrossRef]

40. Gładyś, K.; Dardzińska, J.; Guzek, M.; Adrych, K.; Małgorzewicz, S. Celiac Dietary Adherence Test and Standardized Dietician Evaluation in Assessment of Adherence to a Gluten-Free Diet in Patients with Celiac Disease. *Nutrients* **2020**, *12*, 2300. [CrossRef]

41. Biagi, F.; Andrealli, A.; Bianchi, P.I.; Marchese, A.; Klersy, C.; Corazza, G.R. A gluten-free diet score to evaluate dietary compliance in patients with coeliac disease. *Br. J. Nutr.* **2009**, *102*, 882–887. [CrossRef] [PubMed]

42. Biagi, F.; Bianchi, P.I.; Marchese, A.; Trotta, L.; Vattiato, C.; Balduzzi, D.; Brusco, G.; Andrealli, A.; Cisarò, F.; Astegiano, M.; et al. A score that verifies adherence to a gluten-free diet: A cross-sectional, multicentre validation in real clinical life. *Br. J. Nutr.* **2012**, *108*, 1884–1888. [CrossRef] [PubMed]

43. Wessels, M.M.S.; Te Lintelo, M.; Vriezinga, S.L.; Putter, H.; Hopman, E.G.; Mearin, M.L. Assessment of dietary compliance in celiac children using a standardized dietary interview. *Clin. Nutr.* **2018**, *37*, 1000–1004. [CrossRef] [PubMed]

44. Garg, A.; Gupta, R. Predictors of compliance to gluten-free diet in children with celiac disease. *Int. Sch. Res. Not.* **2014**, *2014*, 248402. [CrossRef] [PubMed]

45. Comino, I.; Fernández-Bañares, F.; Esteve, M.; Ortigosa, L.; Castillejo, G.; Fombuena, B.; Ribes-Koninckx, C.; Sierra, C.; Rodríguez-Herrera, A.; Salazar, J.C.; et al. Fecal gluten peptides reveal limitations of serological tests and food questionnaires for monitoring gluten-free diet in celiac disease patients. *Am. J. Gastroenterol.* **2016**, *111*, 1456–1465. [CrossRef] [PubMed]

46. Moreno, M.L.; Cebolla, Á.; Muñoz-Suano, A.; Carrillo-Carrion, C.; Comino, I.; Pizarro, Á.; León, F.; Rodríguez-Herrera, A.; Sousa, C. Detection of gluten immunogenic peptides in the urine of patients with coeliac disease reveals transgressions in the gluten-free diet and incomplete mucosal healing. *Gut* **2017**, *66*, 250–257. [CrossRef]

47. Czaja-Bulsa, G.; Bulsa, M. Adherence to gluten-free diet in children with celiac disease. *Nutrients* **2018**, *10*, 1424. [CrossRef]
48. Rimarova, K.; Dorko, E.; Diabelkova, J.; Sulinova, Z.; Makovicky, P.; Bakova, J.; Uhrin, T.; Jenca, A.; Jencova, J.; Petrasova, A.; et al. Compliance with gluten-free diet in a selected group of celiac children in the Slovak Republic. *Cent. Eur. J. Public Health* **2018**, *26*, S19–S24. [CrossRef] [PubMed]
49. Comino, I.; Segura, V.; Ortigosa, L.; Espín, B.; Castillejo, G.; Garrote, J.A.; Sierra, C.; Millán, A.; Ribes-Koninckx, C.; Román, E.; et al. Prospective longitudinal study: Use of faecal gluten immunogenic peptides to monitor children diagnosed with coeliac disease during transition to a gluten-free diet. *Aliment. Pharmacol. Ther.* **2019**, *49*, 1484–1492. [CrossRef]
50. Fernández-Miaja, M.; Díaz-Martín, J.J.; Jiménez-Treviño, S.; Suárez-González, M.; Bousoño-García, C. Estudio de la adherencia a la dieta sin gluten en pacientes celiacos [Study of adherence to the gluten-free diet in coeliac patients]. *An. Pediatr.* **2020**, *94*, 377–384. [CrossRef]
51. Roca, M.; Donat, E.; Masip, E.; Crespo-Escobar, P.; Cañada-Martínez, A.J.; Polo, B.; Ribes-Koninckx, C. Analysis of gluten immunogenic peptides in feces to assess adherence to the gluten-free diet in pediatric celiac patients. *Eur. J. Nutr.* **2020**, *60*, 2131–2140. [CrossRef]
52. Sbravati, F.; Pagano, S.; Retetangos, C.; Bolasco, G.; Labriola, F.; Filardi, M.C.; Grondona, A.G.; Alvisi, P. Adherence to gluten- free diet in a celiac pediatric population referred to the general pediatrician after remission. *J. Pediatr. Gastroenterol. Nutr.* **2020**, *71*, 78–82. [CrossRef]
53. Rodrigues, M.; Yonamine, G.H.; Fernandes Satiro, C.A. Rate and determinants of non-adherence to a gluten-free diet and nutritional status assessment in children and adolescents with celiac disease in a tertiary Brazilian referral center: A cross-sectional and retrospective study. *BMC Gastroenterol.* **2018**, *18*, 15.
54. Zingone, F.; Massa, S.; Malamisura, B.; Pisano, P.; Ciacci, C. Coeliac disease: Factors affecting the transition and a practical tool for the transition to adult healthcare. *United Eur. Gastroenterol. J.* **2018**, *6*, 1356–1362. [CrossRef] [PubMed]
55. Johansson, K.; Norström, F.; Nordyke, K.; Myleus, A. Celiac dietary adherence test simplifies determining adherence to a gluten-free diet in swedish adolescents. *J. Pediatr. Gastroenterol. Nutr.* **2019**, *69*, 575–580. [CrossRef]
56. Galli, G.; Esposito, G.; Lahner, E.; Pilozzi, E.; Corleto, V.D.; Di Giulio, E.; Aloe Spiriti, M.A.; Annibali, B. Histological recovery and gluten-free diet adherence: A prospective 1-year follow-up study of adult patients with coeliac disease. *Aliment. Pharmacol. Ther.* **2014**, *40*, 639–647. [CrossRef]
57. Villafuerte-Galvez, J.; Vanga, R.R.; Dennis, M.; Hansen, J.; Leffler, D.A.; Kelly, C.P.; Mukherjee, R. Factors governing long-term adherence to a gluten-free diet in adult patients with coeliac disease. *Aliment. Pharmacol. Ther.* **2015**, *42*, 753–760. [CrossRef]
58. Ramírez-Cervantes, K.L.; Romero-López, A.V.; Núñez-Álvarez, C.A.; Uscanga-Domínguez, L.F. Adherence to a gluten-free diet in mexican subjects with gluten-related ¡disorders: A high prevalence of inadvertent gluten intake. *Rev. Investig. Clin.* **2016**, *68*, 229–234.
59. Silvester, J.; Weiten, D.; Graff, L.; Walker, J.R.; Duerksen, D.R. Living gluten-free: Adherence, knowledge, lifestyle adaptations and feelings towards a gluten-free diet. *J. Hum. Nutr. Diet.* **2016**, *29*, 374–382. [CrossRef]
60. Silvester, J.A.; Graff, L.A.; Rigaux, L.; Walker, J.R.; Duerksen, D.R. Symptomatic suspected gluten exposure is common among patients with coeliac disease on a gluten-free diet. *Aliment. Pharmacol. Ther.* **2016**, *44*, 612–619. [CrossRef]
61. Muhammad, H.; Reeves, S.; Ishaq, S.; Mayberry, J.; Jeanes, Y.M. Adherence to a gluten free diet is associated with receiving gluten free foods on prescription and understanding food labelling. *Nutrients* **2017**, *9*, 705. [CrossRef] [PubMed]
62. Halmos, E.P.; Deng, M.; Knowles, S.R.; Sainsbury, K.; Mullan, B.; Tye-Din, J.A. Food knowledge and psychological state predict adherence to a gluten-free diet in a survey of 5310 Australians and New Zealanders with coeliac disease. *Aliment. Pharmacol. Ther.* **2018**, *48*, 78–86. [CrossRef] [PubMed]
63. Tovoli, F.; Negrini, G.; Sansone, V.; Faggiano, C.; Catenaro, T.; Bolondi, L.; Granito, A. Celiac disease diagnosed through screening programs in at-risk adults is not associated with worse adherence to the gluten-free diet and might protect from osteopenia/osteoporosis. *Nutrients* **2018**, *10*, 1940. [CrossRef]
64. Costa, A.F.; Sugai, E.; Temprano, M.P.; Niveloni, S.I.; Vázquez, H.; Moreno, M.L.; Domínguez-Flores, M.R.; Muñoz-Suano, A.; Smecuol, E.; Stefanolo, J.P.; et al. Gluten immunogenic peptide excretion detects dietary transgressions in treated celiac disease patients. *World J. Gastroenterol.* **2019**, *25*, 1409–1420. [CrossRef]
65. Paganizza, S.; Zanotti, R.; D'Odorico, A.; Scapolo, P.; Canova, C. Is adherence to a gluten-free diet by adult patients with celiac disease influenced by their knowledge of the gluten content of foods? *Gastroenterol. Nurs.* **2019**, *42*, 55–64. [CrossRef]
66. Dana, Z.Y.; Lena, B.; Vered, R.; Haim, S.; Efrat, B. Factors associated with non adherence to a gluten free diet in adult with celiac disease: A survey assessed by BIAGI score. *Clin. Res. Hepatol. Gastroenterol.* **2020**, *44*, 762–767. [CrossRef]
67. Fueyo-Díaz, R.; Magallón-Botaya, R.; Gascón-Santos, S.; Asensio-Martínez, Á.; PalaciosNavarro, G.; Sebastián-Domingo, J.J. The effect of self-efficacy expectations in the adherence to a gluten free diet in celiac disease. *Psychol. Health* **2020**, *35*, 734–749. [CrossRef]
68. Silvester, J.A.; Comino, I.; Kelly, C.P.; Sousa, C.; Duerksen, D.R.; the DOGGIE BAG study group. Most patients with celiac disease on gluten-free diets consume measurable amounts of gluten. *Gastroenterology* **2020**, *158*, 1497–1499.e1. [CrossRef] [PubMed]
69. Silvester, J.A.; Comino, I.; Rigaux, L.N.; Segura, V.; Green, K.H.; Cebolla, Á.; Weiten, D.; Dominguez, R.; Leffler, D.A.; Leon, F.; et al. Exposure sources, amounts and time course of gluten ingestion and excretion in patients with coeliac disease on a gluten-free diet. *Aliment. Pharmacol. Ther.* **2020**, *52*, 1469–1479. [CrossRef]

70. Stefanolo, J.P.; Tálamo, M.; Dodds, S.; de la Paz Temprano, M.; Costa, A.F.; Moreno, M.L.; Pinto-Sánchez, M.I.; Smecuol, E.; Vázquez, H.; Gonzalez, A.; et al. Real-World Gluten Exposure in patients with celiac disease on gluten-free diets, determined from gliadin immunogenic peptides in urine and fecal samples. *Clin. Gastroenterol. Hepatol.* **2020**, *19*, 484–491.e1. [CrossRef]

71. Sharkey, L.M.; Corbett, G.; Currie, E.; Lee, J.; Sweeney, N.; Woodward, J.M. Optimising delivery of care in coeliac disease -comparison of the benefits of repeat biopsy and serological follow-up. *Aliment. Pharmacol. Ther.* **2013**, *38*, 1278–1291. [CrossRef] [PubMed]

72. Taraghikhah, N.; Ashtari, S.; Asri, N.; Shahbazkhani, B.; Al-Dulaimi, D.; Rostami-Nejad, M.; Rezaei-Tavirani, M.; Razzaghi, M.R.; Zali, M.R. An updated overview of spectrum of gluten-related disorders: Clinical and diagnostic aspects. *BMC Gastroenterol.* **2020**, *20*, 258. [CrossRef] [PubMed]

73. Monzani, A.; Rapa, A.; Fonio, P.; Tognato, E.; Panigati, L.; Oderda, G. Use of deamidated gliadin peptide antibodies to monitor diet compliance in childhood celiac disease. *J. Pediatr. Gastroenterol. Nutr.* **2011**, *53*, 55–60. [CrossRef] [PubMed]

74. Zanchi, C.; Ventura, A.; Martelossi, S.; Di Leo, G.; Di Toro, N.; Not, T. Rapid anti-transglutaminase assay and patient interview for monitoring dietary compliance in celiac disease. *Scand. J. Gastroenterol.* **2013**, *48*, 764–766. [CrossRef] [PubMed]

75. Leffler, D.A.; Edwards-George, J.B.; Dennis, M.; Cook, E.F.; Schuppan, D.; Kelly, C.P. A prospective comparative study of five measures of gluten-free diet adherence in adults with coeliac disease. *Aliment. Pharmacol. Ther.* **2007**, *26*, 1227–1235. [CrossRef]

76. Mehta, P.; Pan, Z.; Riley, M.D.; Liu, E. Adherence to a Gluten-free Diet: Assessment by Dietician Interview and Serology. *J. Pediatr. Gastroenterol. Nutr.* **2018**, *66*, e67–e70. [CrossRef]

77. Dickey, W.; Hughes, D.F.; McMillan, S.A. Disappearance of endomysial antibodies in treated celiac disease does not indicate histological recovery. *Am. J. Gastroenterol.* **2000**, *95*, 712–724. [CrossRef]

78. Kaukinen, K.; Sulkanen, S.; Mäki, M.; Collin, P. IgA-class transglutaminase antibodies in evaluating the efficacy of gluten-free diet in coeliac disease. *Eur. J. Gastroenterol. Hepatol.* **2002**, *14*, 311–315. [CrossRef]

79. Bannister, E.G.; Cameron, D.J.; Ng, J.; Chow, C.W.; Oliver, M.R.; Alex, G.; Catto-Smith, A.G.; Heine, R.G.; Webb, A.; McGrath, K.; et al. Can celiac serology alone be used as a marker of duodenal mucosal recovery in children with celiac disease on a gluten-free diet? *Am. J. Gastroenterol.* **2014**, *109*, 1478–1483. [CrossRef]

80. Morón, B.; Cebolla, A.; Manyani, H.; Alvarez-Maqueda, M.; Megías, M.; del Thomas, M.C.; López, M.C.; Sousa, C. Sensitive detection of cereal fractions that are toxic to celiac disease patients by using monoclonal antibodies to a main immunogenic wheat peptide. *Am. J. Clin. Nutr.* **2008**, *87*, 405–414. [CrossRef]

81. Morón, B.; Bethune, M.T.; Comino, I.; Manyani, H.; Ferragud, M.; López, M.C.; Cebolla, A.; Khosla, C.; Sousa, C. Toward the assessment of foodtoxicity for celiac patients: Characterization of monoclonal antibodies to a main immunogenic gluten peptide. *PLoS ONE* **2008**, *3*, e2294. [CrossRef]

82. Comino, I.; Real, A.; Vivas, S.; Síglez, M.Á.; Caminero, A.; Nistal, E.; Casqueiro, J.; Rodríguez-Herrera, A.; Cebolla, A.; Sousa, C. Monitoring of gluten-free diet compliance in celiac patients by assessment of gliadin 33-mer equivalent epitopes in feces. *Am. J. Clin. Nutr.* **2012**, *95*, 670–677. [CrossRef] [PubMed]

83. Cebolla, Á.; Moreno, M.L.; Coto, L.; Sousa, C. Gluten Immunogenic Peptides as Standard for the Evaluation of Potential Harmful Prolamin Content in Food and Human Specimen. *Nutrients* **2018**, *10*, 1927. [CrossRef]

84. Gerasimidis, K.; Zafeiropoulou, K.; Mackinder, M.; Ijaz, U.Z.; Duncan, H.; Buchanan, E.; Cardigan, T.; Edwards, C.A.; McGrogan, P.; Russell, R.K. Comparison of clinical methods with the faecal gluten immunogenic peptide to assess gluten intake in coeliac disease. *J. Pediatr. Gastroenterol. Nutr.* **2018**, *67*, 356–360.

85. Roca, M.; Donat, E.; Masip, E.; Crespo-Escobar, P.; Fornes-Ferrer, V.; Polo, B.; Ribes-Koninckx, C. Detection and quantification of gluten immunogenic peptides in feces of infants and their relationship with diet. *Rev. Esp. Enferm. Dig.* **2019**, *111*, 106–110. [CrossRef]

86. Porcelli, B.; Ferretti, F.; Cinci, F.; Biviano, I.; Santini, A.; Grande, E.; Quagliarella, F.; Terzuoli, L.; Bacarelli, M.R.; Bizzaro, N.; et al. Fecal gluten immunogenic peptides as indicators of dietary compliance in celiac patients. *Minerva Gastroenterol. Dietol.* **2020**, *66*, 201–207. [CrossRef]

87. Soler, M.; Estevez, M.C.; de Moreno, M.L.; Cebolla, A.; Lechuga, L.M. Label-free SPR detection of gluten peptides in urine for non-invasive celiac disease follow-up. *Biosens. Bioelectron.* **2016**, *79*, 158–164. [CrossRef] [PubMed]

88. Peláez, E.C.; Estevez, M.C.; Domínguez, R.; Sousa, C.; Cebolla, A.; Lechuga, L.M. A compact SPR biosensor device for the rapid and efficient monitoring of gluten-free diet directly in human urine. *Anal. Bioanal. Chem.* **2020**, *412*, 6407–6417. [CrossRef]

89. Hall, N.J.; Rubin, G.P.; Charnock, A. Intentional and inadvertent non-adherence in adult coeliac disease. A cross-sectional survey. *Appetite* **2013**, *68*, 56–62. [CrossRef] [PubMed]

90. Myleus, A.; Reilly, N.R.; Green, P.H. Rate, risk factors, and outcomes of nonadherence in pediatric patients with celiac disease: A systematic review. *Clin. Gastroenterol. Hepatol.* **2020**, *18*, 562–573. [CrossRef]

91. Holmes, G.K.; Moor, F. Coeliac disease in Asians in a single centre in southern Derbyshire. *Frontline Gastroenterol.* **2012**, *3*, 283–287. [CrossRef] [PubMed]

92. Muhammad, H.; Reeves, S.; Ishaq, S.; Mayberry, J.; Jeanes, Y.M. PWE-100 Challenges in adhering to a gluten free diet in different ethnic groups. *Gut* **2018**, *67*, 168.

93. Adam, U.U.; Melgies, M.; Kadir, S.; Henriksen, L.; Lynch, D. Coeliac disease in Caucasian and South Asian patients in the North West of England. *J. Hum. Nutr. Diet.* **2019**, *32*, 525–530. [CrossRef] [PubMed]

94. Casellas, F.; Rodrigo, L.; Lucendo, A.J.; Fernández-Bañares, F.; Molina-Infante, J.; Vivas, S.; Rosinach, M.; Dueñas, C.; López-Vivancos, J. Benefit on health-related quality of life of adherence to gluten-free diet in adult patients with celiac disease. *Rev. Esp. Enferm. Dig.* **2015**, *107*, 196–201.

95. Silvester, J.A.; Weiten, D.; Graff, L.A.; Walker, J.R.; Duerksen, D.R. Is it gluten-free? Relationship between self-reported gluten-free diet adherence and knowledge of gluten content of foods. *Nutrition* **2016**, *32*, 777–783. [CrossRef] [PubMed]

96. Thompson, T.; Lee, A.R.; Grace, T. Gluten contamination of grains, seeds, and flours in the United States: A pilot study. *J. Am. Diet. Assoc.* **2010**, *110*, 937–940. [CrossRef]

97. Codex Standard 118-1979. Available online: http://www.fao.org/fao-who-codexalimentarius/sh-proxy/en/?lnk=1&url=https%253A%252F%252Fworkspace.fao.org%252Fsites%252Fcodex%252FStandards%252FCXS%2B118-1979%252FCXS_118e_2015.pdf (accessed on 7 February 2021).

98. Falcomer, A.L.; Santos Araujo, L.; Farage, P.; Santos Monteiro, J.; Yoshio Nakano, E.; Puppin Zandonadi, R. Gluten contamination in food services and industry: A systematic review. *Crit. Rev. Sci. Nutr.* **2018**, *22*, 1–15. [CrossRef] [PubMed]

99. Syage, J.A.; Kelly, C.P.; Dickason, M.A.; Ramirez, A.C.; Leon, F.; Dominguez, R.; Sealey-Voyksner, J.A. Determination of gluten consumption in celiac disease patients on a gluten-free diet. *Am. J. Clin. Nutr.* **2018**, *107*, 201–207. [CrossRef]

100. Zarkadas, M.; Dubois, S.; MacIsaac, K.; Cantin, I.; Rashid, M.; Roberts, K.C.; La Vieille, S.; Godefroy, S.; Pulido, O.M. Living with coeliac disease and a gluten-free diet: A Canadian perspective. *J. Hum. Nutr. Diet.* **2013**, *26*, 10–23. [CrossRef]

101. Rajpoot, P.; Sharma, A.; Harikrishnan, S.; Baruah, B.J.; Ahuja, V.; Makharia, G.K. Adherence to gluten-free diet and barriers to adherence in patients with celiac disease. *Indian J. Gastroenterol.* **2015**, *34*, 380–386. [CrossRef]

102. Hughey, J.J.; Ray, B.K.; Lee, A.R.; Voorhes, K.N.; Kelly, C.P.; Schuppan, D. Self-reported dietary adherence, disease-specific symptoms, and quality of life are associated with healthcare provider follow-up in celiac disease. *BMC Gastroenterol.* **2017**, *17*, 156. [CrossRef]

103. Addolorato, G.; De Lorenzi, G.; Abenavoli, L.; Leggio, L.; Capristo, E.; Gasbarrini, G. Psychological support counselling improves gluten-free diet compliance in coeliac patients with affective disorders. *Aliment. Pharmacol. Ther.* **2004**, *20*, 777–782. [CrossRef]

104. Sainsbury, K.; Mullan, B.; Sharpe, L. A randomized controlled trial of an online intervention to improve gluten-free diet adherence in celiac disease. *Am. J. Gastroenterol.* **2013**, *108*, 811–817. [CrossRef] [PubMed]

105. Dowd, A.J.; Warbeck, C.B.; Tang, K.T.Y.; Fung, T.; Culos-Reed, S.N. My Healthy Gut: Findings from a pilot randomized controlled trial on adherence to a gluten-free diet and quality of life among adults with celiac disease or gluten intolerance. *Digit. Health* **2020**, *6*, 2055207620903627.

106. Wolf, J.; Petroff, D.; Richter, T.; Auth, M.K.H.; Uhlig, H.H.; Laass, M.W.; Lauenstein, P.; Krahl, A.; Händel, N.; de Laffolie, J.; et al. Validation of antibody-based strategies for diagnosis of pediatric celiac disease without biopsy. *Gastroenterology* **2017**, *153*, 410–419.e17. [CrossRef]

107. White, L.E.; Bannerman, E.; Gillett, P.M. Coeliac disease and the gluten-free diet: A review of the burdens; factors associated with adherence and impact on health-related quality of life, with specific focus on adolescence. *J. Hum. Nutr. Diet.* **2016**, *29*, 593–606. [CrossRef]

108. Singh, J.; Whelan, K. Limited availability and higher cost of gluten-free foods. *J. Hum. Nutr. Diet.* **2011**, *24*, 479–486. [CrossRef]

109. Lee, A.R.; Wolf, R.L.; Lebwohl, B.; Ciaccio, E.J.; Green, P.H. Persistent economic burden of the gluten free diet. *Nutrients* **2019**, *11*, 399. [CrossRef] [PubMed]

110. Barratt, S.M.; Leeds, J.S.; Sanders, D.S. Quality of life in coeliac disease is determined by perceived degree of difficulty adhering to a gluten-free diet, not the level of dietary adherence ultimately achieved. *J. Gastrointestin. Liver Dis.* **2011**, *20*, 241–245.

111. Missbach, B.; Schwingshackl, L.; Billmann, A.; Mystek, A.; Hickelsberger, M.; Bauer, G.; König, J. Gluten-free food database: The nutritional quality and cost of packed gluten-free foods. *PeerJ* **2015**, *3*, e1337. [CrossRef]

112. Capacci, S.; Mazzocchi, M.; Lucci, A.C. There is no such thing as a (gluten-) free lunch: Higher food prices and the cost for coeliac consumers. *Econ. Human Biol.* **2018**, *30*, 84–91. [CrossRef] [PubMed]

113. Gorgitano, M.T.; Sodano, V. Gluten-free products: From dietary necessity to premium price extraction tool. *Nutrients* **2019**, *11*, 1997. [CrossRef]

114. Panagiotu, S.; Kontogianni, M.D. The economic burden of gluten-free products and gluten-free diet: A cost estimation analysis in Greece. *J. Hum. Nutr. Diet.* **2017**, *30*, 746–752. [CrossRef]

115. Paul, S.P.; Stanton, L.K.; Adams, H.L.; Basude, D. Coeliac disease in children: The need to improve awareness in resource-limited settings. *Sudan J. Paediatr.* **2019**, *19*, 6–13. [CrossRef]

116. O´Shea, N.; Arendt, E.; Gallagher, E. State of the art in gluten-free research. *J. Food Sci.* **2014**, *79*, R1067–R1076. [CrossRef]

117. Tanner, G.J.; Blundell, M.J.; Colgrave, M.L.; Howitt, C.A. Creation of the first ultra-low gluten barley (*Hordeum vulgare* L.) for coeliac and gluten-intolerant populations. *Plant. Biotechnol. J.* **2016**, *14*, 1139–1150. [CrossRef] [PubMed]

118. Knorr, V.; Wieser, H.; Koehler, P. Production of gluten-free beer by peptidase treatment. *Eur. Food. Res. Technol.* **2016**, *242*, 1129–1140. [CrossRef]

119. Gutowski, E.D.; Weiten, D.; Green, K.H.; Rigaux, L.N.; Bernstein, C.N.; Graff, L.A.; Walker, J.R.; Duerksen, D.R.; Silvester, J.A. Can individuals with celiac disease identify gluten-free foods correctly? *Clin. Nutr. ESPEN* **2020**, *36*, 82–90. [CrossRef] [PubMed]

120. Ljungman, G.; Myrdal, U. Compliance in teenagers with coeliac disease: A Swedish follow-up study. *Acta Paediatr.* **1993**, *82*, 235–238. [CrossRef] [PubMed]

121. Vilppula, A.; Kaukinen, K.; Luostarinen, L.; Krekelä, I.; Patrikainen, H.; Valve, R.; Luostarinen, M.; Laurila, K.; Mäki, M.; Collin, P. Clinical benefit of gluten-free diet in screen-detected older celiac disease patients. *BMC Gastroenterol.* **2011**, *11*, 136. [CrossRef]
122. Casella, S.; Zanini, B.; Lanzarotto, F.; Villanacci, V.; Ricci, C.; Lanzini, A. Celiac disease in elderly adults: Clinical, serological, and histological characteristics and the effect of a gluten-free diet. *J. Am. Geriatr. Soc.* **2012**, *60*, 1064–1069. [CrossRef]
123. Olsson, C.; Lyon, P.; Hörnell, A.; Ivarsson, A.; Matsson Snyder, Y. Food that makes you different: The stigma experienced by adolescents with celiac disease. *Qual. Health Res.* **2009**, *19*, 976–984. [CrossRef]
124. Samasca, G.; Lerner, A.; Girbovan, A.; Sur, G.; Lupan, I.; Makovicky, P.; Matthias, T.; Freeman, H.J. Challenges in gluten-free diet in coeliac disease: Prague consensus. *Eur. J. Clin. Investig.* **2017**, *47*, 394–397. [CrossRef]
125. Sample, D.; Turner, J. Improving gluten free diet adherence by youth with celiac disease. *Int. J. Adolesc. Med. Health* **2019**. Available online: /j/ijamh.ahead-of-print/ijamh-2019-0026/ijamh-2019-0026.xml (accessed on 2 February 2021). [CrossRef] [PubMed]
126. Sainsbury, K.; Halmos, E.P.; Knowles, S.; Mullan, B.; Tye-Din, J.A. Maintenance of a gluten free diet in coeliac disease: The roles of self-regulation, habit, psychological resources, motivation, support, and goal priority. *Appetite* **2018**, *125*, 356–366. [CrossRef] [PubMed]
127. Sainsbury, K.; Marques, M.M. The relationship between gluten free diet adherence and depssive symptoms in adults with celiac disease: A systematic review with meta-analysis. *Appetite* **2018**, *120*, 578–588. [CrossRef] [PubMed]
128. Barberis, N.; Quattropani, M.C.; Cuzzocrea, F. Relationship between motivation, adherence to diet, anxiety symptoms, depression symptoms and quality of life in individuals with celiac disease. *J. Psychosom. Res.* **2019**, *124*, 109787. [CrossRef]
129. Dowd, A.J.; Jung, M.E. Self-compassion directly and indirectly predicts dietary adherence and quality of life among adults with celiac disease. *Appetite* **2017**, *113*, 293–300. [CrossRef]

Review

Food Safety and Cross-Contamination of Gluten-Free Products: A Narrative Review

Herbert Wieser [1,†], Verónica Segura [2], Ángela Ruiz-Carnicer [2], Carolina Sousa [2] and Isabel Comino [2,*]

[1] Independent Researcher, 85354 Freising, Germany; h.wieser2@gmx.de
[2] Department of Microbiology and Parasitology, Faculty of Pharmacy, University of Seville, 41012 Seville, Spain; vsegura@us.es (V.S.); acarnicer@us.es (Á.R.-C.); csoumar@us.es (C.S.)
* Correspondence: icomino@us.es; Tel.: +34-954-55-64-52
† Retired.

Abstract: A gluten-free diet (GFD) is currently the only effective treatment for celiac disease (CD); an individual's daily intake of gluten should not exceed 10 mg. However, it is difficult to maintain a strict oral diet for life and at least one-third of patients with CD are exposed to gluten, despite their best efforts at dietary modifications. It has been demonstrated that both natural and certified gluten-free foods can be heavily contaminated with gluten well above the commonly accepted threshold of 20 mg/kg. Moreover, meals from food services such as restaurants, workplaces, and schools remain a significant risk for inadvertent gluten exposure. Other possible sources of gluten are non-certified oat products, numerous composite foods, medications, and cosmetics that unexpectedly contain "hidden" vital gluten, a proteinaceous by-product of wheat starch production. A number of immunochemical assays are commercially available worldwide to detect gluten. Each method has specific features, such as format, sample extraction buffers, extraction time and temperature, characteristics of the antibodies, recognition epitope, and the reference material used for calibration. Due to these differences and a lack of official reference material, the results of gluten quantitation may deviate systematically. In conclusion, incorrect gluten quantitation, improper product labeling, and poor consumer awareness, which results in the inadvertent intake of relatively high amounts of gluten, can be factors that compromise the health of patients with CD.

Keywords: celiac disease; gluten cross-contaminations; dietary adherence; gluten-free diet; vital gluten; oat; hidden gluten

Citation: Wieser, H.; Segura, V.; Ruiz-Carnicer, Á.; Sousa, C.; Comino, I. Food Safety and Cross-Contamination of Gluten-Free Products: A Narrative Review. *Nutrients* **2021**, *13*, 2244. https://doi.org/10.3390/nu13072244

Academic Editor: Maria Cappello

Received: 31 May 2021
Accepted: 26 June 2021
Published: 29 June 2021

Publisher's Note: MDPI stays neutral with regard to jurisdictional claims in published maps and institutional affiliations.

1. Introduction

The consumption of gluten proteins drives adverse reactions in predisposed individuals who suffer from celiac disease (CD), wheat allergies, non-celiac gluten sensitivity, dermatitis herpetiformis, or gluten ataxia [1–3]. CD is one of the most frequent hypersensitivities, affecting around 1% of the world's population [4]. It is an immune-mediated systemic disorder caused by ingestion of gluten in genetically susceptible individuals and is based on a variable combination of intestinal and extra-intestinal signs and symptoms that are specific to CD antibodies, HLA-DQ2/8 haplotypes, and enteropathy [5,6].

A strict gluten-free diet (GFD) is currently the only safe and efficient therapy for patients with CD [7,8], which implies that all gluten-containing foods and meals, produced from wheat, rye, barley, and some varieties of oats, must be completely excluded from the diet. Nevertheless, such a diet is difficult to follow because of the unintended contamination of "gluten-free" products, improper labeling, social constraints, and ubiquity of gluten proteins in raw or cooked food and pharmaceuticals. Thus, accidental gluten encounters are likely. Most patients with CD can safely tolerate a daily cross-contamination of approximately 10 mg gluten, or 500 g of food containing 20 mg/kg of gluten. However, there is a tremendous degree of variability within this population, and some patients may have worsening histological changes with very low daily gluten exposure [9,10].

People needing to follow a GFD may consume gluten-free foods from two categories. First, they are allowed to eat a wide range of naturally gluten-free foods such as meat, fish, milk products, vegetables, nuts, and fruits. Second, patients may consume dietetic gluten-free products, i.e., alternatives to traditional gluten-containing foods, which are labeled as gluten-free. These are made from cereals that do not contain gluten such as rice, corn, sorghum, and millet, or pseudocereals such as amaranth, buckwheat, and quinoa [11]. Recent advances in the formulation of cereal-based gluten-free products by utilizing alternate ingredients and processing techniques have been summarized by Rai et al. [12]. Definitions, thresholds, and labeling of dietetic gluten-free foods have been specified in international and national regulations. According to the "Codex Alimentarius Standard for Gluten-Free Foods" [13], gluten-free dietetic foods labeled gluten-free should not exceed 20 mg gluten per kg food when sold to consumers [13].

Due to gluten contamination, many inherently gluten-free products (derived from corn, rice, millet, etc.,) cannot be consumed by patients with CD. These products, if misbranded as "gluten-free" and used by the patients with CD, could result in a recurrence of symptoms. Contamination of gluten-free foods with gluten-containing material can occur at many stages of food production, from the fields, farms, mills, and factories, as well as handcraft enterprises, restaurants, and households [14,15]. Until the 1970s, sensitive and accurate quantitation of gluten contamination was not possible, and patients with CD were constantly at risk of inadvertent intake of high amounts of gluten [16]. The picture has clearly improved in recent decades, most likely due to the development of immunochemical methods such as the Enzyme-Linked ImmunoSorbent Assay (ELISA), for gluten detection and the worldwide implementation of the 20 mg/kg maximum threshold of gluten contamination established by the "Codex Alimentarius 118–179" in 2008 [13,17,18]. However, patients with CD are still confronted with foods that are contaminated by gluten above the threshold of 20 mg/kg.

The major aim of this review is to provide insight into the frequent occurrence of gluten in naturally or certified gluten-free foods, as well as the safety of oats as part a GFD and the problem with hidden gluten. In addition, we examine the immunochemical and non-immunochemical methods currently available for the detection of gluten.

2. Methods

PubMed database searches were performed for articles published until March 2021. The search terms used included "c(o)eliac disease, gluten-free diet", "gluten, contamination", "gluten, oats", "gluten, quantitation", and "gluten, quantification". References of included full-text articles were scrutinized for additional studies. We included published articles and review reporting on gluten contamination in gluten-free foods. We excluded publications that did not focus on the aim of this review. Only English publications were selected during the search. Case reports, commentaries, conference papers, and letters were excluded. Retrieved manuscripts were reviewed by the authors, and the data were extracted and described.

3. Gluten Contamination in Gluten-Free Foods

3.1. Naturally and Certified Gluten-Free Foods

In the 1980s and 1990s, newly developed ELISAs, mainly optimized for the detection of wheat gliadins, enabled the sensitive detection of gluten contamination in gluten-free products. At that time, the analytic results demonstrated that raw materials, naturally gluten-free by origin and used for the production of gluten-free products such as rice, buckwheat, corn, or millet flours, were contaminated with wheat up to 3000 mg gliadin/kg [19]. Later, the levels of gluten contamination in gluten-free products were found to be distinctly lower; however, contamination is still a problem, as shown in the following examples (Table 1).

Table 1. Studies on the rate of gluten contamination in natural and certified gluten-free foods, as well as food service products.

Type of Products	Study Country	*n*	Percentage of Food Containing >20 mg/kg of Gluten	References
Gluten-free-labeled products	Ireland	260	10%	McIntosh et al., 2011 [20]
	Italy, Spain, Germany, and Norway	205	0.5%	Gibert et al., 2013 [21]
	United States	78	21%	Lee et al., 2014 [22]
	Brazil	20	16%	Oliveira et al., 2014 [23]
	United States	275	10%	Sharma et al., 2015 [24]
	Spain	3141	12%	Bustamante et al., 2017 [25]
	Turkey	200	17.5%	Atasoy et al., 2017 [26]
	Italy	56	0%	Bianchi et al., 2018 [27]
	Brazil	180	3%	Farage et al., 2019 [28]
	United States	5624	32%	Lerner et al., 2019 [29]
	Indian	160	36%	Raju et al., 2020 [30]
Naturally gluten-free products	United States	22	32%	Thomson et al., 2010 [31]
	Canada	640	9.5%	Koerner et al., 2013b [32]
	United States	186	19%	Sharma et al., 2015 [24]
	Indian	160	10%	Raju et al., 2020 [30]
Naturally or certified gluten-free products	Lebanon	173	6%	Hassan et al., 2017 [33]
	Italy	200	9%	Verma et al., 2017 [34]
	Brazil	130	22%	Farage et al., 2017 [35]

Twenty-two naturally gluten-free grains, seeds, and flours were purchased in the United States and tested in duplicate for gluten contamination [31]. Thirteen samples (59%) contained less than the limit of quantitation (5 mg/kg), and nine samples (41%) contained gluten levels ranging from 8.5 to 2925 mg/kg. Seven samples (32%) exceeded 20 mg/kg and could not be considered gluten-free. In another market survey in the United States, different gluten-free-labeled foods (*n* = 275) and non-labeled foods (without wheat/rye/barley on the ingredient label; *n* = 186) were analyzed for gluten [24]. A total of 10% of gluten-free-labeled foods had gluten contents >20 mg/kg. Among the non-labeled naturally gluten-free foods, 19% had >20 mg/kg of gluten, of which 10% had >100 mg/kg. The investigation of 78 different certified gluten-free foods, offered in the United States, revealed that 61% of the samples contained less gluten than the limit of quantitation (10 mg/kg), and 18% contained between 10 and 20 mg/kg [22]. However, 21% had gluten levels above 20 mg/kg, ranging from 21 to 61 mg/kg. In particular, five of eight labeled breakfast cereal samples showed gluten contents >20 mg/kg.

A large Canadian investigation of naturally gluten-free ingredients, such as flours and starches, showed that 61 of the 640 samples (9.5%) were contaminated with gluten above 20 mg/kg [32]. The largest and most consistent mean contamination came from soy (902 mg/kg), millet (272 mg/kg), and buckwheat (153 mg/kg). An examination of gluten-free products from 25 bakeries in Brasilia revealed that 28 of 130 samples (22%) were contaminated with gluten above 20 mg/kg [35]. This finding was even more concerning considering that 16 bakeries (64%) sold at least one product contaminated with gluten. Only nine establishments (36%) had no gluten-contaminated products in their assortments.

A total of 200 commercially available naturally or certified gluten-free products were randomly collected from different Italian supermarkets and analyzed [34]. Gluten levels were lower than 10 mg/kg in 173 products (87%), between 10 and 20 mg/kg in 9 samples (4.5%), and higher than 20 mg/kg in 18 samples (9%). Contaminated foodstuffs (gluten >20 mg/kg) most commonly belonged to oat-, buckwheat-, and lentil-based items. Naturally gluten-free products were at a significantly higher risk of contamination than products certified as gluten-free. To study the evolution of gluten contamination over time, a total of 3141 cereal-based gluten-free foodstuffs, sold in Spain from 1998 to 2016, were consecutively analyzed [25]. Products were divided into eight categories: flours, breakfast cereals/bars, bakery, pasta, breads, dough, snacks, and yeast. Overall, gluten exceeding 20 mg/kg was detected in 371 samples (12%), with breakfast cereals/bars being the most contaminated group (Table 1). Data obtained on the analyzed products demonstrated that cereal-based gluten-free foods have become safer over time, but gluten contamination remains a problem. One of the few pleasing findings regarding gluten contamination was the reported selection of European foods labeled as gluten-free [21]. A total of 205 representative products among six food groups (bread, pasta, pastry, biscuits, pizza, and breakfast cereals), purchased from markets in Italy, Spain, Germany, and Norway, were investigated. The gluten content ranged between <5 and 28 mg/kg, and only one sample (0.5%) had a gluten concentration above 20 mg/kg.

Miscellaneous gluten-free foods were investigated for gluten contamination to evaluate the situation in Turkey [26]. A total of 200 samples from eight product categories (snack, pasta, bread, cookie, cracker, farina, traditional, and others), manufactured using seven ingredient categories (cereal mixture, buckwheat, corn, rice, locust bean, potato, and others), were analyzed. A significant proportion of the samples (17.5%) were contaminated with gluten, and therefore unacceptable in terms of being called gluten-free. The results pointed to buckwheat as the main cause of this contamination. To evaluate gluten contamination in Lebanon, 173 gluten-free food samples were analyzed over a 2-year period [33]. In 10 samples (6%), the quantity of gluten exceeded 20 mg/kg (Table 1). Eight of the contaminated samples were locally manufactured and based on wheat starch. To assess the gluten content of labeled and naturally gluten-free grain products from markets in Southern India, different "gluten-free" breakfast products, flours, and batters (n = 160) were evaluated [30]. Nearly 36% of the products made from naturally gluten-free grains and 10% of gluten-free-labeled products were found to contain >20 mg/kg gluten.

The ingestion of purified wheat starch as a constituent of gluten-free products is considered safe in many countries, but uncertainties about residual gluten amounts remain. Due to the generally low ratio of gliadins to glutenins in starch, gluten levels determined by ELISA are likely to be underestimated. In comparative analyses of gluten content, eight gluten-free starch samples were analyzed by R5 ELISA (gluten = gliadins × 2) and a chromatographic control method (gluten = gliadins + glutenins) [36]. According to ELISA testing (12–30 mg/kg), only two samples were not gluten-free (21 and 30 mg/kg). In contrast, all eight samples had gluten contents >20 mg/kg (38–69 mg/kg) when gliadins and glutenins were accounted for in chromatographic analysis.

In a recent systematic review, 24 cross-sectional studies were analyzed to evaluate the prevalence of gluten contamination in gluten-free industrial and non-industrial products. The authors evaluated the methodological quality of the included studies using criteria from a Meta-analysis of Statistics Assessment and Review Instrument (MASTARI). In total, 95.83% (n = 23) of the studies presented positive results for contamination (contained gluten above 20 mg/kg). In industrial food products, studies showed a contamination prevalence of 13.2% (95% CI: 10.8–15.7%). In non-industrial food products, studies showed a contamination prevalence of 41.5% (95% CI: 16.6–66.4%). Despite the non-industrial products presenting a higher contamination prevalence than the industrial products, the difference was not significant (p = 0.072). The findings indicated cross-contamination in industrial and non-industrial products [37].

In conclusion, most studies on gluten contents in naturally or certified gluten-free foods revealed relatively high rates of contamination, ranging from 0.5% to 36% of the analyzed samples (Table 1). Contaminated naturally gluten-free products appear to be a higher health risk than certified products for patients with CD. Altogether, both naturally and labeled gluten-free foods do not guaranty safety for patients with CD, and gluten contamination is an important cause of inadvertent non-adherence to a GFD.

3.2. Products from Food Services

Eating at restaurants, workplaces, schools, and home (own or other people's) remains a distinctive risk for inadvertent gluten exposure. In a systematic review, 24 international studies were used to investigate gluten contamination (>20 mg/kg) in gluten-free products from food services and industries [37]. The statistical meta-analysis resulted in a mean contamination prevalence of 42% (17–66%) in certified products offered by food services.

Furthermore, a mean contamination of 13% (11–16%) was detected in industrial food products labeled as gluten-free. The examination of gluten-free-labeled foods, offered in a number of restaurants across the United States, resulted in surprisingly high rates of gluten contamination [29]. A total of 5624 tests were performed by 804 users equipped with a portable gluten detection device (Nima, Nima Labs, Inc., San Francisco, CA, USA). Data were collected during an 18-month period and sorted by food items and regions. Gluten above 20 mg/kg was detected in 32% of products labeled as gluten-free (Table 1). Rates of gluten detection differed by meal, with 27.2% at breakfast and 34.0% at dinner. Gluten-free labeled pizza and pasta were most likely to test positive for gluten, with gluten detected in 53.2% of pizza samples and 50.8% of pasta samples.

The evaluation of gluten content in gluten-free food on request in restaurants in Ireland revealed that the majority of attempts to purchase a gluten-free meal were successful [20]. However, some 10% of all samples contained gluten above 20 mg/kg: 2.7% between 21 and 100 mg/kg and 7.7% above 100 mg/kg, and two unsatisfactory samples were purchased from so-called celiac-friendly restaurants (Table 1). In a study from Brazil, common beans were collected from different self-service restaurants in Brasilia and later analyzed for gluten content [23]. The results revealed that 16% of the samples were contaminated with gluten above 20 mg/kg and almost 45% of the restaurants had gluten contamination in beans on at least one of the days tested (Table 1).

To determine the rate of gluten contamination in typical Brazilian lunch meals, traditionally gluten-free, a total of 180 dishes were purchased from 60 food services in the Federal District Brazil [28]. They were visited at lunch time, and the dishes were chosen randomly. Three different food items were collected for gluten analysis in each food service. Fortunately, only 3% of dish samples were contaminated with gluten (Table 1), and only 7% of food services displayed at least one contaminated food. Thus, traditional Brazilian dishes, made from naturally gluten-free materials, appeared to be safe for patients with CD. Another positive example was found in Italy: all pizzas and cooked dough bases ($n = 56$), produced at certified take-away pizza restaurants in the Turin metropolitan area, were gluten-free (<20 mg/kg) [27]. Thus, attention to and compliance with good manufacturing practices, a requisite for obtaining gluten-free certification for restaurants in Italy, have a positive effect on the production of gluten-free products.

A quantitative assessment of gluten cross-contact in the school environment for children with CD measured the gluten transfer from school activities to gluten-free foods that a child may eat afterwards [38]. Five experiments were used to identify potential gluten transfer to gluten-free bread in classrooms using sensory tables: Play-Doh, baking projects, papier-mâché, dry pasta, and cooked pasta. After activities, slices of gluten-free bread were rubbed on participant's hands and table surfaces and gluten levels were determined. The potential for gluten exposure was found to be high (>20 mg/kg) for papier-mâché, baking projects, and cooked pasta.

Meals may be contaminated not only in food services but also at home. Gluten-free meals should always be prepared, stored, and handled separately from gluten-containing meals. If separate areas are not available, preparing a gluten-free meal before other meals is recommended. However, the need for extra cooking is frequently seen as a problem for maintaining a GFD.

In conclusion, gluten-free products from food services hold a considerable risk for gluten contamination. Patients with CD are advised to check allergen lists, according to Codex Standard 1- 985, and/or to ask staff for information. Food services should make efforts to minimize the risk of cross-contamination in food (Table 1). This would create a more reliable environment for patients with CD who need to eat when away from home. Furthermore, future research should focus on identifying inappropriate procedures that cause gluten-contamination and should propose new strategies to overcome this issue.

3.3. Oats

The necessity of excluding oats from the diet of patients with CD is still a matter of discussion. An update of the ongoing debate on oats, and the pros and cons of using oats in a GFD, was reviewed by Cohen et al. [39] and Hoffmanova et al. [40]. Most clinical studies have reported that moderate amounts of pure oats are well-tolerated by most patients with CD, and only a small number of patients (probably less than 1%) experience harmful effects from oat consumption. Therefore, in many countries, oats are recommended to be included as part of a GFD. The high contents of beneficial compounds such as dietary fiber, unsaturated fatty acids, and antioxidants make oats an attractive component of a GFD. However, oat products can only be tolerated if they are free from wheat, rye, and barley. Pure oats must meet the legislative criteria for gluten-free foods, i.e., the content of gluten from wheat, rye, and barley in the end-product must be less than 20 mg/kg.

The fact that oats are often processed on the same production line as wheat, rye, or barley is a major cause of gluten contamination. Therefore, it is not surprising that commercial oat supply can be heavily contaminated with gluten-containing grains. However, recognition and measurement of gluten contamination in oat products with ELISA kits are still a problem. At AOAC (Association of Official Analytical Collaboration) International, a stakeholder panel convened and agreed upon standard requirements for the quantitation of total wheat, rye, and barley gluten in oat products by ELISA [41]. The defined method acceptance criteria were 5–15 mg/kg of gluten as the analytical range, limits of detection and quantitation below 5 mg/kg of gluten, and 50–200% recovery. The rather wide recovery range was chosen due to the lack of homogeneity inherent in oat samples and different ELISA antibody responses to gluten proteins from wheat, rye, and barley.

If oats are used in a GFD, it is recommended that contamination with wheat, rye, and barley be assessed by a stepwise "test-all-positives" methodology [42]. Oat groats are split into 75 g samples and ground. Afterwards, a 15 g portion is analyzed using a sandwich ELISA. A result of >20 mg/kg disqualifies the production lot, while a result of <20 mg/kg triggers complete analysis of the remaining 60 g of ground sample, which is analyzed in 15 g portions. If all five 15 g tests are <20 mg/kg, the lot can be approved.

One of the first studies on gluten contamination in oats, conducted in Spain in 2006, evaluated 108 oat samples (e.g., rolled oats, oat flakes, and flours) collected from Europe, the United States, and Canada [43]. Three quarters of the samples were contaminated with more than 20 mg/kg of gluten, with a variation of up to 8000 mg/kg. A pilot study on gluten contamination in grains, seeds, and flours in the United States, which included rolled or steel-cut oat samples into the investigation [31], found 9 out of 12 containers, representing four different lots of each of three separate brands (Quaker, Country Choice, and McCann's), had gluten levels ranging from 23 to 1807 mg/kg. Another study, performed in 2011, demonstrated that approximately 88% of 133 Canadian commercial oat samples were contaminated with gluten above 20 mg/kg, and there were no differences between the oat types tested [44]. Among grain-based food products purchased from markets in Southern India, 85% of the oat samples were contaminated with gluten in amounts up to 1830 mg/kg [30].

However, differences in the type of oat grain, oat purity, study design, as well as the specifications for gluten-free products in different countries, are some reasons why the current studies have not clearly established whether oats can be safely consumed by all patients with CD. These apparent contradictions might be explained by the fact that the oat varieties used in the diverse studies were different in terms of their prolamin genes, protein amino acid sequences, and the immunoreactivities of their toxic prolamins [45–48]. Even so, some pure oats cultivars have significant reactivity with the most used monoclonal antibodies R5, G12 [48–51]. Some celiac T-cell activating sequences from oats have been identified [52–54], and some oat varieties have elicited early inflammatory events typical of CD [47]. Despite this evidence, it is still commonly believed that there is no reactivity in pure oats.

In conclusion, while the inclusion of oats in a GFD might be beneficial due to their nutritional and health benefits, the source of the oats used and the cultivar selected are important factors to be considered. It is extremely important to remember that in vitro studies have shown that the immunogenicity of oats varies depending on the cultivar used. In any case, it seems that lack of reactivity with immune assays (R5, G12) may guarantee the absence of toxic gluten regardless of source-from oats itself or from wheat or barley contamination. These factors must also be taken into account when developing food safety regulations, labeling oat-containing products as gluten-free, and designing clinical trials to study the effect of oats in patients with CD for evidence of adverse reactions.

3.4. Hidden Gluten

Patients with CD should be aware of numerous composite foods, medications, and cosmetics that contain "hidden" sources of so-called vital gluten (VG), a by-product of wheat starch production. Therefore, patients are advised to check the ingredients labeled on prepacked products or to obtain information about unpacked foods. Prepacked foods should conform to the regulations of the Codex Standard 1-1985. To protect sensitive consumers from harmful allergic symptoms, the "General Standard for the Labeling of Prepacked Foods" states that the following foods and ingredients, which contain proteins known to cause allergies and other types of hypersensitivities, should always be declared (the top eight food allergens): cereals containing gluten, crustaceans, eggs, fish, peanuts, soybeans, milk, and tree nuts.

However, clear food labeling is not a requirement in all countries. Moreover, a number of patients with CD are not motivated to study the label when they buy prepacked foodstuffs made from naturally gluten-free materials. Likewise, patients usually do not ask for information on unpacked foods that consist of naturally gluten-free components. Thus, hidden VG is one of the main initiators of inadvertent breaks with a GFD. VG is typically added to food in the range of 1–3 g/100 g of dry mass. The water-binding and thickening properties of VG are used to improve the quality of ice cream, coffee creamer, instant pudding, soups, sauces, ketchup, marinades, and dressings. Due to the outbreak of bovine spongiform encephalopathy (BSE) and the subsequent efforts to replace gelatine, VG has found new applications in the production of some special foods such as chewing gum, chew candies, and fruit chews [55].

The properties of VG help to bind vitamin/mineral components to fortified corn flakes, puffed rice, or grain berries. Another application is the coating of dry roasted nuts with VG, which enables the adhesion of salt and other seasonings. VG is also applied as an additive in the production of soy sauce. The binding, film-forming, and thermosetting characteristics of hydrated VG are the basis for various applications in the manufacture of meat products. It is effective in binding meat chunks to form special products such as textured meat, canned hams, and poultry rolls. VG is also useful as a protein binder in sausages and other meat emulsion products. The incorporation of about 2% VG into surimi enhances gel strength and reduces the development of an undesirable rubbery texture after frozen storage.

An increasing number of people opting for a vegetarian or vegan diet has increased the demand for substitutes of animal products that are often produced with the help of VG. Gelatin, used as a thickening and gelling agent, is frequently replaced by VG. The viscoelastic properties of hydrated VGs are exploited for the production of synthetic cheese, with the characteristic texture and sensory properties of natural cheese. The production of seafood analogues is another field of VG utilization.

Gluten is introduced into numerous medications, mostly through the use of wheat starch as a filling agent. Conventional wheat starch contains approximately 3000–4000 mg gluten/kg and, thus, can cause significant gluten contamination in medical products. Apart from medications, wheat starch plays a role in the production of dialysis solutions, enteral nutrition, and even as a substitute for blood plasma. In 2011, the US Food and Drug Administration (FDA) solicited information and public comments on the use of gluten in drug products, but did not include gluten labeling of drugs. Thus, patients with CD do not know whether a product is gluten-free or not, unless it is labeled as such. Therefore, all drug products should be made gluten-free, because there are many alternatives to gluten-containing materials, such as starch, that can be used as excipients during their formulation [56].

VG is frequently added to oral hygiene and cosmetic products such as toothpastes, mouthwashes, and lipsticks. Of 66 items collected from an Italian market, 62 samples were found to be gluten-free (<20 mg/kg), while three toothpastes (21–35 mg/kg) and one lipstick (27 mg/kg) showed a gluten level above 20 mg/kg [57].

In conclusion, hidden gluten in naturally gluten-free foods and drugs is a major contributor to inadvertent gluten intake. In particular, gluten being added to certain products, inadequate labeling, and poor knowledge on the part of consumers are important factors that compromise the health of patients with CD.

4. Analytical Methods to Detect Gluten

Currently, patients with CD are confronted with uncertainties in gluten analysis and, accordingly, may run the risk of inadvertent gluten intake due to inaccurately determined gluten levels. Many methods have been developed for the detection of prolamins, including polymerase chain reaction (PCR), liquid chromatography–mass spectrometry (LC–MS), and immunological methods based on anti-gluten peptide antibodies. The use of LC–MS is difficult because of its cost and technical performance, as well as the complexity of the sample which contains many different peptides. Accordingly, immunoassays such as ELISAs and lateral flow devices (LFD) have been the methods of choice in the food industry to certify gluten-free food because of their combination of specificity, sensitivity, simplicity, and cost effectiveness (Table 2). In recent years, methods have been developed for use by celiac patients themselves. Specifically, Nima™ and EZ Glutent™ LFD have been developed to integrate food processing and gluten detection in a portable device that is available for consumer use [58,59].

Table 2. Analytical techniques for the detection and quantification of gluten.

	Strengths	Weak Points
ELISA immunoenzyme assay	Simple to perform, fast, inex-pensive, high sensitivity, does not produce cross-reactions.	False negatives can occur when proteins are denatured by changes in pressure, temperature, or salt concentration.
POLYMERASE CHAIN REACTION (PCR)	Very high sensitivity in the detection of DNA, allows one to identify the species from which the gluten comes, useful to identify the origin of cross contamination.	Time and qualified personnel are required in the analysis, indirect technique to detect gluten (does not quantify the presence of gluten).
WESTERN BLOT	Highly specific and sensitive, suitable for determining the gluten content in raw and processed foods.	Slow method, requires adequate training and specialization of analysts.
MASS SPECTROMETRY (LC–MS)	Speed, reproducibility, precision.	Complex instrumentation, expensive equipment, not a quantitative technique, etc.
CHROMATOGRAPHY	High capacity for the separation of different peptides.	Time consuming, difficult to automate for many samples.
IMMUNOCHROMATOGRAPHIC STRIPS	Very simple, fast method, visual interpretation.	It does not show the concentration of gluten in the sample.

A number of gluten-specific ELISA, LFD, and PCR kits are commercially available worldwide (Table 3). Each kit has specific features, such as format (sandwich, competitive and LDF), sample extraction buffers, extraction time and temperature, characteristics of the antibodies, and target analysts, as well as the reference material used for calibration. Due to these differences and a lack of official reference material, the results of gluten quantitation may deviate systematically; a number of publications have highlighted that routine ELISA methods do not provide equivalent results for gluten content [60–62].

Table 3. Gluten-specific methods commercially available worldwide. ELISA, Enzyme-Linked ImmunoSorbent Assay; PCR, polymerase chain reaction.

Format	Test Kit	Manufacturer	Target	Antibody
ELISA competitive	AgraQuant ELISA Gluten G12	Romer Labs	QPQLPY	G12 monoclonal
	RIDASCREEN Gliadin Competitive	R-Biopharm, AG	QQPFP, QQQFP, LQPFP, QLPFP	R5 monoclonal
	GlutenTox® ELISA Competitive	Hygiena	QPQLPY	G12 monoclonal
ELISA sandwich	Veratox for Gliadin, 8480	Neogen Corp.	Gluten	USDA monoclonal
	Veratox for Gliadin R5, 8510	Neogen Corp.	QQPFP, QQQFP, LQPFP, QLPFP	R5 monoclonal
	MonoTrace Gluten ELISA Kit GLU-EK-96	BioFront Technologies	Gluten	Set of gluten-specific monoclonal antibodies
	RIDASCREEN®FAST Gliadin sensitive	R-Biopharm, AG	QQPFP, QQQFP, LQPFP, QLPFP	R5 monoclonal
	RIDASCREEN®FAST Gliadin	R-Biopharm, AG	QQPFP, QQQFP, LQPFP, QLPFP	R5 monoclonal
	AllergenControl ™ Gluten Sandwich	Microbiologique Inc.	Gliadin	2D4
	Wheat Protein ELISA (MIoBS)	Morinaga Institute of Biological Sciences, Inc.	Gliadin	Polyclonal
	AllerTek Gluten	ELISA Technologies, Inc.	HMW glutenin	Skerritt monoclonal
	GlutenTox® ELISA Rapid	Hygiena	QPQLPY	G12/A1 monoclonal
	Gluten-Check ELISA kit	Biocheck (UK)	QQPFP, QQQFP, LQPFP, QLPFP	R5 monoclonal
Lateral flow device (LFD)	AgraStrit Gluten G12	Romer Labs	QPQLPY	G12 monoclonal
	RIDA®QUICK Gliadin	R-Biopharm, AG	QQPFP, QQQFP, LQPFP, QLPFP	R5 monoclonal
	GlutenTox® Sticks Plus	Hygiena	QPQLPY	G12/A1 monoclonal
	GlutenTox® Pro	Hygiena	QPQLPY	G12/A1 monoclonal
	Nima Gluten sensor	Nima Labs, Inc.	Gluten	Nima antibody
	EZ Glutent™	ELISA Technologies, Inc.	Gluten	anti-omega gliadin antibody
PCR test	SureFood® ALLERGEN Gluten	R-Biopharm, AG	QQPFP, QQQFP, LQPFP, QLPFP	
	SureFood® ALLERGEN 4plex Cereals	R-Biopharm, AG	QQPFP, QQQFP, LQPFP, QLPFP	

Assays show high variability in specificity that corresponds to the type of cereal, the species, and the variety, as well as the composition of gluten protein types. For example, the evaluation of the five most frequently used ELISA kits showed high variability towards different cereals and gluten protein types [63,64]. Similarly, the determination of gluten in different cultivars of common wheat, spelt, durum wheat, emmer, and einkorn, using three ELISA kits, showed clear differences between kits [65]. The comparison of five ELISA kits containing two polyclonal antibodies and three monoclonal antibodies revealed that wheat prolamins (gliadins) were detected accurately by all tested antibodies, but high variability was observed for rye and barley prolamins [64]. The gluten content (sum of prolamins and glutelins) was either overestimated up to six times (rye) or underestimated up to seven times (barley). Avenins, the gluten proteins of oats, remain a challenge in terms of detection and quantitation because most antibodies used in ELISA do not react with avenins, except monoclonal antibodies G12 and R5 [45–48].

Further problems have been found in determining the gluten content of foods with different matrices. A set of 14 ELISA kits for gluten detection was used to analyze gluten levels in a series of relevant food matrices that varied in complexity [60]. The results demonstrated that there was no single ELISA kit that could accurately detect and quantify gluten in all different matrices. Additional difficulties may be caused by food processing that impairs the detection of gluten, such as heat treatment, extrusion, or fermentation. Accurate quantitation of gluten by antibody-based methods in fermented foods such as beer, baby food, or soy sauce is a particular challenge. The reduced recovery of gluten by ELISA after enzymatic partial hydrolysis of gluten proteins is a well-studied effect. For example, the quantitation of a peptic/tryptic gliadin hydrolysate by a competitive ELISA resulted in 56% recovery compared to the starting gliadin material [66], although toxicity to patients with CD was sustained after peptic/tryptic digestion [67]. It is uncertain whether ELISA, even in a competitive format, is an appropriate method for identifying partially hydrolyzed gluten. Potential diversity in the generation of sequences, relative abundance, and extension of the resultant gluten peptides is almost limitless [68,69]. The estimation of gluten equivalence in hydrolyzed gluten samples is thus a challenge. Firstly, peptides may have only one epitope per molecule. The best approach to measure the immunogenicity level of a beer, for example, is to use two antibodies that are capable of recognizing the peptides of gluten that comprise most of the immunotoxic response of these proteins. These antibodies must recognize sequences that do not overlap one another. Therefore, the difference in estimations between different antibody-based methods could be appreciated more in hydrolyzed food or beverages because of differential resistance of the corresponding epitopes observed [70]. Thus, the concept of gluten content should ideally be changed to gluten immunogenic peptides in beer and other hydrolyzed food, as gluten proteins are actually hydrolyzed and peptides are what remain. A potential risk is that absorption is faster because it may not need digestion in the stomach and intestine to have immunogenic peptides that may be rapidly absorbed.

Moreover, the reactivity and number of immunogenic peptide sequences may vary among different wheat [71–73] and barley [74] varieties. All flours from hexaploid wheats (common wheat and spelt) studied by Schalk et al. [73] contained the immunogenic 33-mer peptide. In contrast, the 33-mer was absent (<limit of detection) from tetra- and diploid species (durum wheat, emmer, einkorn), most likely because of the absence of the D-genome, which encodes α-gliadins. In Comino et al. [75], eight different barley cultivars were analyzed by G12 ELISA, revealing 25-fold differences in reactivity between the most and the least reactive barley cultivars. Three of those cultivars were analyzed by T-cell activation, and the hierarchy of immunogenicity with T-cells isolated from peripheral blood was consistent with the reactivity of the barley kernels.

Most ELISA methods are based on quantifying the prolamin fraction and not the glutelin fraction. To account for this bias, the determined prolamin content is usually multiplied by a factor of two to obtain the gluten content, assuming a prolamin/glutelin ratio of one (Codex Standard 118–179). However, the true ratios are highly variable, ranging from 0.2 in wheat starch to 13.9 in einkorn flour [76]. Consequently, the gluten content of wheat starch tends to be underestimated when the prolamin $\times$ 2 for calculation is applied, as shown by Scherf et al. [77]. Comparative analysis of gluten content in eight starch samples, labeled as gluten-free by R5 ELISA (prolamin $\times$ 2) and a chromatographic control method (sum of prolamins and glutelins), revealed highly different mean values (15 vs. 54 mg/kg). Moreover, gluten analysis of 30 wheat starch samples (14 declared as gluten-free) with seven commercial ELISA kits resulted in up to six different values per sample [77].

In conclusion, an ideal antibody for gluten analysis should not only be a reliable indicator of the presence of prolamins from cereal species known to be toxic to patients with CD, but also should recognize the specific intramolecular regions responsible for such immunotoxicity; as such, it would not underestimate the potential immunogenicity of certain hydrolytic materials [61]. Thus, the reactivity of monoclonal antibodies used in the detection of gluten content could provide different estimations that should be verified with real immunogenicity in human samples. Moreover, ELISA testing is still the most useful method for gluten quantitation and, despite the variability between tests, it provides acceptable results regarding the raw materials used in gluten-free food production. Problems exist in analyzing gluten contamination in wheat starch and processed foods, e.g., heat-treated or fermented foods, and these require further research and development.

5. Conclusions

Most studies on the gluten contents of naturally or certified gluten-free foods reveal relatively high rates of contamination, and contaminated naturally gluten-free products appear to be a higher health risk than certified products for patients with CD. Thus, both naturally and labeled gluten-free foods do not guarantee safety for patients with CD, and gluten contamination is an important cause of inadvertent non-adherence to a GFD. Oats could be included in a GFD, provided that the absence of toxic gluten from the oats themselves, or from contamination by wheat, barley, or rye, is guaranteed. Additionally, gluten-free products from food services represent a considerable risk for gluten contamination. Patients with CD should be aware of numerous composite foods, medications, and cosmetics that contain "hidden" gluten that is used as an additive to improve the properties of gluten-free foods. Many methods have been developed for the detection of gluten proteins, including ELISA, PCR, LFD, and LC/MS. ELISA testing is still the most useful method for gluten quantitation and, despite the variability between tests, it provides acceptable results. Problems exist in analyzing gluten contamination in wheat starch and processed foods, e.g., heat-treated or fermented foods, and these require further research and development.

Author Contributions: Conceptualization, H.W. and C.S.; data curation, H.W. and C.S.; investigation, H.W. and C.S.; methodology, H.W. and C.S.; resources, H.W. and C.S.; writing—original draft, H.W., V.S., Á.R.-C., C.S. and I.C.; writing—review and editing, H.W., V.S., Á.R.-C., C.S. and I.C. All authors have read and agreed to the published version of the manuscript.

Funding: This work was supported by grants from Ministry of Economy, Knowledge, Business and University (project RTC-2016-5441-1).

Conflicts of Interest: The authors have no potential conflicts of interest to declare.

References

1. Taraghikhah, N.; Ashtari, S.; Asri, N.; Shahbazkhani, B.; Al-Dulaimi, D.; Rostami-Nejad, M.; Rezaei-Tavirani, M.; Razzaghi, M.R.; Zali, M.R. An updated overview of spectrum of gluten-related disorders: Clinical and diagnostic aspects. *BMC Gastroenterol.* **2020**, *20*, 258. [CrossRef] [PubMed]
2. Ludvigsson, J.F.; Leffler, D.A.; Bai, J.C.; Biagi, F.; Fasano, A.; Green, P.H.; Hadjivassiliou, M.; Kaukinen, K.; Kelly, C.P.; Leonard, J.N.; et al. The Oslo definitions for coeliac disease and related terms. *Gut* **2013**, *62*, 43–52. [CrossRef] [PubMed]
3. Antiga, E.; Maglie, R.; Quintarelli, L.; Verdelli, A.; Bonciani, D.; Bonciolini, V.; Caproni, M. Dermatitis herpetiformis: Novel perspectives. *Front. Immunol.* **2019**, *10*, 1290. [CrossRef]
4. Rej, A.; Aziz, I.; Sanders, D.S. Coeliac disease and noncoeliac wheat or gluten sensitivity. *J. Intern. Med.* **2020**, *288*, 537–549. [CrossRef]
5. Herrera, M.J.; Hermoso, M.A.; Quera, R. An update on the pathogenesis of celiac disease. *Rev. Med. Chile* **2009**, *137*, 1617–1626.
6. Fasano, A.; Sapone, A.; Zevallos, V.; Schuppan, D. Nonceliac gluten and wheat sensitivity. *Gastroenterology* **2015**, *148*, 1195–1204. [CrossRef]
7. Caio, G.; Volta, U.; Sapone, A.; Leffler, D.A.; De Giorgio, R.; Catassi, C.; Fasano, A. Celiac disease: A comprehensive current review. *BMC Med.* **2019**, *17*, 142. [CrossRef]
8. Lindfors, K.; Ciacci, C.; Kurppa, K.; Lundin, K.E.; Makharia, G.K.; Mearin, M.L.; Murray, J.A.; Verdu, E.F.; Kaukinen, K. Coeliac disease. *Nat. Rev. Dis. Primers* **2019**, *5*, 3. [CrossRef]
9. Hollon, J.R.; Cureton, P.A.; Martin, M.L.; Puppa, E.L.; Fasano, A. Trace gluten contamination may play a role in mucosal and clinical recovery in a subgroup of diet-adherent non-response celiac disease patients. *BMC Gastroenterol.* **2013**, *13*, 40. [CrossRef] [PubMed]
10. Itzlinger, A.; Branchi, F.; Elli, L.; Schumann, M. Gluten-free diet in celiac disease—Forever and for all? *Nutrients* **2018**, *10*, 1796. [CrossRef] [PubMed]
11. Rosell, C.M.; Barro, F.; Sousa, C.; Mena, M.C. Cereals for developing gluten-free products and analytical tools for gluten detection. *J. Cereal Sci.* **2014**, *59*, 354–364. [CrossRef]
12. Rai, S.; Kaur, A.; Chopra, C.S. Gluten-free products for celiac susceptible people. *Front. Nutr.* **2018**, *5*, 116. [CrossRef]
13. Codex Standard 118-1979. Available online: http://www.fao.org/fao-who-codexalimentarius/sh-proxy/en/?lnk=1&url=https%253A%252F%252Fworkspace.fao.org%252Fsites%252Fcodex%252FStandards%252FCXS%2B118-1979%252FCXS_118e_2015.pdf (accessed on 2 February 2021).
14. Francavilla, R.; Cristofori, F.; Stella, M.; Borrelli, G.; Naspi, G.; Castellaneta, S. Treatment of celiac disease: From gluten-free diet to novel therapies. *Minerva Pediatr.* **2014**, *66*, 501–516. [PubMed]
15. Leonard, M.M.; Cureton, P.; Fasano, A. Indications and use of the gluten contamination elimination diet for patients with non-responsive celiac disease. *Nutrients* **2017**, *9*, 1129. [CrossRef] [PubMed]
16. Shehab, D.I. Celiac disease. *Egypt J. Intern. Med.* **2013**, *25*, 53–62.

17. Koerner, T.B.; Abbott, M.; Godefroy, S.B.; Popping, B.; Yeung, J.M.; Diaz-Amigo, C.; Roberts, J.; Taylor, S.L.; Baumert, J.L.; Ulberth, F.; et al. Validation procedures for quantitative gluten ELISA methods: AOAC allergen community guidance and best practices. *J. AOAC Int.* **2013**, *96*, 1033–1040. [CrossRef]
18. Segura, V.; Díaz, J.; Ruiz-Carnicer, Á.; Muñoz-Suano, A.; Carrillo-Carrión, C.; Sousa, C.; Cebolla, Á.; Comino, I. Rapid, Effective and Versatile Extraction of Gluten in Food with Application on Different Immunological Methods. *Foods* **2021**, *10*, 652. [CrossRef]
19. Deutsch, H.; Poms, R.; Heeres, H.; van der Kamp, J.W. Labeling and regulatory issues. In *Gluten-Free Cereal Products and Beverages*; Arendt, E.K., dal Bello, F., Eds.; Academic Press: Cambridge, MA, USA; Elsevier: Amsterdam, The Netherlands, 2008; pp. 29–46.
20. McIntosh, J.; Flanagan, A.; Madden, N.; Mulcahy, M.; Dargan, L.; Walker, M.; Burns, D.T. Awareness of coeliac disease and the gluten status of 'gluten-free' food obtained on request in catering outlets in Ireland. *Int. J. Food Sci. Technol.* **2011**, *46*, 1569–1574. [CrossRef]
21. Gibert, A.; Kruizinga, A.G.; Neuhold, S.; Houben, G.F.; Canela, M.A.; Fasano, A.; Catassi, C. Might gluten traces in wheat substitutes pose a risk in patients with celiac disease? A population-based probabilistic approach to risk estimation. *Am. J. Clin. Nutr.* **2013**, *97*, 109–116. [CrossRef]
22. Lee, H.J.; Anderson, Z.; Ryu, D. Gluten contamination in foods labeled as "gluten-free" in the United States. *J. Food Prot.* **2014**, *77*, 1830–1833. [CrossRef]
23. Oliveira, O.M.; Zandonadi, R.P.; Gandolfi, L.; de Almeida, D.C.; Almeida, L.M.; Pratesi, R. Evaluation of the presence of gluten in beans served at self-service restaurants: A problem for celiac disease carriers. *J. Culin. Sci. Technol.* **2014**, *12*, 22–33. [CrossRef]
24. Sharma, G.M.; Pereira, M.; Williams, K.M. Gluten detection in foods available in the United States—A market survey. *Food Chem.* **2015**, *169*, 120–126. [CrossRef] [PubMed]
25. Bustamante, M.A.; Fernandez-Gil, M.P.; Churruca, I.; Miranda, J.; Lasa, A.; Navarro, V.; Simon, E. Evolution of gluten content in cereal-based gluten-free products: An overview from 1998 to 2016. *Nutrients* **2017**, *9*, 21. [CrossRef]
26. Atasoy, G.; Gokhisar, O.K.; Turhan, M. Gluten contamination in manufactured gluten-free food in Turkey. *Food Addit. Contam. Part. A Chem. Anal. Control. Expo. Risk Assess.* **2020**, *37*, 363–373. [CrossRef] [PubMed]
27. Bianchi, D.M.; Maurella, C.; Gallina, S.; Gorrasi, I.S.; Caramelli, M.; Decastelli, L. Analysis of gluten content in gluten-free pizza from certified take-away pizza restaurants. *Foods* **2018**, *7*, 180. [CrossRef] [PubMed]
28. Farage, P.; Zandonadi, R.P.; Gandolfi, L.; Pratesi, R.; Falcomer, A.L.; Arujo, L.S.; Yoshio Nakano, E.; Cortez-Ginani, V. Accidental gluten contamination in traditional lunch meals from food services in Brasilia, Brazil. *Nutrients* **2019**, *11*, 1924. [CrossRef]
29. Lerner, B.A.; Phan Vo, L.T.; Yates, S.; Rundle, A.G.; Green, P.H.; Lebwohl, B. Detection of gluten in gluten-free labeled restaurant food: Analysis of crowd-sourced data. *Am. J. Gastroenterol.* **2019**, *114*, 792–797. [CrossRef]
30. Raju, N.; Joshi, A.K.; Vahini, R.; Deepika, T.; Bhaskarachari, K.; Devindra, S. Gluten contamination in labelled and naturally gluten-free grain products in Southern India. *Food Addit. Contam. Part A Chem. Anal. Control. Expo. Risk Assess.* **2020**, *37*, 531–538. [CrossRef]
31. Thompson, T.; Lee, A.R.; Grace, T. Gluten contamination of grains, seeds, and flours in the United States: A pilot study. *J. Am. Diet. Assoc.* **2010**, *110*, 937–940. [CrossRef]
32. Koerner, T.B.; Cleroux, C.; Poirier, C.; Cantin, I.; La Vieille, S.; Hayward, S.; Dubois, S. Gluten contamination of naturally gluten-free flours and starches used by Canadians with celiac disease. *Food Addit. Contam. Part A Chem. Anal. Control. Expo. Risk Assess.* **2013**, *30*, 2017–2021. [CrossRef]
33. Hassan, H.; Elaridi, J.; Bassil, M. Evaluation of gluten in gluten-free-labeled foods and assessment of exposure level to gluten among celiac patients in Lebanon. *Int. J. Food Sci. Nutr.* **2017**, *68*, 881–886. [CrossRef]
34. Verma, A.K.; Gatti, S.; Galeazzi, T.; Monachesi, C.; Padella, L.; Baldo, G.D.; Annibali, R.; Lionetti, E.; Catassi, C. Gluten contamination in naturally or labeled gluten-free products marketed in Italy. *Nutrients* **2017**, *9*, 115. [CrossRef] [PubMed]
35. Farage, P.; de Medeiros Nobrega, Y.K.; Pratesi, R.; Gandolfi, L.; Assuncao, P.; Zandonadi, R.P. Gluten contamination in gluten-free bakery products: A risk for coeliac disease patients. *Public Health Nutr.* **2017**, *20*, 413–416. [CrossRef] [PubMed]
36. Scherf, K.A.; Wieser, H.; Koehler, P. Improved quantitation of gluten in wheat starch for celiac disease patients by gel-permeation high-performance liquid chromatography with fluorescence detection (GP-HPLC-FLD). *J. Agric. Food Chem.* **2016**, *64*, 7622–7631. [CrossRef] [PubMed]
37. Falcomer, A.L.; Santos-Araujo, L.; Farage, P.; Santos-Monteiro, J.; Yoshio-Nakano, E.; Puppin-Zandonadi, R. Gluten contamination in food services and industry: A systematic review. *Crit. Rev. Sci. Nutr.* **2018**, *22*, 1–15. [CrossRef] [PubMed]
38. Weisbrod, V.M.; Silvester, J.A.; Raber, C.; Suslovic, W.; Coburn, S.S.; Raber, B.; McMahon, J.; Damast, A.; Kramer, Z.; Kerzner, B. A quantitative assessment of gluten cross-contact in the school environment for children with celiac disease. *J. Pediatr. Gastroenterol. Nutr.* **2020**, *70*, 289–294. [CrossRef]
39. Cohen, I.; Day, A.S.; Shaoul, R. To be or not to be? An update on the ongoing debate on oats for patients with celiac disease. *Front. Pediatr.* **2019**, *7*, 384. [CrossRef]
40. Hoffmanova, I.; Sanchez, D.; Szczepankova, A.; Tlaskalova-Hogenova, H. The pros and cons of using oat in a gluten-free diet for celiac patients. *Nutrients* **2019**, *11*, 2345. [CrossRef]
41. Boison, J.; Allred, L.; Almy, D.; Anderson, L.; Baumert, J.; Bhandari, S.; Cebolla, A.; Chen, Y.; Crowley, E.; Diaz-Amigo, C.; et al. Standard method performance requirements (SMPRs®) 2017.21: Quantitation of wheat, rye and barley gluten in oats. *J. AOAC* **2018**, *101*, 1238–1242. [CrossRef]

42. Chen, Y.; Fritz, R.D.; Kock, L.; Garg, D.; Davis, R.M.; Kasturi, P. A stepwise test-all-positives methodology to assess gluten-kernel contamination at the serving-size level in gluten-free (GF) oat production. *Food Chem.* **2018**, *240*, 391–395. [CrossRef]
43. Hernando, A.; Mujico, J.R.; Juanas, D.; Méndez, E. Confirmation of the cereal type in oat products highly contaminated with gluten. *J. Am. Diet. Assoc.* **2006**, *106*, 665. [CrossRef] [PubMed]
44. Koerner, T.B.; Cléroux, C.; Poirier, C.; Cantin, I.; Alimkulov, A.; Elamparo, H. Gluten contamination in the Canadian commercial oat supply *Food Addit. Contam. Part A Chem. Anal. Control. Expo. Risk Assess.* **2011**, *28*, 705–710. [CrossRef]
45. Silano, M.; Dessì, M.; De Vincenzi, M.; Cornell, H. In Vitro tests indicate that certain varieties of oats may be harmful to patients with coeliac disease. *J Gastroenterol. Hepatol.* **2007**, *22*, 528–531. [CrossRef]
46. Silano, M.; Di Benedetto, R.; Maialetti, F.; De Vincenzi, A.; Calcaterra, R.; Cornell, H.J.; De Vincenzi, M. Avenins from different cultivars of oats elicit response by coeliac peripheral lymphocytes. *Scand. J. Gastroenterol.* **2007**, *42*, 1302e5. [CrossRef] [PubMed]
47. Comino, I.; Real, A.; de Lorenzo, L.; Cornell, H.; López-Casado, M.Á.; Barro, F.; Lorite, P.; Torres, M.I.; Cebolla, A.; Sousa, C. Diversity in oat potential immunogenicity: Basis for the selection of oat varieties with no toxicity in coeliac disease. *Gut* **2011**, *60*, 915–922. [CrossRef]
48. Silano, M.; Pozo, E.P.; Uberti, F.; Manferdelli, S.; Del Pinto, T.; Felli, C.; Budelli, A.; Vincentini, O.; Restani, P. Diversity of oat varieties in eliciting the early inflammatory events in celiac disease. *Eur. J. Nutr.* **2014**, *53*, 1177–1186. [CrossRef] [PubMed]
49. Mujico, J.R.; Mitea, C.; Gilissen, L.J.; de Ru, A.; van Veelen, P.; Smulders, M.J.; Koning, F. Natural variation in avenin epitopes among oat varieties: Implications for celiac disease. *J. Cereal Sci.* **2011**, *54*, 8–12. [CrossRef]
50. Benoit, L.; Masiri, J.; Del Blanco, I.A.; Meshgi, M.; Gendel, S.M.; Samadpour, M. Assessment of Avenins from Different Oat Varieties Using R5-Based Sandwich ELISA. *J. Agric. Food Chem.* **2017**, *65*, 1467–1472. [CrossRef] [PubMed]
51. Gimenez, M.J.; Real, A.; Garcia-Molina, M.D.; Sousa, C.; Barro, F. Characterization of celiac disease related oat proteins: Bases for the development of high quality oat varieties suitable for celiac patients. *Sci. Rep.* **2017**, *7*, 42588. [CrossRef]
52. Arentz-Hansen, H.; Fleckenstein, B.; Molberg, Ø.; Scott, H.; Koning, F.; Jung, G.; Roepstorff, P.; Lundin, K.E.; Sollid, L.M. The molecular basis for oat intolerance in patients with celiac disease. *PLoS Med.* **2004**, *1*, 84–92. [CrossRef]
53. Real, A.; Comino, I.; de Lorenzo, L.; Merchán, F.; Gil-Humanes, J.; Giménez, M.J.; López-Casado, M.Á.; Torres, M.I.; Cebolla, Á.; Sousa, C.; et al. Molecular and immunological characterization of gluten proteins isolated from oat cultivars that differ in toxicity for celiac disease. *PLoS ONE* **2012**, *7*, e48365. [CrossRef] [PubMed]
54. Comino, I.; Bernardo, D.; Bancel, E.; Moreno, M.L.; Sánchez, B.; Barro, F.; Šuligoj, T.; Ciclitira, P.J.; Cebolla, Á.; Knight, S.C.; et al. Identification and molecular characterization of oat peptides implicated on coeliac immune response. *Food Nutr. Res.* **2016**, *60*, 30324. [CrossRef] [PubMed]
55. Wieser, H.; Koehler, P.; Scherf, K.A. *Wheat—An Exceptional Cro*; Woodhead Publishing (Elsevier): Daxford, UK, 2020.
56. Shah, A.V.; Serajuddin, A.T.; Mangione, R.A. Making all medications gluten free. *J. Pharm. Sci.* **2018**, *107*, 1263–1268. [CrossRef]
57. Verma, A.K.; Lionetti, E.; Gatti, S.; Franceschini, E.; Catassi, G.N.; Catassi, C. Contribution of oral hygiene and cosmetics on contamination of gluten-free diet: Do celiac customers need to worry about? *J. Pediatr. Gastroenterol. Nutr.* **2019**, *68*, 26–29. [CrossRef] [PubMed]
58. Allred, L.K.; Park, E.S. EZ Gluten for the qualitative detection of gluten in foods, beverages, and environmental surfaces. *J. AOAC Int.* **2012**, *95*, 1106–1117. [CrossRef] [PubMed]
59. Zhang, J.; Portela, S.B.; Horrell, J.B.; Leung, A.; Weitmann, D.R.; Artiuch, J.B.; Wilson, S.M.; Cipriani, M.; Slakey, L.K.; Burt, A.M.; et al. An integrated, accurate, rapid and economical handheld consumer gluten detector. *Food Chem.* **2019**, *275*, 446–456. [CrossRef]
60. Bruins Slot, I.D.; Bremer, M.G.; van der Fels-Klerx, I.; Hamer, R.J. Evaluating the performance of gluten ELISA test kits: The numbers do not tell the tale. *Cereal Chem.* **2015**, *92*, 513–521. [CrossRef]
61. Mena, M.C.; Sousa, C. Analytical tools for gluten detection, Policies and regulation. In *Advances in the Understanding of Gluten Related Pathology and the Evolution of Gluten-Free Foods*; Arranz, E., Fernández-Bañares, F., Rosell, C.M., Rodrigo, L., Peña, A.S., Eds.; OmniaScience: Barcelona, Spain, 2015; pp. 527–564.
62. Rzychon, M.; Brohée, M.; Cordeiro, F.; Haraszi, R.; Ulberth, F.; O'Connor, G. The feasibility of harmonizing gluten ELISA measurements. *Food Chem.* **2017**, *234*, 144–154. [CrossRef]
63. Lexhaller, B.; Tompos, C.; Scherf, K.A. Comparative analysis of prolamin and glutelin fractions from wheat, rye, and barley with five sandwich ELISA test kits. *Anal. Bioanal. Chem.* **2016**, *408*, 6093–6104. [CrossRef]
64. Lexhaller, B.; Tompos, C.; Scherf, K.A. Fundamental study on reactivities of gluten protein types from wheat, rye and barley with five sandwich ELISA test kits. *Food Chem.* **2017**, *237*, 320–330. [CrossRef]
65. Schopf, M.; Scherf, K.A. Wheat cultivar and species influence variability of gluten ELISA analyses based on polyclonal and monoclonal antibodies R5 and G12. *J. Cereal Sci.* **2018**, *83*, 32–41. [CrossRef]
66. Rallabhandi, P.; Sharma, G.M.; Pereira, M.; Williams, K.M. Immunological characterization of the gluten fractions and their hydrolysates from wheat, rye and barley. *J. Agric. Food Chem.* **2015**, *63*, 1825–1832. [CrossRef]
67. Frazer, A.C.; Fletcher, R.F.; Ross, C.A.; Shaw, B.; Sammons, H.G.; Schneider, R. Gluten-induced enteropathy: The effect of partially digested gluten. *Lancet* **1959**, *2*, 252–255. [CrossRef]
68. Picariello, G.; Mamone, G.; Cutignano, A.; Fontana, A.; Zurlo, L.; Addeo, F.; Ferranti, P. Proteomics, peptidomics, and immunogenic potential of wheat beer (weissbier). *J. Agric. Food Chem.* **2015**, *63*, 3579–3586. [CrossRef] [PubMed]

69. Real, A.; Comino, I.; Moreno, M.D.L.; López-Casado, M.Á.; Lorite, P.; Isabel Torres, M.; Cebolla, Á.; Sousa, C. Identification and in vitro reactivity of celiac immunoactive peptides in an apparent gluten-free beer. *PLoS ONE* **2014**, *9*, 12–14. [CrossRef] [PubMed]
70. Wei, G.; Tian, N.; Siezen, R.; Schuppan, D.; Helmerhorst, E.J. Identification of food-grade subtilisins as gluten-degrading enzymes to treat celiac disease. *Am. J. Physiol. Gastrointest. Liver Physiol.* **2016**, *311*, G571–G580. [CrossRef] [PubMed]
71. Spaenij-Dekking, L.; Kooy-Winkelaar, Y.; Van Veelen, P.; Drijfhout, J.W.; Jonker, H.; Van Soest, L.; Smulders, M.J.M.; Bosch, D.; Gilissen, L.J.W.J.; Koning, F. Natural variation in toxicity of wheat: Potential for selection of nontoxic varieties for celiac disease patients. *Gastroenterology* **2005**, *129*, 797–806. [CrossRef]
72. Gregorini, A.; Colomba, M.; Julia Ellis, H.; Ciclitira, P.J. Immunogenicity characterization of two ancient wheat α-gliadin peptides related to coeliac disease. *Nutrients* **2009**, *1*, 276–290. [CrossRef] [PubMed]
73. Schalk, K.; Lexhaller, B.; Koehler, P.; Scherf, K.A. Isolation and characterization of gluten protein types from wheat, rye, barley and oats for use as reference materials. *PLoS ONE* **2017**, *12*, e0172819. [CrossRef]
74. Mamone, G.; Ferranti, P.; Rossi, M.; Roepstorff, P.; Fierro, O.; Malorni, A.; Addeo, F. Identification of a peptide from alpha-gliadin resistant to digestive enzymes: Implications for celiac disease. *J. Chromatogr. B Analyt. Technol. Biomed. Life Sci.* **2007**, *855*, 236–241. [CrossRef]
75. Comino, I.; Real, A.; Gil-Humanes, J.; Pistón, F.; de Lorenzo, L.; Moreno, M.L.; López-Casado, M.Á.; Lorite, P.; Cebolla, A.; Torres, M.I.; et al. Significant differences in coeliac immunotoxicity of barley varieties. *Mol. Nutr. Food Res.* **2012**, *56*, 1697–1707. [CrossRef] [PubMed]
76. Wieser, H.; Koehler, P. Is the calculation of the gluten content by multiplying the prolamin content by a factor of 2 valid? *Eur. Food Res. Technol.* **2009**, *229*, 9–13. [CrossRef]
77. Scherf, K.A. Gluten analysis of wheat starches with seven commercial ELISA test kits—Up to six different values. *Food Anal. Methods* **2017**, *10*, 234–246. [CrossRef]

Article

Rationale for Timing of Follow-Up Visits to Assess Gluten-Free Diet in Celiac Disease Patients Based on Data Mining

Alfonso Rodríguez-Herrera [1], Joaquín Reyes-Andrade [2] and Cristina Rubio-Escudero [3,*]

[1] Pediatrics, Saint Luke's Hospital, University College Dublin, Kilkenny R95 FY71, Ireland;
     a.rodriguezherrera@hse.ie
[2] Unidad de Gastroenterología y Nutrición, Instituto Hispalense de Pediatría, 41013 Seville, Spain;
     joaquinreyes@ihppediatria.com
[3] Department of Computer Languages and Systems, University of Seville, 41013 Seville, Spain
*   Correspondence: crubioescudero@us.es; Tel.: +34-955421018

Abstract: The assessment of compliance of gluten-free diet (GFD) is a keystone in the supervision of celiac disease (CD) patients. Few data are available documenting evidence-based follow-up frequency for CD patients. In this work we aim at creating a criterion for timing of clinical follow-up for CD patients using data mining. We have applied data mining to a dataset with 188 CD patients on GFD (75% of them are children below 14 years old), evaluating the presence of gluten immunogenic peptides (GIP) in stools as an adherence to diet marker. The variables considered are gender, age, years following GFD and adherence to the GFD by fecal GIP. The results identify patients on GFD for more than two years (41.5% of the patients) as more prone to poor compliance and so needing more frequent follow-up than patients with less than 2 years on GFD. This is against the usual clinical practice of following less patients on long term GFD, as they are supposed to perform better. Our results support different timing follow-up frequency taking into consideration the number of years on GFD, age and gender. Patients on long term GFD should have a more frequent monitoring as they show a higher level of gluten exposure. A gender perspective should also be considered as non-compliance is partially linked to gender in our results: Males tend to get more gluten exposure, at least in the cultural context where our study was carried out. Children tend to perform better than teenagers or adults.

Keywords: celiac disease; data mining gluten free diet; gluten proteins; immunogenicity; evidence-based practice; case management; treatment adherence and compliance

Citation: Rodríguez-Herrera, A.; Reyes-Andrade, J.; Rubio-Escudero, C. Rationale for Timing of Follow-Up Visits to Assess Gluten-Free Diet in Celiac Disease Patients Based on Data Mining. *Nutrients* **2021**, *13*, 357. https://doi.org/10.3390/nu13020357

Academic Editor: Giacomo Caio

Received: 17 December 2020

Accepted: 16 January 2021

Published: 25 January 2021

Publisher's Note: MDPI stays neutral with regard to jurisdictional claims in published maps and institutional affiliations.

1. Introduction

Celiac disease (CD) is a chronic systemic immune-mediated condition that occurs in a genetically susceptible host, produced by the ingestion of nutritional gluten, the major protein component in wheat and other related cereals [1]. It is one of the most common disorders, involving around 1% of the general population and can occur at any age [2]. CD is characterized by the presence of a wide variety of CD-specific antibodies, enteropathy, gluten-dependent clinical expressions, and HLA-DQ2 or HLA-DQ8 haplotypes [3–5].

A lifetime gluten-free diet (GFD) is nowadays the only treatment for CD. Non exposure to gluten is believed to achieve mucosal recovery, resolve symptoms, and avoid the difficulties associated to non-treated CD [6]. Even though following a GFD might seem easy, it becomes a challenge in the gluten-rich Western diet. Indeed, it is increasingly recognized that many CD patients on a presumably GFD may have ongoing symptoms and/or persistent villous atrophy. Therefore, adherence to the GFD needs to be assessed to guarantee potential effects on the patient's health condition and quality of life [5].

There is no consensus regarding the best means for assessing compliance or the optimal frequency of monitoring the GFD. Despite the availability of diverse traditional GFD adherence markers, such as dietary tests or serology, none of them are an accurate evaluation method of the dietary obedience [7,8]. As a result, finding gluten immunogenic peptides (GIP) in human urine and stools have appeared as novel markers for direct verification of GFD compliance [9–11]. GIP show the capacity to resist to gastrointestinal absorption and accounts for immunogenic reaction in T cells of patients with CD. Differently to traditional methods for the monitoring of GFD obedience, which only measures the consequences of GFD non-adherence, this non-intrusive method allows for a direct and quantitative evaluation of gluten exposure [11]. Using this new methodology, GIP were detected in 30–60% of CD patients on a GFD and for whom no gluten exposure was identified by dietary questionnaire or serological tests [1].

It is generally recommended that CD patients have careful long-term follow-up. Silvester et al. [12], conclude that the existing guidelines regarding CD patients follow-up proposed very different recommendations and many were not evidence-based. This study was based on gastroenterological societies and associations guidelines and recommendations by specialists obtained from MEDLINE and other Internet search engines. Javorsky, et al. [13] searched the PubMed database for works related to evidence-based guidelines on follow-up intervals for the 5 topmost chronic conditions according to the highest amount of patient attendance in 2010 in the USA (back problems, arthritis, hypertension, mental disorders, chronic obstructive pulmonary disease/asthma), with some guidelines attempting to recommend specific follow-up intervals, but not being evidence-based. They did not propose intervals based on clinical data or failed to reveal on what timing the visits were based. However, both works conclude that time frequency of visits intervals is relevant. Therefore, prospective studies appear as necessary to create cost-effective, rational, and risk-stratified guidelines for long-term follow-up of these patients [12].

Data mining can be defined as the automatic analysis of data sources to identify models representing knowledge [14]. Clinical data mining is concerned with the application of data mining techniques to clinical data [15], which in turn allows the creation of models of knowledge and aids clinical decision making [16].

In this work, we aimed at providing grounds for evidence-based follow-up frequency suggestions for CD patients, obtained by applying clinical data mining to a dataset extracted from a cohort of 188 CD patients (75% of them are children below 14 years old), whose GFD compliance was assessed. The presence of GIP in stools was used as a distinctive biomarker of GFD adherence in this series. Other variables considered were gender, age and length of ongoing GFD.

2. Materials and Methods

This work is based on the analysis of a retrospective dataset previously collected in a partially blinded nonrandomized, multicenter study including 188 CD patients (75% of them are children below 14 years old) following a GFD recruited between 2012 and 2014 at 13 Spanish hospitals [1]. The trial registration number is NCT02711397. This study was authorized by the ethics committee of each involved institution and informed written consent was acquired from participants over 18 years old and from parents or legal keepers for participants below 18 years old. The group under study was composed of celiac patients on GFD for at least 1 year before being included in the study. Inclusion criteria restricted enrollment to those who had an HLA-DQ2 or HLA-DQ8 haplotype test and a histologically nonstandard duodenal biopsy (grade Marsh IIIB or IIIC) at the time of diagnosis, as well as positive serum anti-endomysium IgA antibodies and/or anti-tissue transglutaminase (anti-tTG) IgA antibodies.

Adherence to GFD was evaluated by GIP detection. The concentration of GIP in feces was assessed with sandwich enzyme-linked immunosorbent assay (ELISA) [17] using the iVYDAL In Vitro Diagnostics iVYLISA GIP-S Kit (Biomedal S.L., Seville, Spain). Patients were also measured on a four-day food record dietitian review and celiac serology (tissue transglutaminase and deamidated gliadin peptide antibodies). Information regarding the date of CD diagnosis, duration of the GFD, and demographic and clinical data were also retrieved.

Data Mining Methods

Data Mining comprises two main tasks: Prediction (supervised learning) and description (unsupervised learning) [18]. Prediction attempts to predict some or several unknown variables from other known ones. The description, however, tries to look for patterns that describe the data in a way that humans can understand.

Within the scope of prediction there are two fundamental tasks: Classification and regression. Classification tries to assign a target variable that belongs to a dataset [19] while regression aims to predict continuous values [20].

We can find a great variety of classification algorithms in the literature [19]. This work has applied the C4.5 algorithm, which according to Wu et al. [21] is one of the top 10 data mining algorithms. This algorithm is one of the best-known ones capable of building decision trees. It was implemented by Quinlan in [22] and is an extension to the ID3 [23] algorithm also implemented by him.

Decision trees can be defined as a classification method that, given a dataset, recursively divides it into subsets using decisions specified at each branch or node in the tree. As we can see in the results shown in Figure 1, the parts of the tree are a root node (made up of all data), inner nodes (branches), and end nodes (leaves). A register from a dataset is classified by successively dividing, following the decision structure defined in the tree, and the target label is assigned to each register according to the node of the leaf on which the register is situated [24,25].

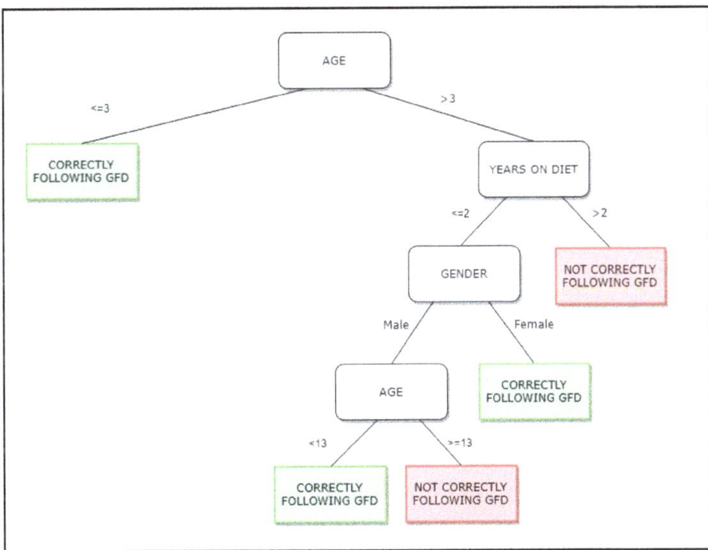

Figure 1. Decision tree obtained by use of the C4.5 algorithm. GFD stands for Gluten Free Diet.

In Figure 1 we show the tree obtained with the dataset under study. Each register stores information related to the variables under study: Gender of the patient, years following GFD diet, age of the subject when collecting the sample and results positive or negative of the fecal GIP. According to the tree, if the patient is 3 years old or below, the GFD is correctly being followed, but if the age is over 3 years old and has been more than 2 years on GFD, the GFD diet is not correctly followed.

The algorithm C4.5 is described below. For a set S registers, C4.5 creates the initial tree using the divide-and-conquer strategy in this way [21,26]:

- Case 1. All the registers in S belong to the same target label or S is not big enough. Then the tree is created with only one leaf, with the target label more frequent S.

- Case 2. In other cases, select a test base on a single variable with two or more outcomes. This test becomes the root of the tree, and one branch is created for each outcome. Then, split S into subsets S1, S2 ... depending to the outcome for each register, and apply the same procedure recursively to each subset.

3. Results and Discussion

Data mining techniques are becoming very popular in clinical data analysis, as a complement to the classically used statistical analysis. Furthermore, data mining is proving to be extremely useful when the volume of data increases [27]. In this era of computer-aided health care, the management of follow-up visits and frequency with an evidence-based approach has the power to decrease costs and improve the population access to the health system [13].

The dataset collected includes four variables. The first, gender, indicates the gender of the patient, the second, years, reports the years that the patient has been on GFD, the third, age, the age of the patient when the sample was collected and finally, results, represents the result of the fecal GIP as positive or negative. This test provides information on whether fecal gluten peptides have been found, so that we can know for sure whether, or not, the subject has followed medical recommendations about not taking gluten [1].

Initially an exploratory analysis of the data was carried out to get an overall vision of the distribution of each of the four variables (see Figure 2).

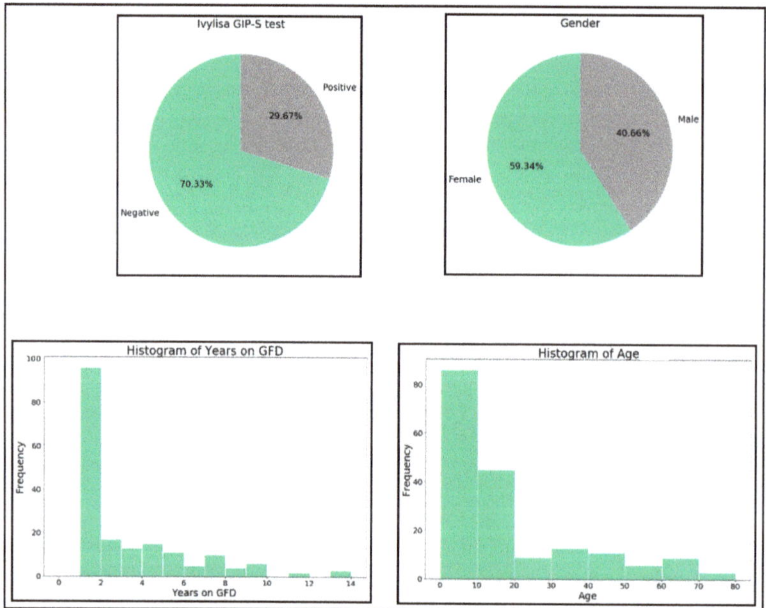

Figure 2. Distribution of the four variables.

Data showed to be unbalanced in regards to GIP (70.23% negatives and 29.67% positives) and gender (59.34% females and 40.66% males). Regarding years on GFD, most of the samples correspond to short term GFD followers. The age of the samples is in the interval (0, 20) for most of the samples.

Data where then analyzed using the C4.5 algorithm. It was executed with different sets of parameters in order to obtain the best resulting tree in terms of area under the curve (AUC). AUC ranks in the (0, 1) interval, with 1 being the best value. It tells us how capable a model is of distinguishing a target variable, positive or negative result for the fecal GIP in our case. The resulting tree can be seen in Figure 1.

This tree is the one best representing the dataset, with an AUC value of 0.7. According to this tree, patients are more adherent following GFD as usually children below 3 years age, women with less than 2 years on GFD, and men up to 13 years old with less than 2 years on GFD. Patients not correctly following the GFD can be characterized as CD patients over 3 years old, with more than 2 years of GFD, and men with less than 2 years of GFD but more than 13 years old. This decision tree identifies patients on the GFD for a longer time, as more prone to poor compliance and perhaps needing more frequent follow-up. These results concur with the results obtained previously by Comino et.al. in [1], in which they identified 13 years old as an age point for increasing dietary transgressions, as well as gender as a determining factor for these transgressions (male at certain ages are more prone to not correctly follow GFD).

This is against the most usual clinical practice of following less the cohort of patients on long term GFD, as they are supposed to perform better.

Current Recommendations for Frequency of Follow-Up in CD

The current clinical practice guidelines on CD do not offer a detailed background with regard to recommendations about how often patients are met for follow-up. These recommendations are based simply on suggestions of periodic visits, usually, or on an annual basis [28–33]. Despite the efforts already made to prevent or diagnose the CD early [34], there is no mention of clinical practice guidelines performing a more thorough control in adolescent patients, despite teenage being a known factor of increased risk to be exposed to gluten intake. The rationale for follow-up frequency in chronic diseases is crucial to maximize the quality of patient care. CD is a chronic disease increasing in frequency in different geographic areas [2]. In CD, non-exposure to gluten is the only "medication". Norris et al. [35] highlighted that compliance is related to how individuals think about their personal need for a treatment in relation to their fears about the potential adverse effects. Reminders or repeated interactions with health personnel may improve compliance by building a therapeutic relationship. Hall et al. describe such process on lengthy therapies such as the one used on physical rehabilitation [36].

Some studies considering the differences linked to gender in CD have been published. Lee et al., in a study carried out by Columbia University, [37] describe gender differences as being highly significant in quality of life perceived. As examples, eating out is a problem for 20% of men and 65% of women, traveling for 18% of men and 64% of women, family life for 18% of men and 49% of women, and diet obedience, regarding the professional career, is a problem for 15% of men and 26% of women. It may be linked to a different level of awareness about the impact of gluten exposure. Despite these significant gender-specific differences, there is no differentiation on the follow up pathways [37]. Does this difference have an actual impact on long term outcomes?

None of the published guidelines consider this gender perspective. It seems sensible to assume that a better avoidance of gluten exposure will render better health outcomes. From our understanding, this is the first time a research work applies data mining to determine follow-up frequency for celiac disease. Although there have been many studies on advances in diagnosis and treatment, the volume of research on patient follow-up is significantly smaller. Scrutiny of performance of medical care can be improved by use of better data analysis. The classic methods of follow-up, serology and dietary surveys, do not present the accuracy needed to measure long term compliance. But in spite of these, most centers keep on relying on it for their decision-making process during follow up without tailoring their care to the actual profile of risk of gluten exposure. The frequency of follow-up has not been analyzed in depth and has been based on general recommendations, without individualization. Appropriate follow-up frequency must be established based on healthcare outcomes. The idea that "one size fits all" proves to be incorrect for follow-up strategies.

4. Conclusions

GFD treatment is very difficult to satisfy, in spite of all efforts for adherence to it, since gluten is present in most of the food we intake. The general population does not need to adhere to GFD, making the coexistence with celiac population a risk. In this work, we have been able to characterize the patients who are more adherent and those who do not correctly follow the GFD based on the four variables studied (gender, age, years on GFD, and fecal GIP).

The cohort of patients on long term GFD should have a more frequent monitoring as they tend to show higher levels of gluten exposure upon longer time on GFD. Males tend to get more gluten exposure when compared with females, at least in the cultural context where our study was carried out.

Data mining techniques applied to records could improve the identification of celiac patients who regularly transgress (voluntarily or involuntarily) whilst following a GFD. It would help to avoid more serious consequences due to persistent exposure to gluten. Timing of follow-up frequency should be different for patients newly diagnosed than for patients on the GFD for a longer period. A gender perspective should be considered as the risk off non-compliance is partially linked to gender in our results. CD management can greatly benefit from evidence-based timing of follow-up visits.

Author Contributions: Conceptualization, A.R.-H. and C.R.-E.; methodology, A.R.-H. and C.R.-E.; software, C.R.-E.; validation, A.R.-H. and J.R.-A.; formal analysis, C.R.-E.; investigation, A.R.-H. and J.R.-A.; resources, A.R.-H.; data curation, C.R.-E.; writing—original draft preparation, A.R.-H. and C.R.-E.; writing—review and editing, J.R.-A.; visualization, C.R.-E.; supervision, A.R.-H. and C.R.-E.; project administration, A.R.-H., and C.R.-E.; funding acquisition, A.R.-H., and C.R.-E. All authors have read and agreed to the published version of the manuscript.

Funding: The authors want to thank the financial support given by the Saint Luke's Hospital, Kilkenny, Republic of Ireland. They also want to thank the support by the Spanish Ministry of Economy and Competitivity (TIN2017-88209-C2-2-R) and the Junta de Andalucía Project US-1263341.

Institutional Review Board Statement: The study was conducted according to the guidelines of the Declaration of Helsinki, and approved by the Institutional Review Board (or Ethics Committee) of Hospital Universitario Virgen Macarena, Seville (Spain), Protocol code Deliac 01-2012, date of approval 25.5.2012.

Informed Consent Statement: Informed consent was obtained from all subjects involved in the study.

Data Availability Statement: Data is available upon request.

Conflicts of Interest: The authors declare no conflict of interest.

References

1. Comino, I.; Fernández-Bañares, F.; Esteve, M.; Ortigosa, L.; Castillejo, G.; Fambuena, B.; Ribes-Koninckx, C.; Sierra, C.; Rodríguez-Herrera, A.; Salazar, J.C.; et al. Fecal gluten peptides reveal limitations of serological tests and food questionnaires for monitoring gluten-free diet in celiac disease patients. *Am. J. Gastroenterol.* **2016**, *111*, 1456–1465. [CrossRef] [PubMed]
2. Catassi, C.; Gatti, S.; Fasano, A. The New Epidemiology of Celiac Disease. *J. Pediatr. Gastroenterol. Nutr.* **2014**, *59*, S7–S9. [CrossRef] [PubMed]
3. Lundin, K.E.A.; Qiao, S.-W.; Snir, O.; Sollid, L.M. Coeliac disease—From genetic and immunological studies to clinical applications. *Scand. J. Gastroenterol.* **2015**, *50*, 708–717. [CrossRef] [PubMed]
4. Sollid, L.M.; Jabri, B. Triggers and drivers of autoimmunity: Lessons from coeliac disease. *Nat. Rev. Immunol.* **2013**, *13*, 294–302. [CrossRef]
5. Vriezinga, S.L.; Schweizer, J.J.; Koning, F.; Mearin, M.L. Coeliac disease and gluten-related disorders in childhood. *Nat. Rev. Gastroenterol. Hepatol.* **2015**, *12*, 527–536. [CrossRef]
6. Ludvigsson, J.F.; Bai, J.C.; Biagi, F.; Card, T.R.; Ciacci, C.; Ciclitira, P.J.; Green, P.H.R.; Hadjivassiliou, M.; Holdoway, A.; Van Heel, D.A.; et al. Diagnosis and management of adult coeliac disease: Guidelines from the British Society of Gastroenterology. *Gut* **2014**, *63*, 1210–1228. [CrossRef]
7. Silvester, J.A.; Kurada, S.; Szwajcer, A.; Kelly, C.P.; Leffler, D.A.; Duerksen, D.R. Tests for serum transglutaminase and endomysial antibodies do not detect most patients with celiac disease and persistent villous atrophy on gluten-free diets: A meta-analysis. *Gastroenterology* **2017**, *1533*, 689–701. [CrossRef]

8. de Lourdes Moreno, M.; Cebolla, Á.; Muñoz-Suano, A.; Carrillo-Carrion, C.; Comino, I.; Pizarro, Á.; Sousa, C. Detection of gluten immunogenic peptides in the urine of patients with coeliac disease reveals transgressions in the gluten-free diet and incomplete mucosal healing. *Gut* **2017**, *662*, 250–257. [CrossRef]
9. Morón, B.; Cebolla, Á.; Manyani, H.; Álvarez-Maqueda, M.; Megías, M.; Thomas, M.D.C.; López, M.C.; Sousa, C. Sensitive detection of cereal fractions that are toxic to celiac disease patients by using monoclonal antibodies to a main immunogenic wheat peptide. *Am. J. Clin. Nutr.* **2008**, *87*, 405–414. [CrossRef]
10. Morón, B.; Bethune, M.T.; Comino, I.; Manyani, H.; Ferragud, M.; López, M.C.; Cebolla, Á.; Khosla, C.; Sousa, C. Toward the Assessment of Food Toxicity for Celiac Patients: Characterization of Monoclonal Antibodies to a Main Immunogenic Gluten Peptide. *PLoS ONE* **2008**, *3*, e2294. [CrossRef]
11. Comino, I.; Real, A.; Vivas, S.; Síglez, M.Á.; Caminero, A.; Nistal, E.; Casqueiro, J.; Rodríguez-Herrera, A.; Cebolla, Á.; Sousa, C. Monitoring of gluten-free diet compliance in celiac patients by assessment of gliadin 33-mer equivalent epitopes in feces. *Am. J. Clin. Nutr.* **2012**, *95*, 670–677. [CrossRef]
12. A Silvester, J.; Rashid, M. Long-term Follow-Up of Individuals with Celiac Disease: An Evaluation of Current Practice Guidelines. *Can. J. Gastroenterol.* **2007**, *21*, 557–564. [CrossRef]
13. Javorsky, E.; Robinson, A.; Kimball, A.B. Evidence-based guidelines to determine follow-up intervals: A call for action. *Am. J. Manag. Care* **2014**, *201*, 17.
14. Han, J.; Pei, J.; Kamber, M. *Data Mining: Concepts and Techniques*; Elsevier: Amsterdam, The Netherlands, 2011.
15. Iavindrasana, J.; Cohen, G.; Depeursinge, A.; Müller, H.; Meyer, R.; Geissbuhler, A. Clinical data mining: A review. *Yearb. Med. Inform.* **2009**, *1801*, 121–133.
16. Syage, J.A.; Murray, J.A.; Green, P.H.; Khosla, C.; Adelman, D.C.; Sealey-Voyksner, J.A. Oral Latiglutenase Improves Chronic Symptoms in Seropositive Celiac Disease Patients. *Gastroenterology* **2017**, *152*, S163. [CrossRef]
17. Kaw, C.; Hefle, S.; Taylor, S.L. Sandwich Enzyme-Linked Immunosorbent Assay (ELISA) for the Detection of Lupine Residues in Foods. *J. Food Sci.* **2008**, *73*, T135–T140. [CrossRef]
18. Fayyad, U.; Piatetsky-Shapiro, G.; Smyth, P. From data mining to knowledge discovery in databases. *AI Mag.* **1996**, *173*, 37.
19. Phyu, T.N. Survey of classification techniques in data mining. In Proceedings of the International MultiConference of Engineers and Computer Scientists, Hong Kong, China, 18–20 March 2009; pp. 18–20.
20. Goebel, M.; Gruenwald, L. A survey of data mining and knowledge discovery software tools. *ACM SIGKDD Explor. Newsl.* **1999**, *11*, 20–33. [CrossRef]
21. Wu, X.; Kumar, V.; Quinlan, J.R.; Ghosh, J.; Yang, Q.; Motoda, H.; Hou, Z.H. Top 10 algorithms in data mining. *Knowl. Inf. Syst.* **2008**, *141*, 1–37. [CrossRef]
22. Quinlan, J.R. Bagging, boosting, and C4. 5. *AAAI Conf. Artif. Intell.* **1996**, *1*, 725–730.
23. Quinlan, J.R. Induction of decision trees. *Mach. Learn.* **1986**, *11*, 81–106. [CrossRef]
24. Friedl, M.; Brodley, C. Decision tree classification of land cover from remotely sensed data. *Remote Sens. Environ.* **1997**, *61*, 399–409. [CrossRef]
25. Play Tennis. Simple Dataset with Decisions about Playing Tennis. Available online: https://www.kaggle.com/fredericobreno/play-tennis (accessed on 16 January 2021).
26. Muniyandi, A.P.; Rajeswari, R.; Rajaram, R.K. Network Anomaly Detection by Cascading K-Means Clustering and C4.5 Decision Tree algorithm. *Procedia Eng.* **2012**, *30*, 174–182. [CrossRef]
27. Asencio–Cortés, G.; Morales–Esteban, A.; Shang, X.; Martínez–Álvarez, F. Earthquake prediction in California using regression algorithms and cloud-based big data infrastructure. *Comput. Geosci.* **2018**, *115*, 198–210. [CrossRef]
28. Husby, S.; Koletzko, S.; Korponay-Szabó, I.; Mearin, M.L.; Phillips, A.; Shamir, R.; Troncone, R.; Giersiepen, K.; Branski, D.; Catassi, C.; et al. European Society for Pediatric Gastroenterology, Hepatology, and Nutrition Guidelines for the Diagnosis of Coeliac Disease. *J. Pediatr. Gastroenterol. Nutr.* **2012**, *54*, 136–160. [CrossRef]
29. Hill, I.D.; Dirks, M.H.; Liptak, G.S.; Colletti, R.B.; Fasano, A.; Guandalini, S.; Seidman, E.G. Guideline for the diagnosis and treatment of celiac disease in children: Recommendations of the North American Society for Pediatric Gastroenterology, Hepatology and Nutrition. *J. Pediatric Gastroenterol. Nutr.* **2005**, *40*, 1–19. [CrossRef]
30. Al-Toma, A.; Volta, U.; Auricchio, R.; Castillejo, G.; Sanders, D.S.; Cellier, C.; Mulder, C.J.; E A Lundin, K. European Society for the Study of Coeliac Disease (ESsCD) guideline for coeliac disease and other gluten-related disorders. *United Eur. Gastroenterol. J.* **2019**, *7*, 583–613. [CrossRef]
31. National Institute for Health and Clinical Excellence. Coeliac Disease: Recognition, Assessment and Management. Available online: https://www.guidelines.co.uk/gastrointestinal/nice-coeliac-disease-guideline/252667.article2015 (accessed on 16 January 2021).
32. Rubio-Tapia, A.; Hill, I.D.; Kelly, C.P.; Calderwood, A.H.; Murray, J.A. American College of Gastroenterology clinical guideline: Diagnosis and management of celiac disease. *Am. J. Gastroenterol.* **2013**, *1085*, 656. [CrossRef]
33. SPAIN Ministry of Health. Working Group of the Protocol for the Early Diagnosis of Celiac Disease. Protocol for the Early Diagnosis of Celiac Disease. Ministry of Health, Social Services and Equality. Canary Islands Health Service Evaluation Service SESCS. Available online: https://www.mscbs.gob.es/profesionales/prestacionesSanitarias/publicaciones/Celiaquia/enfermedadCeliaca.pdf (accessed on 20 August 2019).

34. Cervino, G.; Fiorillo, L.; Laino, L.; Herford, A.S.; Lauritano, F.; Giudice, G.L.; Famà, F.; Santoro, R.; Troiano, G.; Iannello, G.; et al. Oral health impact profile in celiac patients: Analysis of recent findings in a literature review. *Gastroenterol. Res. Pract.* **2018**, 7848735. [CrossRef]
35. Norris, S.L.; Lau, J.; Smith, S.J.; Schmid, C.H.; Engelgau, M.M. Self-management education for adults with type 2 diabetes: A meta-analysis of the effect on glycemic control. *Diabetes Care* **2002**, *257*, 1159–1171. [CrossRef]
36. Hall, A.M.; Ferreira, P.H.; Maher, C.G.; Latimer, J.; Ferreira, M.L. The influence of the therapist-patient relationship on treatment outcome in physical rehabilitation: A systematic review. *J. Phys. Ther.* **2010**, *908*, 1099–1110. [CrossRef] [PubMed]
37. Lee, A.R.; Newman, J.M. Celiac diet: Its impact on quality of life. *J. Am. Diet. Assoc.* **2003**, *103*, 1533–1535. [CrossRef] [PubMed]

Article

Dynamics and Considerations in the Determination of the Excretion of Gluten Immunogenic Peptides in Urine: Individual Variability at Low Gluten Intake

Laura Coto [1,2], Carolina Sousa [3] and Angel Cebolla [1,*]

[1]   Biomedal S.L., 41900 Seville, Spain; laura.coto@biomedal.com
[2]   Human Nutrition and Food Science Doctoral Program, University of Granada, 18011 Granada, Spain
[3]   Department of Microbiology and Parasitology, Faculty of Pharmacy, University of Seville, 41012 Seville, Spain; csoumar@us.es
*   Correspondence: acebolla@biomedal.com; Tel.: +34-955-983-215

Abstract: Background: A lifelong strict gluten-free diet is the only available treatment for celiac disease, but total exclusion of gluten is difficult to achieve. The aim of this study was to determine the range of time and the amount of gluten immunogenic peptides (GIP) excreted in urine after specific gluten ingestions. Methods: 20 healthy participants followed the same diet for 12 days in which 50 mg and 2 g of gluten were ingested and all the urinations were collected. GIP were analyzed by lateral flow immunoassay (LFIA) tests and quantified using an LFIA reader. Results: GIP were detected in 15% and 95% of participants after 50 mg and 2 g gluten intakes, respectively. The higher frequency and concentration of GIP was found between 6 and 9 h after both gluten ingestions. The ranges of detection were 3–12 h (50 mg) and 0–15 h (2 g). Conclusions: An increase in the frequency of urine tests may be a suitable approach to avoid false negative results. The use of the LFIA test in three urine samples collected at different times may show a sensitivity of 19.6% for a gluten ingestion like 50 mg, increasing to 93% after 2 g consumption.

Keywords: gluten immunogenic peptides; gluten excretion urine; gluten-free diet monitoring; celiac disease

Citation: Coto, L.; Sousa, C.; Cebolla, A. Dynamics and Considerations in the Determination of the Excretion of Gluten Immunogenic Peptides in Urine: Individual Variability at Low Gluten Intake. *Nutrients* **2021**, *13*, 2624. https://doi.org/10.3390/nu13082624

Academic Editor: Armin Alaedini

Received: 7 July 2021
Accepted: 26 July 2021
Published: 29 July 2021

Publisher's Note: MDPI stays neutral with regard to jurisdictional claims in published maps and institutional affiliations.

1. Introduction

Celiac disease (CD) is a chronic systemic immune-mediated disease triggered by the ingestion of dietary gluten in genetically predisposed individuals with the human leukocyte antigen, HLA-DQ2, and/or HLA-DQ8 haplotypes [1]. The clinical presentation of CD is extremely variable, ranging from typical gastrointestinal symptomatology to extraintestinal symptoms or have no symptoms at all. Importantly, extraintestinal symptoms comprise a substantial proportion of the clinical manifestations of CD such as dermatitis herpetiformis, arthritis, neurological symptoms, anaemia, osteopenia, osteoporosis, tooth enamel defects, aphthous stomatitis, hypertransaminasemia, etc. [1–3]. The pathogenesis of CD involves structural changes in the small intestinal mucosa and intraepithelial lymphocyte infiltration when gluten immunogenic peptides (GIP) resistant to digestive enzymes cross the epithelial barrier to the lamina propria, leading to the activation of both innate and adaptive immune responses [2,4].

Currently, the only treatment available for CD is a lifelong gluten-free diet (GFD). Strict adherence to the GFD is crucial to reverse the clinical manifestations and to prevent long-term complications [1,3,5–7]. However, a diet with the total exclusion of gluten is challenging for most patients, who need high levels of discipline and motivation [1]. Moreover, GFD is more expensive, less palatable, and imposes social constraints, such as when dining out and traveling [8–10]. Consequently, a substantial number of patients with CD, especially those who are asymptomatic, commit diet transgressions and they are at risk of developing histological lesions and complications as a result of their condition [11].

The reported rates of GFD adherence range between 12% and 90% in adults [11–13] and between 23% and 98% in children [14].

Although it has been described in the literature that a daily ingestion of less than 50 mg appears to be safe for most patients with CD [15], other authors have decreased this level to 30 mg to avoid intestinal mucosal abnormalities [16]. As there is a great diversity in gluten sensitivity among individuals [15], the establishment of a harmless threshold of daily gluten intake for the celiac population remains a troublesome task.

Gluten is an alcohol-soluble mixture of storage proteins, known as prolamins, of cereals such as wheat, rye, and barley [17]. These proteins are fundamental for dough formation in bakery products because of their viscoelasticity; however, their applications in the food industry are broader [18]. Wheat gluten prolamins, called gliadins and glutenins, are characterized by being rich in proline and glutamine amino acids, which make them resistant to hydrolysis by gastric and pancreatic enzymes [17]. As a result, an innumerable diversity of GIP is produced in the gastrointestinal tract, triggering an immune response in individuals with CD. In any case, most of the immunogenicity could be assigned to a limited number of gluten epitopes [19]; among the GIP containing the most active T cell epitopes of CD, the α-gliadin 33-mer peptide has been described as a paradigm of immunodominance [20].

There is limited evidence regarding gluten digestion, metabolism, and excretion mechanisms. As a dietary protein, gluten hydrolyzation occurs mainly in the small intestine by pancreatic enzymes, which break polypeptides into small peptides and amino acids that are transported through the intestinal barrier [21,22]. Furthermore, it has been described that a fraction of longer peptides resistant to the action of the peptidases can also cross the basolateral membrane of the enterocytes and reach portal circulation [21]. Several authors have reported the detection of GIP in the urine of patients with CD and healthy individuals using mass spectrometry and antibody-based methods [11,13,23–28]. Thus, they demonstrated that gluten-derived peptides enter the kidneys, and after the ultrafiltration process they are partially or totally excreted in the urine. It remains unknown if a proportion of these peptides is also reabsorbed and then metabolized or excreted using alternative pathways.

The use of GIP detection in urine has been developed as a direct test for GFD monitoring in contrast to the classical methods, rather than only detecting the consequences of diet transgressions [11,23]. Urine is an advantageous sample for disease monitoring, as it can be collected fully non-invasively, in large amounts, and repeatedly over long periods of time [29]. Urine is a complex matrix of different components, such as water, glucose, proteins, amino acids, and inorganic salts [29]. However, the usual low concentration of protein in urine and its heterogeneity within and between individuals complicate the determination of the specific moment of analyte excretion [29]. The aim of this study was to determine the individual variability and the dynamics and limit of detection (LoD) of urine GIP excretion after two different amounts of low/moderate gluten ingestion (50 mg and 2 g) by monitoring a significant number of participants with minimized diet variations.

2. Materials and Methods

2.1. Study Population

Between January 2020 and March 2020, 20 healthy volunteers were enrolled from circles of relatives of Biomedal S.L. (Seville, Spain) employees in collaboration with the research group of the University of Seville (Seville, Spain). The criteria for inclusion as healthy volunteers were: (1) participants who were >18 years old; (2) not been diagnosed with CD, non-celiac gluten sensitivity, and no food allergies, food intolerances, and other kinds of gastrointestinal diseases; (3) participants who were prepared to follow a strict diet; and (4) to have the determination and abilities for daily urine and stool collection. The exclusion criteria were as follows: (1) participants with associated pathologies or severe psychiatric diseases; and (2) participants who did not collect the samples properly on at least 70% of occasions.

All the subjects provided written informed consent to participate in the study, which was approved by the local ethics committee (n. 2381-N-19).

2.2. Study Design

The study involved all participants over a 19-day period. The first week was the wash-out stage, in which the participants had to follow a strict GFD. Two days before the first gluten ingestion, they were asked to collect one sample each of urine and feces to confirm the absence of dietary gluten (Figure 1). After the wash-out period, participants were provided with equivalent gluten-free lunch and dinner menus and gluten-free bread, which were supplied daily by the research team. The meals were consumed within the prescribed GFD. Two doses of gluten (50 mg and 2 g) were ingested in the morning (9:00) on days 8 and 12, respectively, and one sample of all the ordinary individual urinations and depositions (data published separately) were collected during the whole period (12 days in total). From the beginning to the end of the study, a food-recall questionnaire was used to assess GFD adherence and fluid intake, and the participants had to record the name and the quantity of the dishes that they consumed daily.

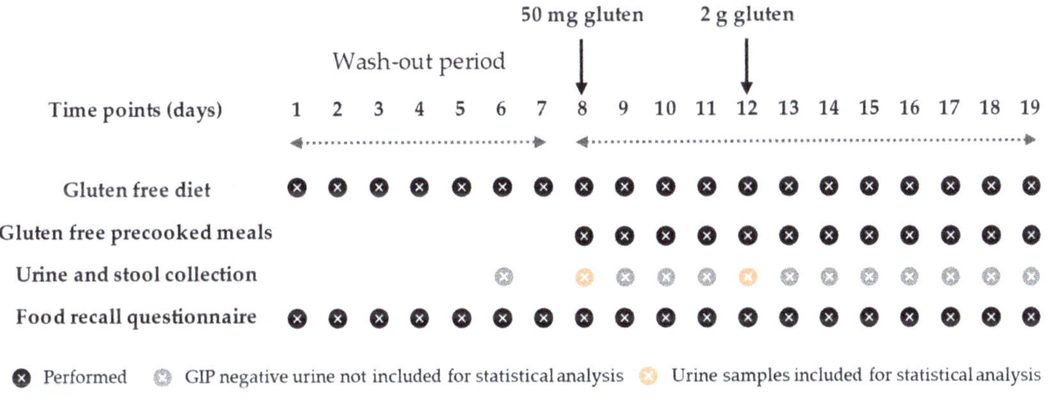

Figure 1. Study timeline.

2.3. Gluten Administration

Gluten ingestions consisted of two doses of 50 mg and 2 g of powdered wheat gluten (El Granero Integral™; Biogran S.L., Madrid, Spain) encapsulated in "000" size gelatin caps (Your Supplements™, Bredbury, Stockport, England). The quantity selection was based on the minimum amount of gluten that, when eaten daily, could provoke histological changes in patients with CD [15] and an amount considered appropriate to observe the dynamic of excretion of GIP in urine. The gelatin caps were analyzed using GlutenTox® ELISA Sandwich kit (Hygiena, Seville, Spain), based on G12 and A1 antibodies, to confirm the absence of gluten. Gluten estimation was calculated by analyzing several samples of maize starch Maizena™ (Unilever, London, England) spiked with the powdered gluten at different concentrations and analyzed using the GlutenTox® ELISA Sandwich kit (Hygiena, Seville, Spain). Considering the results obtained (near 100% recovery), gluten doses for each subject were prepared using the total weight of the powdered gluten: 50 ± 5 mg and 2000 ± 5 mg in 1 and 4 caps, respectively.

An equivalence calculation of the gluten dosages to bread portions was performed using the methodology described by Biagi et al. [30]. The slice of bread was 11 cm × 12 cm and weighted 30 g. Based on the nutritional composition given by the manufacturer, the whole slice contained 2.48 g of gluten. The corresponding amount of gluten in the bread slice was 0.6 g of slice for 50 mg of gluten (Figure 2a) and 24 g for 2 g of gluten (Figure 2b). A battery (AAA) was used as the standard for size comparison.

(a) **(b)**

2.4. Meal Administration

All participants followed the same GFD during the gluten excretion period and were provided with ready-to-eat meals for lunch and dinner in addition to gluten-free certified bread (Beiker™, Dr. Schär, Postal BZ, Italy) to complete meals and for breakfast time. The diet was isocaloric and the ingestion of fresh fruits, unprocessed nuts, and gluten-free beverages was free of choice, depending on the energy requirements and habits of each participant. The meals were ordered from a catering company and were analyzed daily by the ISO17025 certified laboratory services of Biomedal S.L. (Seville, Spain), using GlutenTox® ELISA Sandwich kit (Hygiena, Seville, Spain) to confirm the absence of gluten.

2.5. Urine Collection

Detailed instructions were given to all participants at the beginning of the study. The subjects were provided with all materials for urine collection, including specific plastic screw-capped containers, labels, cool bags, isothermal boxes, and cool packs. The participants were instructed to collect between 30 and 60 mL of each micturition and to write down the date and time of when they pass urine. All urine samples were preserved in isothermal boxes with cool packs at 4–8 °C and dropped off within 48 h of collection. All samples were stored at −20 °C until processing.

2.6. Urine Analysis

GIP qualitative results in urine were measured using a lateral flow immunoassay (LFIA) (iVYCHECK GIP Urine kit, Biomedal S.L., Seville, Spain) following the manufacturer's recommendations. Defrosted urine samples were homogenized and mixed with a conditioning solution. Thereafter, 100 µL of the mixture was added to the immunochromatographic cassette and visual interpretation of the results was carried out after 30 min (recommended time for samples containing a low amount of GIP). A positive result was considered when the test line showed a red color, and the control line showed a green color. A negative result was considered when only the control line showed a green color. The LoD of the technique determined by visual inspection was 2 ng/mL.

The concentration of GIP in urine was also measured in the immunochromatographic strips after 30 min using the iVYCHECK Reader (Biomedal S.L., Seville, Spain). The validity of this method was previously described by Moreno et al. [23]. The reader was calibrated prior to urine analysis using the α-gliadin 33-mer peptide as a standard. The measuring range established for this method was: 1.56–25 ng GIP/mL urine. The results are expressed as ng GIP per mL of urine. Each sample was run in duplicate, and at least two different aliquots of each sample were tested.

2.7. Statistics

The results of the quantitative variables were expressed using the mean (SD) and median (IQR or range), and those of the categorical variables were expressed as absolute (N) and relative (%) frequencies. The goodness-of-fit to normality was calculated using the Shapiro–Wilk test. The Mann–Whitney U test was employed to compare quantitative variables in independent groups and for paired quantitative variables, the Wilcoxon test was used.

Only urines not later than 24 h post gluten ingestion were included for statistical analysis due to later urines from all participants giving a negative result. Ranges of time were established for the study of the dynamics of GIP excretion in intervals of 3 h. All samples from each participant collected in each range were clustered to obtain one result per participant. Any GIP+ sample indicated a total positive result.

Spearman's correlation was used to calculate the association between the liquid consumption after gluten ingestion and the concentration of GIP in urine. Basic probability rules were used to obtain the diagnostic sensitivity of the studied techniques over a predetermined range of time with the different samples collected.

Statistical analyses were performed with IBM SPSS Statistics 25.0 for Windows (IBM Corp, Armonk, NY, USA). Statistical significance was set at $p < 0.05$.

3. Results

3.1. Subjects and Samples

A total of 20 individuals, including 13 (65%) females and 7 (35%) males, completed the study after 10 dropouts from the preselection process due to unforeseen events ($n = 6$) and COVID-19 mobility restrictions ($n = 4$). The median age of participants was 30.5 years (IQR 24.7–34.0) (Figure 3).

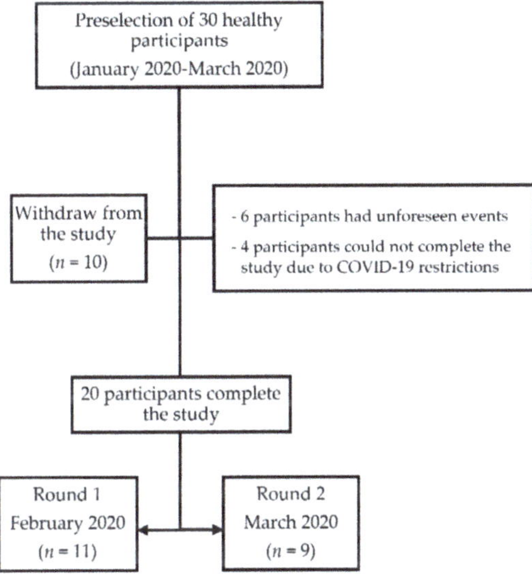

Figure 3. Flowchart of the study participants.

None of the participants were declared to be diagnosed with a relevant disease or had been taking any probiotics or fiber supplements. One participant reported following a special fitness diet before the study. According to the food-recall completed, all participants were compliant with the prescribed GFD and the gluten dose ingestion. The average fluid intake per participant during the study period was 1.5 ± 0.6 L/day.

3.2. GIP Detection in Urine Samples

A total of 290 urine samples were collected from all participants during the 24 h after gluten ingestion, 142 corresponding to the 50 mg gluten dose, and 148 to the 2 g gluten dose. The remaining samples of the study were excluded for statistical analysis as they obtained GIP negative results. The medians of the number of samples collected per participant in the first 24 h were 7 (IQR 5–8) for the 50 mg intake and 7 (IQR 5.5–8.5) for the 2 g intake (Table 1).

GIP were detected in 4/142 (2.8%) of the urine samples up to 24 h after 50 mg gluten ingestion, corresponding to 3/20 (15%) participants. From these participants, GIP were detected in only one sample for two subjects and in two samples for one subject. Regarding the 2 g dose, 33.1% (49/148) of the urine samples were GIP+ during the 24 h of collection, corresponding to 19/20 (95%) of the participants. GIP+ samples were obtained in only one to two urinations for 10/19 participants (52.6%), in three to four urinations for 7/19 participants (36.8%), and in five to six urinations for 2/19 participants (10.5%).

GIP+ samples were found from the first to the fourth collected samples after 50 mg gluten ingestion. The detection of GIP in urine could be extended up to the eight urinations after the 2 g dose, with the third sample being where most participants (16/20; 80%) had GIP+ urine.

3.3. Time Course of GIP Excretion

GIP were detected in urine samples collected in the first 3 h after 2 g gluten intake and between 3–6 h after 50 mg gluten ingestion (Figures 4 and 5). The majority of GIP+ urine samples were found in the range of 6–9 h after both gluten doses (18.8% and 78.8%, respectively) (Figures 4 and 5). As expected, the 2 g ingestion resulted in significant proportions of positive samples for a longer period (3–15 h) with rates between 41.2% and 78.8% (Figure 5). No positive results were found after 12 h and 15 h post ingestion for the 50 mg and the 2 g doses, respectively (Figures 4 and 5).

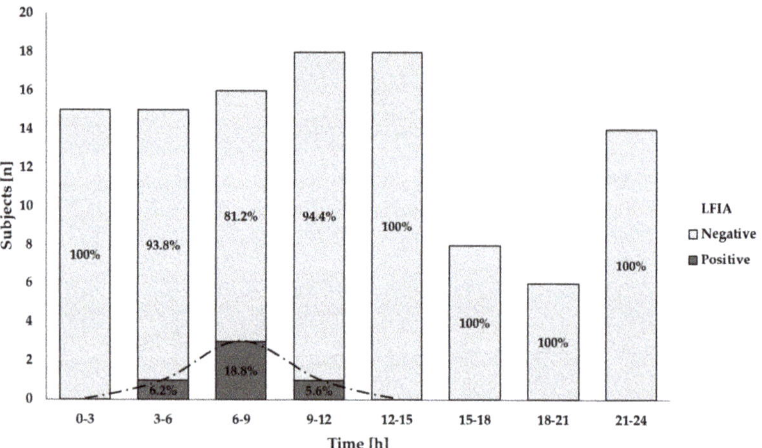

Figure 4. Results of qualitative analysis of GIP excretion in urine after 50 mg of gluten ingestion using a LFIA test. The trend of the GIP detection dynamics is represented by the dashed line.

Table 1. Individual characteristics of GIP excretion in urine within 24 h after 50 mg and 2 g gluten ingestions.

							Urine GIP Excretion in 24 h						
	50 mg Gluten							2 g Gluten					
Participant	Samples	LFIA+	Time Median	Time Range	Peak Max GIP	GIP Median	GIP Range	Samples	LFIA+	Time Median	Time Range	Peak Max GIP	GIP Median	GIP Range
	n	n	h	h	h	ng/mL	ng/mL	n	n	h	h	h	ng/mL	ng/mL
1	7	0						6	1	4.50	0.00 (4.50)	4.50	1.80	0.00 (1.80)
2	7	0						8	6	6.00	5.00 (4.00–9.00)	5.00	9.10	13.23 (2.93–16.17)
3	4	0						6	3	7.88	3.25 (6.25–9.50)	9.50	2.20	0.80 (1.80–2.60)
4	8	0						9	4	8.54	6.50 (5.50–12.00)	7.00	3.50	5.43 (2.30–7.73)
5	6	0						4	1	14.67	0.00 (14.67)	14.67	2.50	0.00 (2.50)
6	4	1	3.33	0.00 (3.33)	3.33	4.40	0.00 (4.40)	4	3	6.33	5.00 (3.83–8.83)	8.83	4.92	5.63 (2.10–7.73)
7	15	0						18	1					
8	10	0						10	5	7.67	3.00 (7.00–10.00)	7.00	4.17	4.60 (2.77–7.37)
9	7	0						7	3	6.00	2.33 (5.00–7.33)	6.00	3.47	2.40 (2.20–4.60)
10	10	0						7	4	7.96	8.75 (6.00–14.75)	6.00	6.50	11.43 (1.77–13.20)
11	6	0						7	2	3.21	3.42 (1.50–4.92)	1.50	1.92	0.37 (1.73–2.10)
12	6	2	8.29	2.42 (7.08–9.50)	7.08	2.69	0.23 (2.57–2.80)	5	2	9.04	5.92 (6.08–12.00)	6.08	4.60	2.47 (3.37–5.83)
13	5	0						7	1	7.83	0.00 (7.83)	7.83	1.95	0.00 (1.95)
14	5	0						3	0					
16	7	0						10	1	8.00	0.00 (8.00)	8.00	3.23	0.00 (3.23)
17	10	0						10	2	7.50	0.00 (7.50)	7.50	2.60	0.00 (2.60)
19	7	0						5	2	6.32	1.97 (5.33–7.30)	5.33	3.07	2.73 (1.70–4.43)
21	8	0						8	4	8.04	7.07 (4.42–11.48)	4.42	3.12	5.53 (1.73–7.27)
22	5	1	7.67	0.00 (7.67)	7.67	2.30	0.00 (2.30)	8	1	5.13	0.00 (5.13)	5.13	1.70	0.00 (1.70)
23	5	0						6	3	7.65	3.50 (6.30–9.80)	7.65	2.43	0.97 (1.60–2.57)
TOTAL	7	0	7.67	0 (3.33–9.50)	7.08	2.57	2.20 (2.20–4.40)	7	2	7.00	13.25 (1.50–14.75)	6.54	2.68	14.83 (1.33–16.17)

GIP: gluten immunogenic peptides; LFIA: lateral flow immunoassay.

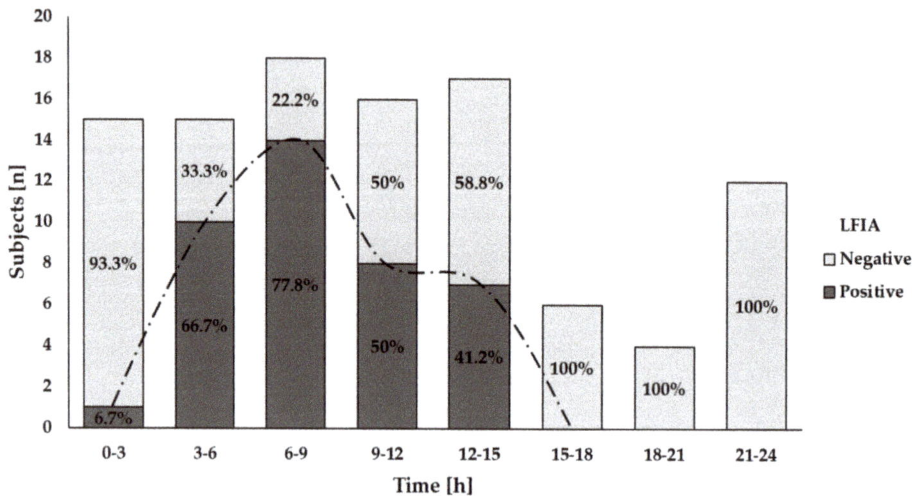

Figure 5. Results of qualitative analysis of GIP excretion in urine after 2 g of gluten ingestion using a LFIA test. The trend of the GIP detection dynamics is represented by the dashed line.

Despite the variability observed among individuals, both gluten ingestions showed a comparable period for initial GIP detection (7.1 h (range 3.3–7.7)) for the 50 mg dose and 5.8 h (range 1.5–14.7) for the 2 g dose) ($p = 0.285$). A longer range of time for detectable GIP per participant was found in the larger gluten dose (0 (range 0–2.4) vs. 3.1 (range 0–8.8)) but without statistical significance ($p = 0.180$).

3.4. GIP Quantification in Urine

In line with the time for GIP detection after gluten ingestion, higher concentrations of GIP were measured in the urine of most participants in the same period (6–9 h) using an LFIA reader. The median of GIP in this period was 0 ng GIP/mL urine (range 0–2.8) for the 50 mg dose and 2.57 ng GIP/mL urine (range 0–13.2) for the 2 g intake (Figure 6, Table 2).

Table 2. Urine GIP detection in 3-hour periods after 50 mg and 2 g of gluten intakes.

	50 mg Gluten			2 g Gluten		
Time	Participants	GIP+ Participants	GIP [ng/mL]	Participants	GIP+ Participants	GIP [ng/mL]
h	*n*	*n*	Median (Range)	*n*	*n*	Median (Range)
0–3	15	0	0.00 (0)	15	1	0.00 (0–2.10)
3–6	15	1	0.00 (0–4.40)	15	10	1.70 (0–16.17)
6–9	16	3	0.00 (0–2.80)	18	14	2.57 (0–13.20)
9–12	18	1	0.00 (0–2.57)	16	8	0.00 (0–4.83)
12–15	18	0	0.00 (0)	17	7	0.00 (0–3.37)
15–18	8	0	0.00 (0)	6	0	0.00 (0)
18–21	6	0	0.00 (0)	4	0	0.00 (0)
21–24	14	0	0.00 (0)	12	0	0.00 (0)

GIP: gluten immunogenic peptides.

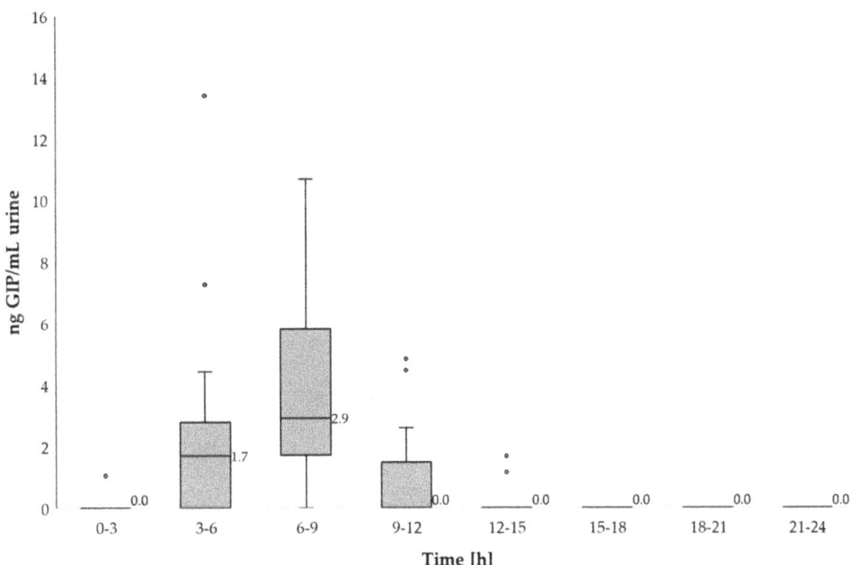

Figure 6. Dynamic of GIP excretion in urine after 2 g of gluten intake using a LFIA reader. Potential outliers are represented as dots.

However, the peak levels of GIP in urine were observed at different time periods; for the 50 mg dose, it was 4.4 ng GIP/mL, detected 3.3 h post ingestion and for the 2 g dose it was 16.17 ng GIP/mL, detected 5 h after gluten ingestion. Considering the period of GIP detection after both gluten doses (3–12 h) the median of GIP was 0 ng GIP/mL urine (range 0–4.4) for the 50 mg dose and 1.73 ng GIP/mL urine (range 0–16.17) for the 2 g intake with statistical differences between ingestions ($p < 0.001$) (Figure 7).

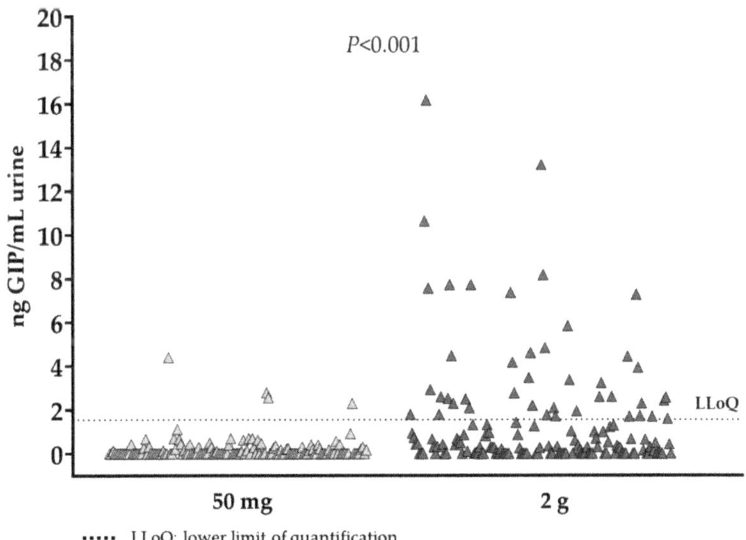

Figure 7. GIP detected in urine samples within 24 h after 50 mg and 2 g gluten ingestions.

We observed a significant negative correlation between the liquid consumption 12 h after gluten consumption and the levels of GIP in urine detected using an LFIA reader in the same period (rho = −0.79, 95% CI [−0.91, −0.53]; $p < 0.001$) (Figure 8). In contrast, no correlations were found between liquid consumption and urination frequency (rho = 0.119, 95% CI [−0.34, 0.53]; $p = 0.62$) and GIP concentrations and urination frequency (rho = −0.004, 95% CI [−0.45, 0.44]; $p = 0.99$).

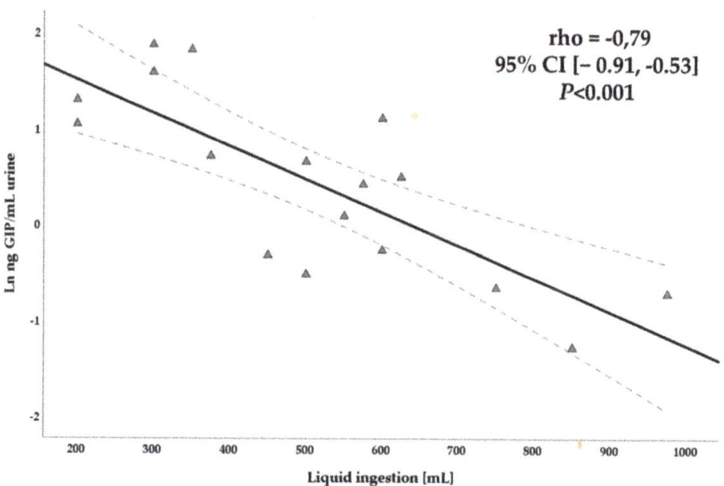

Figure 8. Scatterplot of liquid consumption after 12 h gluten intake and GIP concentration in urine using a LFIA reader.

3.5. Interindividual Variability in GIP Excretion

Despite the limited period of GIP detection in urine after gluten intake, differences in GIP excretion were observed among individuals. Regarding the 50 mg gluten dose, one of three participants reached the peak of GIP concentrations in urine in the first 6 h post ingestion and two of three participants between 6 and 9 h post ingestion. Regarding the 2 g gluten dose, 6/20 (30%) participants reached the peak of GIP detection in the range of 0–6 h, 9/20 (45%) in the range of 6–9 h and 3/20 (15%) after 9 h post gluten ingestion (Figure 9). Although there were differences in GIP concentrations, GIP excretion patterns were similar in the participants with measurable excreted GIP after both gluten intakes. However, one participant (subject 6) obtained the peak of GIP concentration in the 2 g dose approximately 6 h later than the 50 mg dose. In general, no unusual patterns were observed, as most individuals showed a GIP elevation (3–9 h after gluten intake) followed by a decreasing tendency.

When the results were compared between sex, the higher GIP concentrations were seen in the group of females in both gluten intakes, however no statistical significance was observed between females and males in GIP detection (4.40 ng/mL vs. 2.50 ng/mL, respectively, for the 50 mg dose ($p = 0.319$); 2.69 ng/mL vs. 3.12 ng/mL, respectively, for the 2 g dose ($p = 0.162$)). Moreover, similar results were found between groups in terms of initial time of GIP detection (3.3 h vs. 7.38 h, respectively, for the 50 mg dose ($p = 0.221$); 6.22 h vs. 5.59 h, respectively, for the 2 g dose ($p = 0.730$)) and time range of GIP detection (0 h vs. 1.21 h, respectively, for the 50 mg dose ($p = 0.480$); 2.90 h vs. 3.59 h, respectively, for the 2 g dose ($p = 0.688$)).

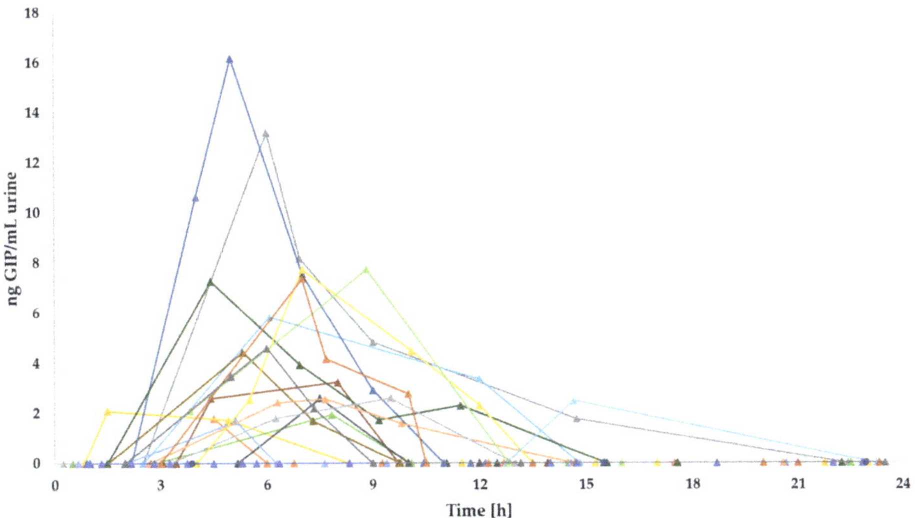

Figure 9. Individual GIP excretion patterns in urine after 2 g gluten ingestion.

3.6. Diagnostic Sensitivity of the LFIA Test

The LFIA test in urine samples demonstrated the capacity for GIP detection after a low gluten ingestion (50 mg). However, this amount of gluten was only detected in 3 out of 20 subjects (15%). In contrast, with a dose of 2 g of gluten, the sensitivity of the test increased to 95% of participants (19/20).

The theoretical probability of finding at least one GIP+ result for a single gluten ingestion was calculated considering the interval of time for GIP detection in both amounts of gluten, 3–12 h after ingestion (Table 3). The diagnostic sensitivity of the LFIA test to detect GIP in a unique urine sample from a small amount of gluten (50 mg) is 7%, which may increase to 13.5% when two samples are collected and to 19.6% in three samples collected. In the hypothetical situation of frequent gluten ingestion (i.e., 2 g), the sensitivity in a single sample may be 59%, reaching rates of 83.2% and 93.1% when two and three samples are collected, respectively.

Table 3. Estimated diagnostic sensitivity of the methods in a specific range of time and with different sample collections.

	LFIA Sensitivity in Urine		
	50 mg Gluten/2 g Gluten		
Time (h)	1 Sample (%)	2 Samples (%)	3 Samples (%)
3–6	6/47.8	11.6/72.8	16.9/85.8
6–9	13/72	24.3/92.2	34.1/97.8
9–12	4/57.1	7.8/81.6	11.5/92.1
3–12	7/59	13.5/83.2	19.6/93.1

4. Discussion

In the present study, we described the dynamics of excretion of GIP in urine samples of healthy subjects who ingested two small doses of gluten under controlled dietary conditions using an immunoassay method based on the anti-33-mer antibodies, G12 and A1 [23,31]. Our results confirmed that the LFIA test could detect a single ingestion of 50 mg of gluten in urine samples collected in a range of 3–10 h post ingestion, with most GIP excreted in a unique sample per participant. Equivalent results were obtained when the gluten ingestion was 40 times higher (2 g). GIP were detected for this gluten dose in the range of 1–15 h, with most of them obtained between 6 and 9 h post gluten ingestion.

Consistent with our data, Moreno et al. [23] were the first to demonstrate that the same LFIA method could detect the ingestion of 25 mg and 50 mg of gluten from processed bread in the urine of healthy volunteers. Their results revealed that GIP from those gluten doses were detectable 3–9 h post ingestion; however, the time of GIP disappearance after a normal gluten-containing diet was extended to 16–34 h compared to this study. They estimated that the time of excretion of gluten-derived peptides ranged from 1 to 2 days. In agreement with these data, other authors found an association between confirmed gluten exposure and GIP presence in urine within 36 h after ingestion in patients with CD [26,27]. Although the results exhibited a high interindividual variability, the interval between gluten consumption and GIP detection in urine was generally consistent, ranging from <4 to >24 h. Our results with more participants ($n = 20$) showed GIP detection in the first 15 h after the 2 g gluten challenge. Thus, depending on the amount of gluten consumed, the period of GIP detection may vary, with a positive trend between gluten consumption and GIP excretion. Nonetheless, it seems that the interval of time between 3 and 9 h post ingestion may be crucial for GIP detection, independent of the magnitude of gluten exposure. In any case, disagreement in the period to excrete all GIP could also vary depending on the type of ingested gluten, for instance, capsulated gluten in this study vs. other alternatives such as cookies, bread, or cereal bars that have been used in other studies [23,26,27].

Regarding the GIP concentration in urine, we found a significant variation in GIP content of samples collected 24 h after ingestion of 50 mg and 2 g gluten ($p < 0.001$). In this study, we observed that the higher the amount of gluten consumed, the more frequent GIP detection and quantification in urine. Nevertheless, interindividual variability was observed, with GIP medians of participants ranging from 2.3–4.4 ng/g for the 50 mg dose and from 1.7–9.1 ng/g for the 2 g dose. The study carried out by Moreno et al. [23] also showed slightly less differences in the maximum GIP content in urine collected after 25 and 50 mg of gluten ingestion (10–15 ng/mL vs. 15–20 ng/mL, respectively), however they pretreat the urine sample with solid phase extraction. Deviations between our results and those from previous studies could be due to the matrix containing the gluten used in the study and the methodology used for GIP quantification.

The correlation between the amount of gluten consumed and the excretion of gluten-derived peptides has been previously described [11,23,26,27]. Generally, urine samples from healthy subjects under a normal gluten-containing diet showed a higher amount of GIP than in diet transgressions made by patients with CD. Indeed, most urine samples from these patients were detectable but were under the limit of quantitation [11,23,26,27]. Moreno et al. [23] reported GIP quantifications ranging from 6.5 to 600 ng/mL and 6.5 to 370 ng/mL in healthy adults and children, respectively, whereas GIP content in urines from patients with CD ranged from 9.27 to 78.12 ng GIP/mL and from 9.33 to 29.78 ng GIP/mL (in adults and children, respectively). Other authors reported significant differences in urine GIP concentrations between patients with CD under a GFD and de novo CD-diagnosed patients (average range 40.26 ng/mL vs. 80.31 ng/mL, $p < 0.001$) [11].

Although gluten consumption and diet composition were controlled in our study, urinary GIP excretion varied among participants. Urine composition can vary between individuals due to differences in biological factors, body size, physical exercise, environmental conditions, and fluid, salt, and high protein ingestion [32]. Moreover, the sample collection timing in relation to exposure, variation in the kinetics of elimination within and between individuals, and physicochemical properties of the urine matrix should be considered [33]. Hydration status plays a crucial role in variations in the urinary flow rate (volume of urine produced per unit time), and therefore in the concentrations of the biomarker in the study [33]. Although the mechanisms of GIP elimination in urine are currently unknown, our results showed a significant inverse correlation between liquid consumption and GIP concentration in urine ($p < 0.001$), as expected for a higher dilution of the urine peptides. However, we did not find a correlation between the urination frequency and GIP concentrations. Thus, it seems that urinary GIP detection may be affected by the amount of liquid ingested. On the other hand, some individuals may not absorb and excrete sufficient GIP in urine to be detected or a fraction of the absorbed GIP might go back to the portal circulation after their pass through the kidneys. Despite one of the participants of the present study obtaining negative results in all the urine samples collected after the 2 g gluten ingestion, GIP were significantly found in the stools coming from the same period, confirming the gluten exposure (data published separately). Another explanation could be that this participant missed the collection of one or more samples due to the tedious methodology employed in the study.

In a real-life scenario, following a strict GFD is a difficult task for patients with CD. Gluten is reported to be present in a significant percentage of foodstuffs [34]. Consequently, the frequency of diet transgressions is considerable, despite the assumed efforts of the patients [11,13,26,27,35]. Furthermore, it was suggested that inadvertent gluten ingestion may be more recurrent than intentional intake, not only when eating out, but also at home [12,26,27]. The main goal of this study was to comprehensively determine the pattern of GIP excretion in urine related to a single ingestion of a low amount of gluten, which is the expected situation for inadvertent-involuntary gluten exposure. This information will be valuable in providing more accurate guidance for the use of the GIP tests in patients with CD. On the basis of these results, future studies with the target population may build an effective protocol for urine sample collection to establish the algorithm of assessment of GFD adherence.

On the other hand, a recent publication with a cohort of 77 participants under a GFD for ≥ 2 years revealed that the urine LFIA test obtained a diagnostic sensitivity of 94.4% and negative predictive value of 96.9% in detecting mucosal damage when urine samples were collected on three different days, two of them over the weekend [11]. Other authors reported a rate of 69.8% of patients with at least one GIP+ in urine when they collected weekly samples on weekends over 4 weeks [13]. Thus, considering the short period of GIP detection in urine (3–12 h) after gluten intake and the variability in GIP excretion within and between individuals in our study, it seems that the increase in the frequency of tests may be a worthwhile approach to reduce the probability of false negative results due to punctual gluten consumptions [11,13]. In this scenario, the use of three LFIA tests in urine collected at different times during weekdays and weekends may reach a sensitivity of 19.6% for very low gluten intakes, such as 50 mg, while this sensitivity could increase to 93.1% with higher gluten exposures, such as 2 g.

Regarding the optimum time for urine collection, several circumstances need to be considered, such as the period with a higher rate of GIP detection: 3–12 h post ingestion, meals with a greater chance of gluten exposure (lunch and dinner), and the best time to obtain a concentrated urine sample in most individuals. Hence, it seems that the last urine in the night, or alternatively the first one in the morning, would meet most of these conditions. Routinely, first-morning urine samples are required for urinary analysis as a representative sample of the average urine of the day and because they have the highest concentration of peptides [33]. However, food ingestion occurs during the day, therefore, depending on the dynamics of excretion, analyte detection among subjects could fluctuate [33]. Alternations of first-morning urine and the last urine in the day may offer more probabilities to reveal a diet transgression made at lunch and dinner times. Moreno et al. [23] suggested the collection of 24-h total urine to increase the probability of GIP detection from a low amount of gluten ingestion; however, the complexity for patients is higher and urine samples with detectable amounts of GIP could be diluted, decreasing the concentration of the final sample.

The main limitations of this study were the inclusion of only healthy volunteers, and that maybe the sample size could be increased. The complexity of the study design, requiring a big effort from the volunteers, and the declaration of the COVID-19 pandemic make the recruitment process a difficult task. Furthermore, the inclusion of patients with CD had ethical concerns. Although it is generally believed that gluten metabolism is similar between patients with CD and healthy subjects, several aspects need to be addressed when the CD population is considered as they may present digestive alterations, intestinal permeability, and differences in the microbiota involved in the gluten degradation process [21,36,37]. In fact, it was described that patients with CD may have a higher proteolytic activity in the intestine leading to a gluten reduction in feces in comparison to healthy subjects and first-degree relatives on normal diet [37]. Moreover, a recent study confirmed that patients with CD consuming wheat excreted in urine a significantly higher diversity of gluten-derived peptides than healthy subjects, however differences the healing of the intestinal epithelia between patients with CD were not contemplated [28]. Thus, since the test for GIP detection in urine is intended for use by people suffering from CD and gluten-related disorders, future studies with these populations with similar gluten consumptions will confirm the compatibility of our results for the definition of clinical practice guidelines for the application of GIP in the monitoring of GFD.

The ability to capture a biomarker in a sample of urine is a noninvasive procedure that is convenient for almost all population [33]. Therefore, urinary GIP detection provides a supplemental tool to evaluate gluten exposure in individuals following a GFD. In conclusion, the results of this study will provide additional knowledge about gluten metabolism and GIP excretion, which could be useful to fine-tune the application of GIP determination in the follow-up of patients with CD and gluten-related disorders.

Author Contributions: Conceptualization, L.C. and A.C.; Data curation, L.C.; Formal analysis, L.C.; Investigation, L.C.; Project administration, A.C.; Resources, A.C.; Supervision, C.S. and A.C.; Validation, C.S. and A.C.; Writing—original draft, L.C.; Writing—review & editing, C.S. and A.C. All authors have read and agreed to the published version of the manuscript.

Funding: This work was supported by grants from Ministerio de Ciencia e Innovación (DI-16-08943) and Junta de Andalucía, Consejería de Economía, Conocimiento, Empresas y Universidad and FEDER funds (AT17_5489_USE). We also thank the generous volunteer subjects who enrolled in the study.

Institutional Review Board Statement: The study was conducted according to the guidelines of the Declaration of Helsinki, and approved by the Ethics Committee of Virgen del Rocío and Virgen Macarena University Hospitals (n. 2381-N-19; 03/02/2020).

Informed Consent Statement: Informed consent was obtained from all subjects involved in the study.

Acknowledgments: This paper will be part of the Laura Coto's doctorate that it is being carried out within the context of "Human Nutrition Program" at the University of Granada. She was supported by a Research Fellowship from the Government of Spain (DI-16-08943).

Conflicts of Interest: Angel Cebolla is the founder and current CEO of Biomedal S.L. (Seville, Spain), Angel Cebolla and Carolina Sousa are inventors of the patent "Detecting gluten peptides in human fluids" (no. WO/2016/005643), and Laura Coto is employee at Biomedal S.L.

References

1. Lindfors, K.; Ciacci, C.; Kurppa, K.; Lundin, K.E.A.; Makharia, G.K.; Mearin, M.L.; Murray, J.A.; Verdu, E.F.; Kaukinen, K. Coeliac Disease. *Nat. Rev. Dis. Prim.* **2019**, *5*, 3. [CrossRef]
2. Fasano, A.; Catassi, C. Celiac Disease. *N. Engl. J. Med.* **2012**, *367*, 2419–2426. [CrossRef]
3. Caio, G.; Volta, U.; Sapone, A.; Leffler, D.A.; De Giorgio, R.; Catassi, C.; Fasano, A. Celiac Disease: A Comprehensive Current Review. *BMC Med.* **2019**, *17*, 142. [CrossRef] [PubMed]
4. Abadie, V.; Sollid, L.M.; Barreiro, L.B.; Jabri, B. Integration of Genetic and Immunological Insights into a Model of Celiac Disease Pathogenesis. *Annu. Rev. Immunol.* **2011**, *29*, 493–525. [CrossRef]
5. Lebwohl, B.; Sanders, D.S.; Green, P.H.R. Seminar Coeliac Disease. *Lancet* **2018**, *391*, 70–81. [CrossRef]
6. Elli, L.; Ferretti, F.; Orlando, S.; Vecchi, M.; Monguzzi, E.; Roncoroni, L.; Schuppan, D. Management of Celiac Disease in Daily Clinical Practice. *Eur. J. Intern. Med.* **2019**, *61*, 15–24. [CrossRef]
7. Therrien, A.; Kelly, C.P.; Silvester, J.A. Celiac Disease. *J. Clin. Gastroenterol.* **2020**, *54*, 8–21. [CrossRef] [PubMed]
8. Troncone, R.; Auricchio, R.; Granata, V. Issues Related to Gluten-Free Diet in Coeliac Disease. *Curr. Opin. Clin. Nutr. Metab. Care* **2008**, *11*, 329–333. [CrossRef]

9.   Hall, N.J.; Rubin, G.; Charnock, A. Systematic Review: Adherence to a Gluten-Free Diet in Adult Patients with Coeliac Disease. *Aliment. Pharmacol. Ther.* **2009**, *30*, 315–330. [CrossRef]

10. Villafuerte-Galvez, J.; Vanga, R.R.; Dennis, M.; Hansen, J.; Leffler, D.A.; Kelly, C.P.; Mukherjee, R. Factors Governing Long-Term Adherence to a Gluten-Free Diet in Adult Patients with Coeliac Disease. *Aliment. Pharmacol. Ther.* **2015**, *42*, 753–760. [CrossRef]

11. Ruiz-Carnicer, Á.; Garzón-Benavides, M.; Fombuena, B.; Segura, V.; García-Fernández, F.; Sobrino-Rodríguez, S.; Gómez-Izquierdo, L.; Montes-Cano, M.A.; Rodríguez-Herrera, A.; Millán, R.; et al. Negative Predictive Value of the Repeated Absence of Gluten Immunogenic Peptides in the Urine of Treated Celiac Patients in Predicting Mucosal Healing: New Proposals for Follow-up in Celiac Disease. *Am. J. Clin. Nutr.* **2020**, *112*, 1240–1251. [CrossRef]

12. Muhammad, H.; Reeves, S.; Jeanes, Y.M. Identifying and Improving Adherence to the Gluten-Free Diet in People with Coeliac Disease. *Proc. Nutr. Soc.* **2019**, *78*, 418–425. [CrossRef]

13. Stefanolo, J.P.; Tálamo, M.; Dodds, S.; de la Paz Temprano, M.; Costa, A.F.; Moreno, M.L.; Pinto-Sánchez, M.I.; Smecuol, E.; Vázquez, H.; Gonzalez, A.; et al. Real-World Gluten Exposure in Patients With Celiac Disease on Gluten-Free Diets, Determined From Gliadin Immunogenic Peptides in Urine and Fecal Samples. *Clin. Gastroenterol. Hepatol.* **2021**, *19*, 484–491.e1. [CrossRef]

14. Myléus, A.; Reilly, N.R.; Green, P.H.R. Rate, Risk Factors, and Outcomes of Nonadherence in Pediatric Patients with Celiac Disease: A Systematic Review. *Clin. Gastroenterol. Hepatol.* **2020**, *18*, 562–573. [CrossRef]

15. Catassi, C.; Fabiani, E.; Iacono, G.; D'Agate, C.; Francavilla, R.; Biagi, F.; Volta, U.; Accomando, S.; Picarelli, A.; De Vitis, I.; et al. A Prospective, Double-Blind, Placebo-Controlled Trial to Establish a Safe Gluten Threshold for Patients with Celiac Disease. *Am. J. Clin. Nutr.* **2007**, *85*, 160–166. [CrossRef] [PubMed]

16. Collin, P.; Thorell, L.; Kaukinen, K.; Mäki, M. The Safe Threshold for Gluten Contamination in Gluten-Free Products. Can Trace Amounts Be Accepted in the Treatment of Coeliac Disease? *Aliment. Pharmacol. Ther.* **2004**, *19*, 1277–1283. [CrossRef] [PubMed]

17. Mena, M.C.; Sousa, C. Analytical Tools for Gluten Detection. Policies and Regulation. In *Advances in the Understanding of Gluten related Pathology and the Evolution of Gluten-Free Foods*; OmniaScience: Terrassa, Spain, 2015; pp. 527–564. [CrossRef]

18. Shewry, P.R.; Halford, N.G.; Belton, P.S.; Tatham, A.S. The Structure and Properties of Gluten: An Elastic Protein from Wheat Grain. *Philos. Trans. R. Soc. B Biol. Sci.* **2002**, *357*, 133–142. [CrossRef] [PubMed]

19. Cebolla, Á.; Moreno, M.; Coto, L.; Sousa, C. Gluten Immunogenic Peptides as Standard for the Evaluation of Potential Harmful Prolamin Content in Food and Human Specimen. *Nutrients* **2018**, *10*, 1927. [CrossRef] [PubMed]

20. Shan, L. Structural Basis for Gluten Intolerance in Celiac Sprue. *Science* **2002**, *297*, 2275–2279. [CrossRef]

21. Caminero, A.; Nistal, E.; Herrán, A.R.; Pérez-Andrés, J.; Vaquero, L.; Vivas, S.; Ruíz de Morales, J.M.; Casqueiro, J. Gluten Metabolism in Humans. In *Wheat and Rice in Disease Prevention and Health*; Elsevier: Amsterdam, The Netherlands, 2014; pp. 157–170. [CrossRef]

22. Bhutia, Y.D.; Ganapathy, V. Protein Digestion and Absorption. In *Physiology of the Gastrointestinal Tract*; Elsevier: Amsterdam, The Netherlands, 2018; Volume 2, pp. 1063–1086. [CrossRef]

23. Moreno, M.D.L.; Cebolla, Á.; Muñoz-Suano, A.; Carrillo-Carrion, C.; Comino, I.; Pizarro, Á.; León, F.; Rodríguez-Herrera, A.; Sousa, C. Detection of Gluten Immunogenic Peptides in the Urine of Patients with Coeliac Disease Reveals Transgressions in the Gluten-Free Diet and Incomplete Mucosal Healing. *Gut* **2017**, *66*, 250–257. [CrossRef] [PubMed]

24. Costa, A.F.; Sugai, E.; de la Paz Temprano, M.; Niveloni, S.I.; Vázquez, H.; Moreno, M.L.; Domínguez-Flores, M.R.; Muñoz-Suano, A.; Smecuol, E.; Stefanolo, J.P.; et al. Gluten Immunogenic Peptide Excretion Detects Dietary Transgressions in Treated Celiac Disease Patients. *World J. Gastroenterol.* **2019**, *25*, 1409–1420. [CrossRef]

25. Lähdeaho, M.-L.; Scheinin, M.; Vuotikka, P.; Taavela, J.; Popp, A.; Laukkarinen, J.; Koffert, J.; Koivurova, O.-P.; Pesu, M.; Kivelä, L.; et al. Safety and Efficacy of AMG 714 in Adults with Coeliac Disease Exposed to Gluten Challenge: A Phase 2a, Randomised, Double-Blind, Placebo-Controlled Study. *Lancet Gastroenterol. Hepatol.* **2019**, *4*, 948–959. [CrossRef]

26. Silvester, J.A.; Comino, I.; Rigaux, L.N.; Segura, V.; Green, K.H.; Cebolla, A.; Weiten, D.; Dominguez, R.; Leffler, D.A.; Leon, F.; et al. Exposure Sources, Amounts and Time Course of Gluten Ingestion and Excretion in Patients with Coeliac Disease on a Gluten-Free Diet. *Aliment. Pharmacol. Ther.* **2020**, *52*, 1469–1479. [CrossRef]

27. Silvester, J.A.; Comino, I.; Kelly, C.P.; Sousa, C.; Duerksen, D.R.; Bernstein, C.N.; Cebolla, A.; Dominguez, M.R.; Graff, L.A.; Green, K.H.; et al. Most Patients With Celiac Disease on Gluten-Free Diets Consume Measurable Amounts of Gluten. *Gastroenterology* **2020**, *158*, 1497–1499.e1. [CrossRef] [PubMed]

28. Palanski, B.A.; Weng, N.; Zhang, L.; Hilmer, A.J.; Fall, L.A.; Swaminathan, K.; Jabri, B.; Sousa, C.; Fernandez-becker, N.Q. An Efficient Urine Peptidomics Workflow Identifies Chemically Defined Dietary Gluten Peptides from Patients with Celiac Disease. *bioRxiv* **2021**, 1–23. [CrossRef]

29. Adachi, J.; Kumar, C.; Zhang, Y.; Olsen, J.V.; Mann, M. The Human Urinary Proteome Contains More than 1500 Proteins, Including a Large Proportion of Membrane Proteins. *Genome Biol.* **2006**, *7*, R80. [CrossRef] [PubMed]

30. Biagi, F.; Andrealli, A.; Bianchi, P.I.; Marchese, A.; Klersy, C.; Corazza, G.R. A Gluten-Free Diet Score to Evaluate Dietary Compliance in Patients with Coeliac Disease. *Br. J. Nutr.* **2009**, *102*, 882–887. [CrossRef]

31. Morón, B.; Bethune, M.T.; Comino, I.; Manyani, H.; Ferragud, M.; López, M.C.; Cebolla, Á.; Khosla, C.; Sousa, C. Toward the Assessment of Food Toxicity for Celiac Patients: Characterization of Monoclonal Antibodies to a Main Immunogenic Gluten Peptide. *PLoS ONE* **2008**, *3*, e2294. [CrossRef]

32. Rose, C.; Parker, A.; Jefferson, B.; Cartmell, E. The Characterization of Feces and Urine: A Review of the Literature to Inform Advanced Treatment Technology. *Crit. Rev. Environ. Sci. Technol.* **2015**, *45*, 1827–1879. [CrossRef]

33. Aylward, L.L.; Hays, S.M.; Smolders, R.; Koch, H.M.; Cocker, J.; Jones, K.; Warren, N.; Levy, L.; Bevan, R. Sources of Varia-bility in Biomarker Concentrations. *J. Toxicol. Environ. Health B Crit. Rev.* **2014**, *17*, 45–61. [CrossRef]
34. Lerner, B.A.; Phan Vo, L.T.; Yates, S.; Rundle, A.G.; Green, P.H.R.; Lebwohl, B. Detection of Gluten in Gluten-Free Labeled Restaurant Food: Analysis of Crowd-Sourced Data. *Am. J. Gastroenterol.* **2019**, *114*, 792–797. [CrossRef]
35. Comino, I.; Fernández-Bañares, F.; Esteve, M.; Ortigosa, L.; Castillejo, G.; Fambuena, B.; Ribes-Koninckx, C.; Sierra, C.; Rodríguez-Herrera, A.; Salazar, J.C.; et al. Fecal Gluten Peptides Reveal Limitations of Serological Tests and Food Questionnaires for Monitoring Gluten-Free Diet in Celiac Disease Patients. *Am. J. Gastroenterol.* **2016**, *111*, 1456–1465. [CrossRef] [PubMed]
36. Ménard, S.; Lebreton, C.; Schumann, M.; Matysiak-Budnik, T.; Dugave, C.; Bouhnik, Y.; Malamut, G.; Cellier, C.; Allez, M.; Crenn, P.; et al. Paracellular versus Transcellular Intestinal Permeability to Gliadin Peptides in Active Celiac Disease. *Am. J. Pathol.* **2012**, *180*, 608–615. [CrossRef] [PubMed]
37. Caminero, A.; Nistal, E.; Herrán, A.R.; Pérez-Andrés, J.; Ferrero, M.A.; Vaquero Ayala, L.; Vivas, S.; Ruiz De Morales, J.M.G.; Albillos, S.M.; Casqueiro, F.J. Differences in Gluten Metabolism among Healthy Volunteers, Coeliac Disease Patients and First-Degree Relatives. *Br. J. Nutr.* **2015**, *114*, 1157–1167. [CrossRef] [PubMed]

 *nutrients*

Review

New Insights into Non-Dietary Treatment in Celiac Disease: Emerging Therapeutic Options

Verónica Segura, Ángela Ruiz-Carnicer, Carolina Sousa and María de Lourdes Moreno *

Departamento de Microbiología y Parasitología, Facultad de Farmacia, Universidad de Sevilla, 41012 Sevilla, Spain; vsegura@us.es (V.S.); acarnicer@us.es (Á.R.-C.); csoumar@us.es (C.S.)
* Correspondence: lmoreno@us.es; Tel.: +34-954556453

Abstract: To date, the only treatment for celiac disease (CD) consists of a strict lifelong gluten-free diet (GFD), which has numerous limitations in patients with CD. For this reason, dietary transgressions are frequent, implying intestinal damage and possible long-term complications. There is an unquestionable need for non-dietary alternatives to avoid damage by involuntary contamination or voluntary dietary transgressions. In recent years, different therapies and treatments for CD have been developed and studied based on the degradation of gluten in the intestinal lumen, regulation of the immune response, modulation of intestinal permeability, and induction of immunological tolerance. In this review, therapeutic lines for CD are evaluated with special emphasis on phase III and II clinical trials, some of which have promising results.

Keywords: celiac disease; gluten-free diet; gluten; gliadin; gluten immunogenic peptides; non-dietary therapies

Citation: Segura, V.; Ruiz-Carnicer, Á.; Sousa, C.; Moreno, M.d.L. New Insights into Non-Dietary Treatment in Celiac Disease: Emerging Therapeutic Options. *Nutrients* **2021**, *13*, 2146. https://doi.org/10.3390/nu13072146

Academic Editor: Giacomo Caio

Received: 27 May 2021
Accepted: 15 June 2021
Published: 23 June 2021

Publisher's Note: MDPI stays neutral with regard to jurisdictional claims in published maps and institutional affiliations.

1. Introduction

Celiac disease (CD) is a chronic immune-mediated enteropathy triggered by exposure to dietary gluten in genetically predisposed individuals [1,2]. The pooled global prevalence of CD has been reported to be approximately 1%, however, the prevalence values for CD varies in South America, Africa, North America, Asia, Europe, and Oceania; the prevalence is higher in female vs. male individuals and is 4–8 times higher among non-Hispanic white people compared with other races. Moreover, there has been an increase in the diagnosis rate in the last 10 years [3–7]. CD is characterized by intestinal and/or extraintestinal manifestations, elevation of specific antibodies such as anti-gliadin and anti-tissue transglutaminase (anti-tTG), and the presence of HLA-DQ2/DQ8 haplotypes [8–11].

Gluten is a complex mixture of seed storage proteins known as prolamins, found in cereals grains such as wheat, barley, rye, oats, and their derivatives. The viscoelastic network generated by gluten enables an excellent aerated structure, contributing to the baking quality of these cereals [3–12]. Gluten proteins are characterized by high proline and glutamine content. Therefore, these proteins are partially degraded to peptides by digestive proteases of the gastrointestinal track that persist in the intestine and potentiate their deamidation through tTG [13].

The prevailing hypothesis of immunopathogenesis is the two-signal model, which establishes that gluten has a dual effect on the duodenum of celiac patients mediated by innate and adaptive immune systems [14,15]. Certain peptides, such as the 19-mer gliadin peptide, trigger an innate immune response mainly characterized by the production of interleukin-15 (IL-15) by epithelial cells and the disruption of the epithelial barrier caused by increased permeability and induction of enterocyte apoptosis [16,17]. Consequently, other peptides such as the 33-mer gliadin can now reach the lamina propria to be deamidated by tTG, providing a negative charge to gliadin peptides that activate the immune-adaptive system. The affinity of the HLA-DQ2/8 peptide is enhanced and expressed on the surface of dendritic cells (DCs) [18–20]. DCs present a gluten antigen to T-cells and drive the progression of the proinflammatory response, thereby contributing to the symptomatology of the disease [21,22].

2. Gluten-Free Diet: Challenge and Gluten Exposure

Currently, the only available treatment for CD is a strict, lifelong gluten-free diet (GFD). Dietary gluten restriction is a safe and effective therapy; however, unintentional gluten exposure on a GFD is common and intermittent. Recent findings suggest that most CD patients can only attain a gluten-reduced diet instead of the recommended strict GFD. Gluten exposure may be more common than realized and is distinct from lapses in an otherwise intentionally strict GFD [23,24].

Among the main causes of gluten exposure in a GFD is the ubiquitous nature of gluten, food cross-contamination, and the limitations and socio-emotional toll [25]. In addition, many of the manufactured gluten-free products tend to be less healthy than their gluten analogues since high amounts of lipids, sugars, and other additives are incorporated in their production to simulate the viscoelastic properties of gluten proteins [26]. Although it is well known that legislation on the labeling of gluten-free products is based on the limitation of 20 parts per million (ppm) of gluten [27], there is no clear consensus on the safe amount of daily gluten intake due to the threshold for triggering symptoms has interindividual variability. Total daily gluten consumption that seems to be safe for most CD patients is <50 mg gluten; nevertheless, little amounts as 10 mg of daily gluten for some CD patients could promote development of intestinal mucosal abnormalities [28].

Several studies based on nutritional questionnaires, serological tests, and evaluating gluten immunogenic peptides in feces and urine, have reported variable gluten exposure rates in patients with CD, reaching up to 69% in adults, 64% in adolescents, and 45% in children (Figure 1) despite their best efforts to avoid it. Studies reporting gluten exposure rates may compromise high rates of ongoing symptoms [29–31] and enteropathy [32–35] in patients with CD, leading to comorbidities such as anemia, severe malabsorption, and various forms of malignancies [36]. Hence, it is important to drive efforts to develop non-dietary adjunctive or alternative therapies for CD treatment [37]. Recently, researchers have attempted to meet the requests of celiac patients seeking therapies aside from GFD. In this review, we summarize the spectrum of potential therapeutic agents to improve CD management and their research status, highlighting several drug candidates in phase II/III clinical trials.

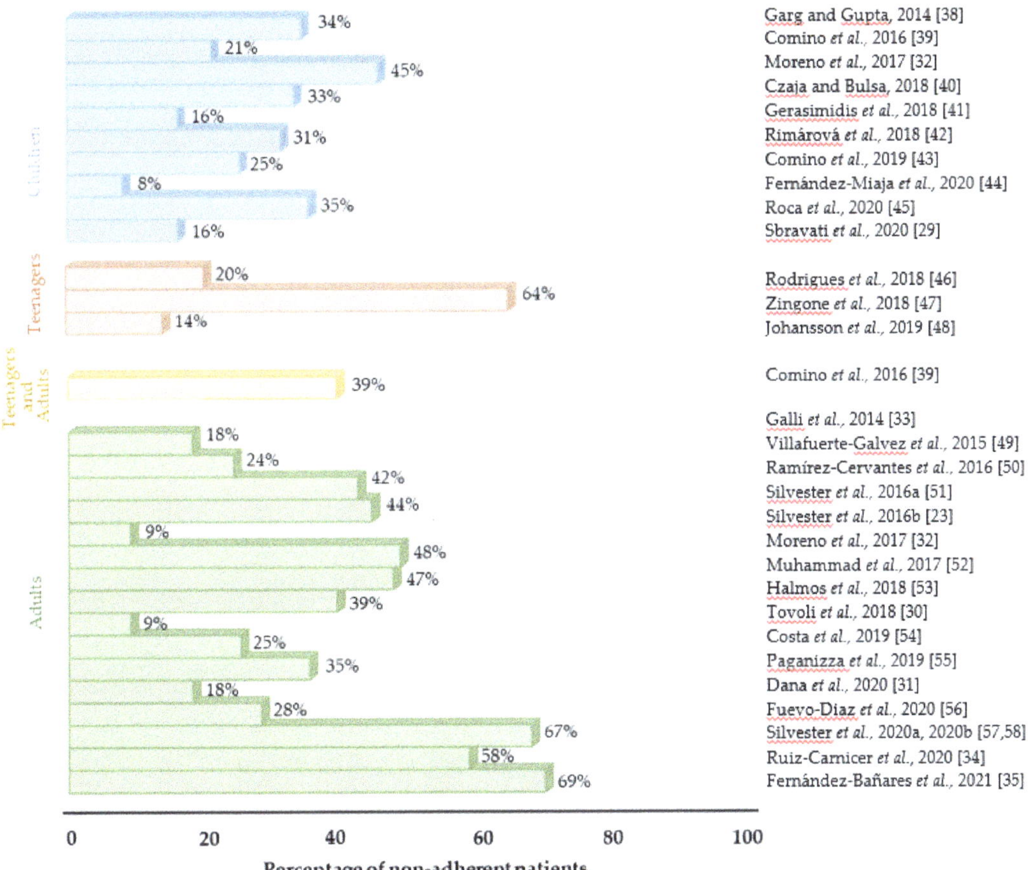

Figure 1. Studies reporting gluten exposure rates in CD patients on a supposed GFD. CD, celiac disease; GFD, gluten-free diet [23,29–58].

3. Potential Alternative or Adjuvant Non-Dietary Treatments for CD

The emerging therapeutic options for CD can be broadly classified into one of the following strategies—(1) removal of toxic gluten peptides before reaching the intestine, (2) regulation of the immunostimulatory effects of toxic gluten peptides, (3) modulation of intestinal permeability, (4) immune modulation and induction of gluten tolerance, and (5) restoration of the imbalance in the gut microbiota (Figure 2).

Many of the sequential steps in CD pathogenesis are well-elucidated; hence, multiple well-defined targets for research and drug development are available (Table 1). Likewise, therapies focused on the regulation of the immunostimulatory effects have been described for other related pathologies, and due to their efficacy, their indications have been extended to CD.

3.1. Removal or Reduction of Toxic Gluten Peptides

Therapies aimed at eliminating or reducing gluten peptides can act in food before marketing, during digestion in the human tract, or masking the antigenic capacity before reaching the intestinal mucosa.

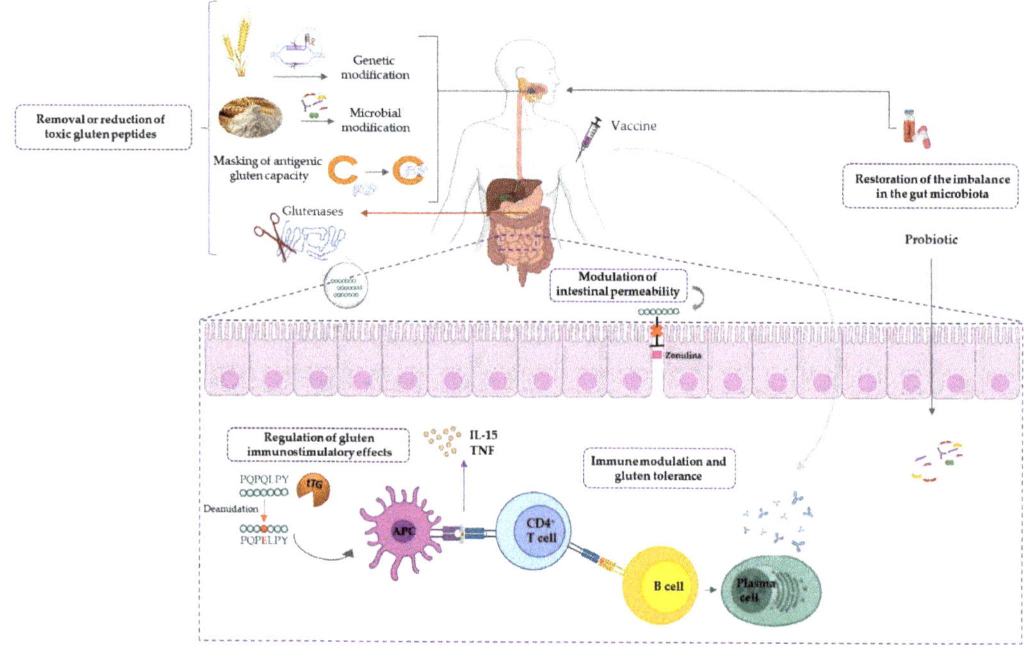

Figure 2. Emerging therapeutic approaches for non-dietary CD treatment. APC, antigen-presenting cell; CD, celiac disease; IL-15, interleukin 15; TNF, tumor necrosis factor; tTG, tissue transglutaminase.

Table 1. Summary of strategies for CD grouped according to their goals.

Strategy	Goal	Therapy	References
Removal of toxic gluten peptides before reaching the intestine	Genetic modification of gluten-containing cereals	Genetically modified wheat flours	[59–61]
	Microbial gluten modification	Pretreatment with probiotic bacteria of the genus *Lactobacillus* (VSL#3)	[62]
		Pretreatment with microbial transglutaminase (m-TG) and N-methyl-lysine	[63]
	Masking of antigenic gluten capacity	Polymeric resins HEMA-co-SS AGY-010	[64,65] [66]
	Luminal gluten detoxification	Prolyl endopeptidases (PEPs): *Flavobacterium meningosepticum* (FM-PEP)	[67,68]
		Prolyl endopeptidases (PEPs): *Myxococcus xanthus* (MX-PEP)	[69]
		Prolyl endopeptidases (PEPs): *Sphingomonas capsulata* (SC-PEP)	[70,71]
		Prolyl endopeptidases (PEPs): *Aspergillus niger* (AN-PEP)	[72]
		Gluten hydrolytic enzyme cocktail: SC-PEP and EPB-2 (ALV003)	[73]
		Gluten hydrolytic enzyme cocktail: FM-PEP and EPB-2	[74]
		Subtilisin derived from *Rothia mucilaginosa* (Sub-A)	[75]
		Cysteine endopeptidase derived from *Hordeum vulgare* (EP-B2)	[21]
		Elastase derived from *Homo sapiens* (CEL-3B)	[22]

Table 1. *Cont.*

Strategy	Goal	Therapy	References
Regulation of the immunostimulatory effects of toxic gluten peptides	Immune response regulation	Inhibition of transglutaminase (ZED 1227)	[76]
		Blocker of HLA DQ binding to T-cells	[77]
		NK lymphocyte activation blocker: NKG2D receptor antagonists	[78]
		Lymphocyte recruitment blocker: Anti-α4 integrin (natalizumab) Anti-integrin α4β7 (vedolizumab) Binding inhibitors CD40-CD40L Binding inhibitors CXCL10- CXCR3 Binding inhibitors CCL25-CCR9	[79]
		Anti-cytokines: Anti-IL-15, PRV-015, CALY-002 (AMG714) Anti-TNF-α (infliximab and adalimumab) Anti-TNF-γ (fontolizumab)	[76,80]
	Inhibition of the proinflammatory cascade	Anti-inflammatories (generic corticosteroids, budesonide, mesalazine)	[81]
Modulation of intestinal permeability	Barrier enhancing therapies	Larazotide acetate (AT-1001 and INN-202)	[82,83]
Immune modulation and induction of tolerance to gluten	Immunomodulation and gluten tolerance	Vaccine Nexvax2	[84,85]
		TAK-101 (CNP-101 and TIMP-GLIA)	[86]
		KAN-101	[87]
		Hookworm infection (*Necator americanus*)	[88]
		Mucosal tolerance due to genetic modification	[89]
Restoration of the imbalance in the gut microbiota	Probiotic supplementation	Microbial therapies	[90,91]

TNF, tumor necrosis factor; IgA, immunoglobulin A; Il-15, interleukin 15; NK, natural killer; PEP, prolyl endopeptidase; P-HEMA-co-SS, poly-hydroxyethylmethacrylateco-styrene sulfonate.

3.1.1. Genetic Modification of Gluten-Containing Cereals

The development of cereals with reduced or absent immunogenic gluten proteins is important for the management of CD. The wheat variants currently used have been reported to be more immunogenic than the ancestral or wild variants such as those belonging to the genera *Tritordeum* or *Triticum* [92,93]. Genetic advances in plants have successfully allowed the production of wheat lines with very low or completely lacking gluten content through the hybridization of wheat species [94]. A recent study described the traditional breeding and characterization of a novel ultralow gluten barley variety in which the gluten content was reduced to below 5 ppm by combining three recessive alleles, which act independently to lower the hordein content in the parental varieties [59].

RNA interference to silence the expression of gluten proteins that contain immunogenic epitopes for CD has been employed as a genetic engineering strategy [95]. This approach has allowed the development of wheat lines that contain very few immunogenic epitopes of CD, and, therefore, could be consumed by patients with non-celiac gluten sensitivity, since it produces no adverse clinical symptoms [96,97]. Currently, several studies are in progress to understand the effects of these new lines in patients with CD.

The use of CRISPR/Cas9 (Clustered Regulatory Interspaced Palindromic Repeats associated protein 9) technology can precisely and efficiently reduce the amount of α-gliadin in the seed kernel, providing bread and durum lines with reduced immunoreactivity for the celiac community [60,94]. However, it is likely that the deleted gliadin genes need to be replaced by non-immunogenic gliadin variants to obtain adequate elasticity. Additionally, governmental regulations for genetic modification of food products require expensive and time-consuming food safety assessments to be met before product marketing [94].

3.1.2. Microbial Gluten Modification

The addition of diverse microorganisms in sourdough for fermentation has been studied because it contains proteases capable of hydrolyzing gluten peptides rich in glutamine and proline residues. Diverse studies using species of the genus *Lactobacillus* have reported that this baking method could obtain safe breads for celiac patients [62,98]. The well-known probiotic preparation VSL#3 comprises eight strains belonging to the genera *Bifidobacterium*, *Lactobacillus*, and *Streptococcus*. This cocktail was assayed during the food processing step and produced tolerable predigested gliadins without α-gliadin peptides p62-75 and 33-mer, but with the palatability of gluten-free products [99]. This study demonstrated the improvement in the symptoms of adult CD patients with irritable bowel syndrome (IBS) [100]. Furthermore, the probiotic preparation was capable of stabilizing intraepithelial junctions, promoting the barrier effect that prevents the entry of toxic peptides into the lamina propria [91,101]. However, individual probiotic strains are inadequate to break down gliadin compared to the group efficacy [101,102].

Another investigated approach in the preclinical phase consists of the pretreatment of flours or sourdoughs with microbial transglutaminase (m-TG) and N-methyl-lysine [103,104]. The use of N-methyl lysine and m-TG derived from *Streptomyces mobaraensis* provoked gluten modification and loss of affinity for the HLA-DQ2 molecule, which leads to less activation of intestinal T lymphocytes [105]. Although the effect of standard bakery concentrations of microbial transglutaminase (m-TG) in wheat bread preparation on the immunoreactivity of sera of CD patients was investigated, its use in food preparation remains a subject of debate [63].

3.1.3. Masking of Antigenic Gluten Capacity

The gluten-binding polymer BL-7010 or copolymer poly-hydroxyethylmethacrylate-co-styrene sulfonate (P-HEMA-co-SS) complex is a non-absorbable synthetic origin blocking agent that binds intraluminal gluten [64]. Therefore, digestive enzymes cannot access the cleavage sites, preventing the degradation of immunogenic peptides that are not absorbed by the intestine and do not induce an immune response. The effect of BL-7010 has been investigated in intestinal biopsy samples from patients with CD [64,65,106]. Attenuation of the immune response and the high safety profile in animal models were observed; however, this phase II therapy was discontinued in 2017.

Recent studies have developed neutralizing anti-gliadin antibodies extracted from egg yolk (AGY-010). IgY antibodies have shown effectiveness in neutralizing and absorbing gliadin, as well as resistance to stomach conditions [66]. This therapy is currently in phase II studies and a study is ongoing to evaluate its efficacy and safety in CD patients [107]. As the use of egg yolk antibodies might be inefficient for large-scale clinical production, parallel recombinant antibody fragments in single-chain format have been produced for the same purpose [108].

3.1.4. Luminal Gluten Detoxification

Oral enzyme therapy is focused on the inactivation of gluten peptides in the human gastrointestinal tract before reaching the intestine. Gluten-degrading enzymes seem to hold the most promise as attractive therapies for helping patients with CD to avoid accidental gluten ingestion and to promote better overall health. A prerequisite is that such enzymes should be active under gastro-duodenal conditions, quickly neutralize the T-cell-activating gluten peptides and be safe for human consumption [67,68,70,109].

Glutenases have been identified in bacteria, fungi, plants, and even insects (Table 2). Although the enzymes studied are endopeptidases, interesting exopeptidases have also been described [110]. Endopeptidases are further subdivided depending on their catalytic mechanism; among them, prolyl endopeptidases (PEPs) are especially effective in hydrolyzing peptide bonds on the carboxyl side of internal proline residues in gluten-derived oligopeptides [69]. The potential synergism between gluten-degrading enzymes that differ in their cleavage specificities and optimum pH values raises the possibility of a mixture that would more effectively eliminate the antigenicity of ingested gluten fractions [111].

Table 2. Summary of glutenases used in enzyme therapy and classified according to origin of isolation, producer organism, and catalytic mechanism. ND, not determined.

Source of Enzymes	Peptidase Type	Organism	Isolated Enzyme	References
Bacterial peptidases	Prolyl endopeptidase	*S. capsulata*	SC-PEP	[68]
		M. xanthus	MX-PEP	[65]
		F.meningosepticum	FM-PEP	[66]
		Chryseobacterium taeanense	PEP 2RA3	[109]
	Subtilisin	*Rothia aeria*	ND	[112]
		R. mucilaginosa	Sub-A	[112]
		Bacillus licheniformis	ND	[113]
	Pseudolysin	*Pseudomonas aeruginosa*	lasB	[114]
	Thermolysin	*Bacillus thermoproteolyticus*	ND	[113]
	Serine peptidase	*Bacillus tequilensis*	ND	[115]
	ND	*Bacillus spp GS 188*	ND	[116]
	Serine carboxyl peptidase	*Actinoallomurus A8*	E40	[117]
Fungal peptidases	Prolyl endopeptidase	*A. niger*	AN-PEP	[72]
	Aspergillopepsin	*A. niger*	ASP	[118]
	Exopeptidase	*Aspergillus oryzae*	AO-DPP-IV	[119]
Plant peptidases	Cysteine endopeptidase	*H. vulgare*	EP-B2	[120]
		Carica papaya	Caricain	[121]
		Triticum aestivum	Triticain-α	[122]
		H. vulgare	HvPap-6 CysProt	[123]
Insect peptidases	Prolyl peptidase	*Rhizopertha dominica*	ND	[123]
	Prolidase	*Tenebrio molitor*	ND	[124]
Human peptidases	Elastase	*Homo sapiens*	CEL3B	[22]
		Homo sapiens	CEL2A	[22]
	Carboxypeptidase	*Homo sapiens*	CBPA1	[22]

Among the bacterial enzymes capable of degrading gluten, PEPs are produced by *F. meningosepticum* [68,69], *S. capsulata* [70,71] and *M. xanthus* [69]. These three enzymes showed high specificity against reference chromogenic substrates and the potential to successfully degrade the immunogenic sequences of gluten. The cysteine endoprotease EP-B2 and PEP from *F. meningosepticum* complement each other in terms of their gluten hydrolytic properties; however, significant efforts have been made to increase their thermostability to be suitable for industrial applications [111].

Fungal PEP from *A. niger*, known as AN-PEP, exhibits post-proline cleavage activity and is highly efficient in degrading gluten [72]. A clinical study with Tolerase G, an AN-PEP-based supplement, reduced the amount of gluten exposed in the duodenum efficiently, despite not completely degrading the gluten [72]. The enzyme preparation consisting of AN-PEP from *A. niger* and DPP-IV from *A. oryzae* (STAN 1) administered orally in celiac patients appeared to be modest because of the non-specificity of AN-PEP and the very limited proteolytic effect of DPP-IV. Therefore, these studies were stalled in phase II in 2017. In the genus *Aspergillus*, another enzyme was detected with gluten-degrading activity, termed aspergillopepsin (ASP) from *A. niger*, although ASP needs to be used as a complementary enzyme because of its incomplete degradation [118]. In this sense, a dietary supplement has been widely used in the food and feed industry containing ASP from *A. niger* and DPP-IV from *A. oryzae*, which successfully degraded small amounts of gluten in vitro [119].

As previously argued, the combination of enzymes appears to be a future direction in enzyme therapy. The enzymatic cocktail, latiglutenase or IMGX-003 (formerly ALV003), consists of a 1:1 combination of cysteine endoprotease from barley EP-B2 (IMGX-001), and PEP from *S. capsulate* SC-PEP (IMGX-002). A phase II gluten challenge to investigate its effect on both mucosal and symptomatic protection in CD patients is in progress. Initial findings with latiglutenase have been shown to mitigate gluten-induced intestinal mucosal injury as well as to reduce the severity and frequency of symptoms in patients with CD [73,125]. Evidence of symptom relief was particularly pronounced in patients with positive serology despite following a GFD [61,126,127].

An engineered synthetic gluten-degrading enzyme, KumaMax, with technological improvements, is being studied. KumaMax showed similar in vitro results to IMGX-003, although it is still under development [128]. The gluten-degrading enzyme subtilisin-A (Sub-A) from *B. licheniformis* was modified by PEGylation and subjected to microencapsulation. The effectiveness was confirmed in vitro and in vivo and showed a significant increase in protection against acid exposure [113].

Investigating the effect of glutenases on the symptoms and biomarkers in CD patients with randomized, placebo-controlled studies is mandatory; however, this is not as straightforward as it might seem.

3.2. Immune Response Regulation

As inflammatory mediators are common in CD and other gastrointestinal pathologies, certain therapies aimed at avoiding chronic gastrointestinal inflammation could be applied in CD.

tTG plays a critical role in the pathogenesis of CD through the deamidation and transamidation of gluten peptides, which leads to an immune response with inflammation of the intestinal mucosa [129,130]. Hence, the inhibition of tTG results in the abolishment of gluten peptide presentation by HLA-DQ2/DQ8, preventing the immune response. Three varieties of tTG-2 inhibitors have been well described, namely, irreversible inhibitors, reversible inhibitors, and competitive amine inhibitors. ZED-1227 is a highly specific orally active irreversible inhibitor with promising preliminary preclinical results. A phase II clinical study with ZED-1227 is ongoing in EU countries in healthy volunteers [76]. Nevertheless, tTG plays a critical role in gut wound healing, and its safety and efficacy require further study [131]. Among competitive inhibitors, cystamine is currently the only competitive commercially available tTG-2 inhibitor despite that it has not been explored for its potential role in CD. Recently, Palansky et al. [132] discovered that disulfiram, an FDA-approved drug for alcohol abuse, is also a tTG inhibitor. This is the first clinically approved compound to show human tTG inhibitory activity, raising further interesting possibilities for the future in terms of tTG inhibition as a therapeutic strategy in CD [133].

Another attractive therapeutic target to prevent the activation of the immune response is the HLA-DQ2 blocker. Gluten-like molecules in which proline residues have been replaced by azidoprolines do not elicit an immune response in T-cells isolated from individuals with CD [8]. Cyclic and dimeric peptides have also been developed that bind DQ2, partially blocking T-cell proliferation and antigen presentation. However, these molecules do not fully block the activation of T-cells; therefore, other nontoxic antagonists with high affinity are currently being studied [129].

Some studies have highlighted the role of IL-15 and the receptor activator NKG2D and other immune soluble factors as targets of CD treatment. IL-15 plays a critical role in the activation of intraepithelial lymphocytes and participates in both innate and adaptive responses. NKG2D is the receptor of T-cells and natural killer cells [134]. The first monoclonal antibody (moAb) studied against the IL-15 receptor was Hu-Mik-Beta-1, and positive results were obtained in refractory CD. However, this therapy was stuck in phase I. Second, PRV-015 (also known as AMG 714) is a fully human moAb that has emerged as a leading investigational candidate for nonresponsive CD (NRCD), in which patients maintain disease activity despite an ongoing GFD. Phase II studies have shown a reduction in inflammation and symptoms in a clinical trial with patients with refractory CD type 2 [80]. Lastly, CALY-002 is a moAb whose safety, tolerability, pharmacokinetics, and pharmacodynamics are being evaluated in phase II studies in both CD and eosinophilic esophagitis [135].

Tumor necrosis factor (TNF)-γ secreted by T-cells in response to gluten is another therapeutic target under study. Fontolizumab was initially developed for inflammatory bowel disease (IBD) treatment and has been proposed for CD, although clinical trials for this indication have not yet been registered. Infliximab and adalimumab moAbs targeting TNF-α have been used in clinical practice for IBD and could be useful in treating CD [76,136].

Among T-cell-targeted therapies aimed at blocking lymphocyte recruitment, natalizumab is an anti-α4 used in Crohn's disease and could be useful in CD, although its side effects are very high [79,137]. Vedolizumab is scheduled to start phase II studies that block α4β7 integrin [138]. In addition, chemokine receptor inhibitors such as CXCR3 and its specific ligands CXCL10 and CXCL11 have also been studied [79]. These molecules are among the main determinants in the recruitment of immune cells to the intestinal lamina propria and are involved in the uptake of lymphocytes in the presence of gliadin peptides. CCL25 and its receptor CCR9 appear to be a therapeutic alternative in the future, although to date it has only been studied in animal models with Crohn's disease [139,140].

Anti-inflammatory drugs such as corticosteroids and budesonide are generally used to treat the symptoms of refractory CD. Likewise, mesalazine has been proposed, although it must be remembered that most of these formulations are prepared to be released in the colon and the inflammation in CD affects the small intestine [66]. Recent studies have shown that mesalazine has a beneficial effect on the molecules and biological mediators of inflammation that occur in the mucosa of celiac patients [81].

3.3. Barrier Enhancing Therapies

Increased intestinal permeability has been implicated in CD due to both transcellular and paracellular epithelial permeability, with apical junctional protein complexes called tight junctions being key components in the latter process [141].

Larazotide acetate, formerly known as AT-1001 or INN-202, is a locally acting octapeptide with a sequence analogous to a portion of *Vibrio cholerae* zonula occludens toxin [141]. In cultured intestinal epithelial monolayers, larazotide acetate enhanced actin rearrangement and prevented the disassembly of tight junctions [142,143]. In addition, larazotide acetate prevents the passage of gluten peptides to the lamina propria by closing the intercellular junctions of the enterocytes, which could help prevent the development of the immune cascade in celiac patients. Therefore, larazotide acetate is the most advanced experimental drug, showing a reduction in symptoms as well as a reduction in anti-tTG antibody levels. Three phase II studies of larazotide acetate have been completed and published in CD patients undergoing a gluten challenge, but only an excellent safety profile and efficacy with low dose have been reported in patients with NRCD. Therefore, larazotide acetate has moved forward to a phase III registration study for this indication [82,83,144].

3.4. Immunomodulation and Gluten Tolerance

Vaccine therapy is the preferred option among alternative treatments to a GFD in patients with CD. It is based on immunization with gluten epitopes, which induces the expansion of regulatory T-cells, restoring oral tolerance to gluten [145]. The Nexvax2 vaccine (ImmusanT, Cambridge, MA, USA) comprises the use of three gluten epitopes chosen based on a study by Tye-Din et al. [145]. This study examined epitopes within wheat, barley, and rye with the ability to induce and stimulate T-cells isolated from the serum of patients with CD on a gluten-containing diet. Nexvax2 is one of several CD drugs that has reached phase II clinical trials [141]. However, although Nexvax2 showed a good safety profile, its efficacy has yet to be demonstrated. Nexvax2 is specific only for individuals with the HLA-DQ2 genotype. Therefore, another vaccine should be investigated in patients with HLA-DQ8 genotyping [84,85].

Biodegradable nanoparticles encapsulated with gliadin proteins TAK-101 (formerly known as CNP-101 and TIMP-GLIA) seem to be a first-in-class agent that induces antigen-specific immune tolerance to CD [141]. TAK-101 binds inflammatory cells to initiate tolerogenic immune reprogramming. According to the clinicaltrials.gov, the phase II developmental trial of TAK-101 for treating patients with CD was estimated to be completed in July 2019, but it is still in the active phase, not the recruiting phase [146].

A new therapy in phase I focuses on restoring normal immune tolerance by targeting specific receptors in the liver, named KAN-101 [141]. The tolerogenic nanoparticles for intravenous injection trigger a cascade of events that drive the re-education of T-cells so that they do not respond to gluten antigens [87].

The administration of *N. americanus* infective larvae in patients with CD interferes with the host immune response due to its survival in the intestine. Studies of duodenal biopsies from CD individuals infected with *N. americanus* and exposed to gluten have shown a reduction in the production of IL-2, IFN-γ, and IL-17. In addition, the absence of histological lesions and even a decrease in anti-tTG antibody levels have been demonstrated [88]. *N. americanus* is currently in phase II clinical trials, although problems with CD patient acceptance for routine clinical use are anticipated [66,147].

Finally, other studies based on the tolerance of the mucosa to genetic modification are in the initial phase of investigations. These studies specifically focused on organoids derived from the human intestine, providing a model to study the response to gluten and the effects of molecules derived from the microbiota in patients with CD [89].

3.5. Restoration of the Imbalance in the Gut Microbiota

The gut microbiota is involved in the initiation and perpetuation of intestinal inflammation in several chronic diseases. Indeed, several studies have identified certain microorganisms in CD patients and healthy subjects. Therefore, alteration of the microbiota could play a significant role in the pathogenesis of CD. Recent studies have focused on the role of the gut microbiota in CD and the complex relationship between its composition, genetic background, GFD, and persistence of clinical symptoms [90,148]. The specific mechanisms by which microorganisms can participate in the development of responses to gluten are broad and include the metabolism of trigger antigen responses, enhancement of the intestinal barrier, and modulation of adaptive and innate immune responses [149].

Recent data have shown that genetics (HLA-DQ-2 or DQ-8) may predispose individuals with CD to dysbiosis [90,148]. Palma et al. [150] studied the effects of following a GFD on the composition of gut microbiota in healthy subjects. A significant decrease of *Bifidobacterium*, *Clostridium lituseburense*, and *Faecalibacterium prausnitzii* and an increase in *Enterobacteriaceae* and *Escherichia coli* counts were found. Therefore, the supplementation with a probiotic to restore the imbalance in the gut microbiota might be a reasonable therapeutic option by downregulating the proinflammatory immune response in CD patients [90]. The design of specific probiotics comprises advanced genomic and metabolomics techniques using the interactions between the human body-microbiota and intra-microbiota, eventually leading to tailored specific probiotic therapies for microbiome regulation and health sustainability.

Probiotics play an important role in preventing the overgrowth of potentially pathogenic bacteria and maintaining the integrity of the gut mucosal barrier. The beneficial effects of probiotics have been previously studied in adult patients with IBS. Oral administration of a probiotic mixture of *Lactobacillus plantarum* 14D-CECT 4528, *Lactobacillus casei*, *Bifidobacterium breve* Bbr8 LMG P-17501, *B. breve* Bl10 LMG P-17500, and *Bifidobacterium animalis* under randomized, double-blind, and placebo-controlled conditions showed the improvement in symptoms of adult CD patients with IBS [100]. In the future, microorganisms or even genetically engineered microorganisms could be used to act as living enzyme machinery as well as vectors for the delivery of endopeptidases capable of digesting gluten in the stomach, thereby allowing celiac patients to have a controlled dietary gluten intake [91,151].

In conclusion, probiotics are not expected to provide a rapid cure for complex diseases such as CD, but rather to alleviate the severity of symptoms [99]. More studies are needed to address how the gut microbiome can modulate or alter the course of the disease. To date, there are no guidelines available that recommend probiotic use in patients with CD. However, the data suggest a strong adjunctive role in the management of symptoms and bacterial overgrowth.

4. Clinical Trials

Clinical endpoints are variables to quantify the potential effect of the treatment or intervention under study and reflect or characterize how a subject "feels, functions, or survives" [152]. Many major disease areas have established clinical trial endpoints because a fair number of registration trials have already been conducted and drugs approved for marketing. As in CD, there are no approved products and little experience, and agreed endpoints are lacking. Certain treatments in CD could control symptoms and prevent worsening of damage, while others are, at least initially, focused primarily on healing and maintenance of healing, with little effect on symptoms. Therefore, different endpoints or endpoint instruments are needed [153]. To date, only larazotide acetate is currently in phase III studies; most of them at phase II and a few phase I trials have explored its efficacy. Some therapies are being evaluated in preclinical trials and are postulated to be promising treatments for CD pathogenesis (Figure 3). We are facing many promising and emerging options for the treatment of CD.

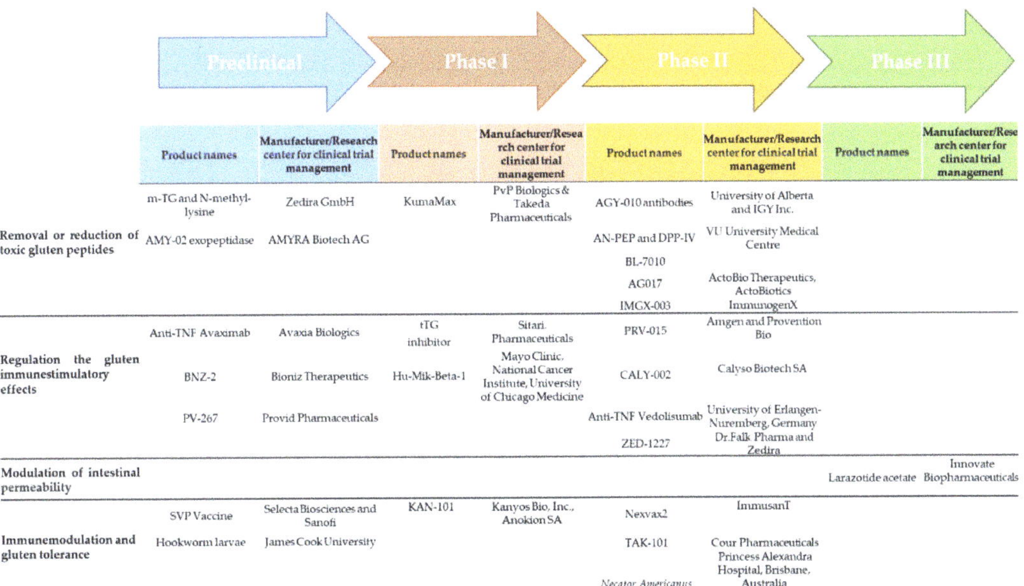

	Preclinical Product names	Preclinical Manufacturer/Research center for clinical trial management	Phase I Product names	Phase I Manufacturer/Research center for clinical trial management	Phase II Product names	Phase II Manufacturer/Research center for clinical trial management	Phase III Product names	Phase III Manufacturer/Research center for clinical trial management
Removal or reduction of toxic gluten peptides	m-TG and N-methyl-lysine	Zedira GmbH	KumaMax	PvP Biologics & Takeda Pharmaceuticals	AGY-010 antibodies	University of Alberta and IGY Inc.		
	AMY-02 exopeptidase	AMYRA Biotech AG			AN-PEP and DPP-IV	VU University Medical Centre		
					BL-7010			
					AG017	ActoBio Therapeutics, ActoBiotics		
					IMGX-003	ImmunogenX		
Regulation the gluten immunestimulatory effects	Anti-TNF Avaximab	Avaxia Biologics	tTG inhibitor	Sitari Pharmaceuticals	PRV-015	Amgen and Provention Bio		
	BNZ-2	Bioniz Therapeutics	Hu-Mik-Beta-1	Mayo Clinic, National Cancer Institute, University of Chicago Medicine	CALY-002	Calyso Biotech SA		
	PV-267	Provid Pharmaceuticals			Anti-TNF Vedolisumab	University of Erlangen-Nuremberg, Germany		
					ZED-1227	Dr.Falk Pharma and Zedira		
Modulation of intestinal permeability							Larazotide acetate	Innovate Biopharmaceuticals
Immunemodulation and gluten tolerance	SVP Vaccine	Selecta Biosciences and Sanofi	KAN-101	Kanyos Bio, Inc., Anokion SA	Nexvax2	ImmusanT		
	Hookworm larvae	James Cook University			TAK-101	Cour Pharmaceuticals		
					Necator Americanus	Princess Alexandra Hospital, Brisbane, Australia		

Figure 3. Clinical and preclinical development pipeline for CD. CD, celiac disease; PEP, prolyl endopeptidases; TNF, tumor necrosis factor; tTG, tissue transglutaminase.

5. Conclusions

Although a GFD has been shown to be safe and effective in most celiac patients, the limitations caused by dietary gluten restriction and high gluten exposure rates raise the need to develop new therapies for CD. Different non-dietary therapeutic strategies are currently in the development phase and in clinical research, which could be a useful option in the medium- or long-term in patients with CD. To date, larazotide acetate is the most advanced experimental drug that has shown a reduction in symptoms as well as anti-tTG antibody titers. Promising PRV-015 immunotherapy requires more assays to establish rational targets for disease prevention. The use of glutenases as food preprocessors has proven to be very effective; however, the use of oral glutenases is perhaps the most accepted strategy for patients with CD and one of the most numerous options in terms of ongoing studies. All efforts are now being made to assess the effectiveness of these enzymes as a supplement to a GFD, highlighting the phase II results of IMGX-003 being very promising. Vaccine therapy has limitations, such as that it can engage only known or previously investigated immunogenic epitopes and effectiveness with the specific HLA-DQ2 genotype. However, if successful, it has the potential to have prolonged benefits on patients.

In addition, other many interesting drugs are in early research stages, such as tTG inhibitors, HLA blockers, and probiotics, although probiotics will probably need to be combined with long-term dietary changes. While several trials are ongoing or underway for CD, there is no consensus on outcome measures in CD patient trials.

Preventing the onset of CD entirely would be the most beneficial and desirable approach; however, recent approaches argue whether ingesting certain amounts of gluten plays a complementary or "adjuvant" role to a GFD and not as a substitute to a GFD in patients with CD. Nevertheless, some of these therapies could also be effective in other gluten-related pathologies in which a minimal amount of gluten is tolerable.

Great efforts are ongoing to determine the effectiveness and the dose limit of gluten ingested in these therapies. It is also obvious that the possibility of using synergistic strategies could increase the maximum safe doses allowed for CD; therefore, this issue will be the next challenge.

Author Contributions: Conceptualization, V.S. and M.d.L.M.; data curation, V.S. and M.d.L.M.; formal analysis, V.S. and M.d.L.M.; investigation, V.S., Á.R.-C., C.S. and M.d.L.M.; methodology, V.S. and M.d.L.M.; resources, V.S. and M.d.L.M.; writing—original draft, V.S., Á.R.-C., C.S. and M.d.L.M.; writing—review and editing, V.S., Á.R.-C., C.S. and M.d.L.M. All authors have read and agreed to the published version of the manuscript.

Funding: This work was supported by grant from Federación de Asociaciones de Celíacos de España (FACE) (SUBN/2019/005).

Conflicts of Interest: The authors have no potential conflict of interest to declare. None of the authors have any conflicts of interest that could affect the performance of the work or the interpretation of the data.

References

1. Lebwohl, B.; Cao, Y.; Zong, G.; Hu, F.B.; Green, P.H.R.; Neugut, A.I.; Rimm, E.B.; Sampson, L.; Dougherty, L.W.; Giovannucci, E.; et al. Long-term gluten consumption in adults without celiac disease and risk of coronary heart disease: Prospective cohort study. *BMJ* **2017**, *357*, j1892. [CrossRef] [PubMed]
2. Al-Toma, A.; Volta, U.; Auricchio, R.; Castillejo, G.; Sanders, D.S.; Cellier, C.; Mulder, C.J.; Lundin, K.E.A. European Society for the Study of Coeliac Disease (ESsCD) guideline for coeliac disease and other gluten-related disorders. *United Eur. Gastroenterol. J.* **2019**, *7*, 583–613. [CrossRef] [PubMed]
3. Lebwohl, B.; Sanders, D.S.; Green, P.H.R. Coeliac disease. *Lancet* **2018**, *391*, 70–81. [CrossRef]
4. Ludvigsson, J.F.; Murray, J.A. Epidemiology of celiac disease. *Gastroenterol. Clin. N. Am.* **2019**, *48*, 1–18. [CrossRef]
5. Singh, P.; Arora, A.; Strand, T.A.; Leffler, D.A.; Catassi, C.; Green, P.H.; Kelly, C.P.; Ahuja, V.; Makharia, G.K. Global prevalence of celiac disease: Systematic review and meta-analysis. *Clin. Gastroenterol. Hepatol.* **2018**, *16*, 823–836.e2. [CrossRef] [PubMed]
6. Glissen, J.R.B.; Singh, P. Coeliac disease. *Paediatr. Int. Child. Health* **2019**, *39*, 23–31. [CrossRef] [PubMed]

7.   Mardini, H.E.; Westgate, P.; Grigorian, A.Y. Racial differences in the prevalence of celiac disease in the US population: National Health and Nutrition Examination Survey (NHANES) 2009–2012. *Dig. Dis. Sci.* **2015**, *60*, 1738–1742. [CrossRef] [PubMed]
8.   Husby, S.; Koletzko, S.; Korponay-Szabo, I.R.; Mearin, M.L.; Phillips, A.; Shamir, R.; Troncone, R.; Giersiepen, K.; Branski, D.; Catassi, C.; et al. European Society for Pediatric Gastroenterology, Hepatology, and Nutrition guidelines for the diagnosis of coeliac disease. *J. Pediatr. Gastroenterol. Nutr.* **2012**, *54*, 136–160. [CrossRef]
9.   Rubio-Tapia, A.; Hill, I.D.; Kelly, C.P.; Calderwood, A.H.; Murray, J.A. American College of Gastroenterology. ACG clinical guidelines: Diagnosis and management of celiac disease. *Am. J. Gastroenterol.* **2013**, *108*, 656–676. [CrossRef]
10. Ludvigsson, J.F.; Bai, J.C.; Biagi, F.; Card, T.R.; Ciacci, C.; Ciclitira, P.J.; Green, P.H.; Hadjivassiliou, M.; Holdoway, A.; van Heel, D.A.; et al. Diagnosis and management of adult coeliac disease: Guidelines from the British Society of Gastroenterology. *Gut* **2014**, *63*, 1210–1228. [CrossRef]
11. Fasano, A.; Sapone, A.; Zevallos, V.; Schuppan, D. Nonceliac gluten sensitivity. *Gastroenterology* **2015**, *148*, 1195–1204. [CrossRef] [PubMed]
12. Mena, M.C.; Sousa, C. Analytical tools for gluten detection. Policies and regulation. In *Advances in the Understanding of Gluten Related Pathology and the Evolution of Gluten-Free Foods*; Arranz, E., Fernández-Bañares, F., Eds.; OmniaScience: Barcelona, Spain, 2015; pp. 527–564.
13. Caio, G.; Volta, U.; Sapone, A.; Leffler, D.A.; De Giorgio, R.; Catassi, C.; Fasano, A. Celiac disease: A comprehensive current review. *BMC Med.* **2019**, *17*, 142. [CrossRef]
14. Sharma, N.; Bhatia, S.; Chunduri, V.; Kaur, S.; Sharma, S.; Kapoor, P.; Kumari, A.; Garg, M. Pathogenesis of celiac disease and other gluten related disorders in wheat and strategies for mitigating them. *Front. Nutr.* **2020**, *7*, 6. [CrossRef] [PubMed]
15. Lindfors, K.; Ciacci, C.; Kurppa, K.; Lundin, K.E.A.; Makharia, G.K.; Mearin, M.L.; Murray, J.A.; Verdu, E.F.; Kaukinen, K. Coeliac disease. *Nat. Rev. Dis. Primers* **2019**, *5*, 3. [CrossRef]
16. Maiuri, L.; Ciacci, C.; Auricchio, S.; Brown, V.; Quaratino, S.; Londei, M. Interleukin 15 mediates epithelialmchanges in celiac disease. *Gastroenterology* **2000**, *119*, 996–1006. [CrossRef]
17. Maiuri, L.; Ciacci, C.; Ricciardelli, I.; Vacca, L.; Raia, V.; Auricchio, S.; Picard, J.; Osman, M.; Quaratino, S.; Londei, M. Association between innate response to gliadin and activation of pathogenic T cells in coeliac disease. *Lancet* **2003**, *362*, 30–37. [CrossRef]
18. Qiao, S.W.; Bergseng, E.; Molberg, O.; Xia, J.; Fleckenstein, B.; Khosla, C.; Sollid, L.M. Antigen presentation to celiac lesion-derived t cells of a 33-mer gliadin peptide naturally formed by gastrointestinal digestion. *J. Immunol.* **2004**, *173*, 1757–1762. [CrossRef]
19. Ráki, M.; Tollefsen, S.; Molberg, Ø.; Lundin, K.E.A.; Sollid, L.M.; Jahnsen, F.L. A Unique dendritic cell subset accumulates in the celiac lesion and efficiently activates gluten-reactive T cells. *Gastroenterology* **2006**, *131*, 428–438. [CrossRef]
20. Tollefsen, S.; Arentz-Hansen, H.; Fleckenstein, B.; Molberg, Ø.; Ráki, M.; Kwok, W.W.; Jung, G.; Lundin, K.E.A.; Sollid, L.M. HLA-DQ2 and -DQ8 signatures of gluten T cell epitopes in celiac disease. *J. Clin. Investig.* **2006**, *116*, 2226–2236. [CrossRef]
21. Martinez, M.; Gómez-Cabellos, S.; Giménez, M.J.; Barro, F.; Diaz, I.; Diaz-Mendozam, M. Plant Proteases: From key enzymes in germination to allies for fighting human gluten-related disorders. *Front. Plant Sci.* **2019**, *10*, 721. [CrossRef] [PubMed]
22. Gutiérrez, S.; Pérez-Andrés, J.; Martínez-Blanco, H.; Ferrero, M.A.; Vaquero, L.; Vivas, S.; Casqueiro, J.; Rodríguez-Aparicio, L.B. The human digestive tract has proteases capable of gluten hydrolysis. *Mol. Metab.* **2017**, *6*, 693–702. [CrossRef]
23. Silvester, J.A.; Graff, L.A.; Rigaux, L.; Walker, J.R.; Duerksen, D.R. Symptomatic suspected gluten exposure is common among patients with coeliac disease on a gluten-free diet. *Aliment. Pharmacol. Ther.* **2016**, *44*, 612–619. [CrossRef]
24. Stefanolo, J.P.; Tálamo, M.; Dodds, S.; Temprano, M.P.; Costa, A.F.; Moreno, M.L.; Pinto-Sánchez, M.I.; Smecuol, E.; Vázquez, H.; Gonzalez, A.; et al. Real-world gluten exposure in patients with celiac disease on gluten-free diets, determined from gliadin immunogenic peptides in urine and fecal samples. *Clin. Gastroenterol. Hepatol.* **2021**, *19*, 484–491.e1. [CrossRef]
25. Wolf, R.L.; Lebwohl, B.; Lee, A.R.; Zybert, P.; Reilly, N.R.; Cadenhead, J.; Amengual, C.; Green, P.H.R. Hypervigilance to a gluten-free diet and decreased quality of life in teenagers and adults with celiac disease. *Dig. Dis. Sci.* **2018**, *63*, 1438–1448. [CrossRef]
26. Khoury, D.E.; Balfour-Ducharme, S.; Joye, I.J. A review on the gluten-free diet: Technological and nutritional challenges. *Nutrients* **2018**, *10*, 1410. [CrossRef] [PubMed]
27. Codex Standard 118-1979. Available online: http://www.fao.org/fao-who-codexalimentarius/sh-proxy/en/?lnk=1&url=https%253A%252F%252Fworkspace.fao.org%252Fsites%252Fcodex%252FStandards%252FCXS%2B118-1979%252FCXS_118e_2015.pdf (accessed on 7 January 2021).
28. Cohen, I.S.; Day, A.S.; Shaoul, R. Gluten in celiac disease-more or less? *Rambam Maimonides Med. J.* **2019**, *10*, e0007. [CrossRef] [PubMed]
29. Sbravati, F.; Pagano, S.; Retetangos, C.; Spisni, E.; Bolasco, G.; Labriola, F.; Filardi, M.C.; Grondona, A.G.; Alvisi, P. Adherence to gluten-free diet in a celiac pediatric population referred to the general pediatrician after remission. *J. Pediatr. Gastroenterol. Nutr.* **2020**, *71*, 78–82. [CrossRef] [PubMed]
30. Tovoli, F.; Negrini, G.; Sansone, V.; Faggiano, C.; Catenaro, T.; Bolondi, L.; Granito, A. Celiac disease diagnosed through screening programs in at-risk adults is not associated with worse adherence to the gluten-free diet and might protect from osteopenia/osteoporosis. *Nutrients* **2018**, *10*, 1940. [CrossRef] [PubMed]
31. Dana, Z.Y.; Lena, B.; Vered, R.; Haim, S.; Efrat, B. Factors associated with non adherence to a gluten free diet in adult with celiac disease: A survey assessed by BIAGI score. *Clin. Res. Hepatol. Gastroenterol.* **2020**, *44*, 762–767. [CrossRef] [PubMed]

32. Moreno, M.L.; Cebolla, Á.; Muñoz-Suano, A.; Carrillo-Carrion, C.; Comino, I.; Pizarro, Á.; León, F.; Rodríguez-Herrera, A.; Sousa, C. Detection of gluten immunogenic peptides in the urine of patients with coeliac disease reveals transgressions in the gluten-free diet and incomplete mucosal healing. *Gut* **2017**, *66*, 250–257. [PubMed]

33. Galli, G.; Esposito, G.; Lahner, E.; Pilozzi, E.; Corleto, V.D.; Di-Giulio, E.; Spiriti, M.A.; Annibale, B. Histological recovery and gluten-free diet adherence: A prospective 1-year follow-up study of adult patients with coeliac disease. *Aliment. Pharmacol. Ther.* **2014**, *40*, 639–647. [CrossRef]

34. Ruiz-Carnicer, A.; Garzón-Benavides, M.; Fombuena, B.; Segura, V.; García-Fernández, F.; Sobrino-Rodríguez, S.; Gómez-Izquierdo, L.; Montes-Cano, M.A.; Rodríguez-Herrera, A.; Millán, R.; et al. Negative predictive value of the repeated absence of gluten immunogenic peptides in the urine of treated celiac patients in predicting mucosal healing: New proposals for follow-up in celiac disease. *Am. J. Clin. Nutr.* **2020**, *112*, 1240–1251. [CrossRef]

35. Fernández-Bañares, F.; Beltrán, B.; Salas, A.; Comino, I.; Ballester-Clau, R.; Ferrer, C.; Molina-Infante, J.; Rosinach, M.; Modolell, I.; Rodríguez-Moranta, F.; et al. Persistent villous atrophy in de novo adult patients with celiac disease and strict control of gluten-free diet adherence: A multicenter prospective study (CADER Study). *Am. J. Gastoenterol.* **2021**, *116*, 1036–1043. [CrossRef]

36. Abu-Janb, N.; Jaana, M. Facilitators and barriers to adherence to gluten-free diet among adults with celiac disease: A systematic review. *J. Hum. Nutr. Diet* **2020**, *33*, 786–810. [CrossRef] [PubMed]

37. Caio, G.; Ciccocioppo, R.; Zoli, G.; Giorgio, R.D.; Volta, U. Therapeutic options for coeliac disease: What else beyond gluten-free diet? *Digest. Liver Dis.* **2020**, *52*, 130–137. [CrossRef] [PubMed]

38. Garg, A.; Gupta, R. Predictors of compliance to gluten-free diet in children with celiac disease. *Int. Sch. Res. Not.* **2014**, *2014*, 248402. [CrossRef] [PubMed]

39. Comino, I.; Fernández-Bañares, F.; Esteve, M.; Ortigosa, L.; Castillejo, G.; Fombuena, B.; Ribes-Koninckx, C.; Sierra, C.; Rodríguez-Herrera, A.; Salazar, J.C.; et al. Fecal gluten peptides reveal limitations of serological tests and food questionnaires for monitoring gluten-free diet in celiac disease patients. *Am. J. Gastroenterol.* **2016**, *111*, 1456–1465. [CrossRef] [PubMed]

40. Czaja-Bulsa, G.; Bulsa, M. Adherence to gluten-free diet in children with celiac disease. *Nutrients* **2018**, *10*, 1424. [CrossRef]

41. Gerasimidis, K.; Zafeiropoulou, K.; Mackinder, M.; Ijaz, U.Z.; Duncan, H.; Buchanan, E.; Cardigan, T.; Edwards, C.A.; McGrogan, P.; Russell, R.K. Comparison of clinical methods with the faecal gluten immunogenic peptide to assess gluten intake in coeliac disease. *J. Pediatr. Gastroenterol. Nutr.* **2018**, *67*, 356–360. [CrossRef]

42. Rimárová, K.; Dorko, E.; Diabelková, J.; Sulinová, Z.; Makovický, P.; Baková, J.; Uhrin, T.; Jenča, A.; Jenčová, J.; Petrášová, A.; et al. Compliance with gluten-free diet in a selected group of celiac children in the Slovak Republic. *Cent. Eur. J. Public Health* **2018**, *26*, S19–S24. [CrossRef]

43. Comino, I.; Segura, V.; Ortigosa, L.; Espín, B.; Castillejo, G.; Garrote, J.A.; Sierra, C.; Millán, A.; Ribes-Koninckx, C.; Román, E.; et al. Prospective longitudinal study: Use of faecal gluten immunogenic peptides to monitor children diagnosed with coeliac disease during transition to a gluten-free diet. *Aliment. Pharmacol. Ther.* **2019**, *49*, 1484–1492. [CrossRef]

44. Fernández-Miaja, M.; Díaz-Martín, J.J.; Jiménez-Treviño, S.; Suárez-González, M.; Bousoño-García, C. Estudio de la adherencia a la dieta sin gluten en pacientes celiacos [Study of adherence to the gluten-free diet in coeliac patients]. *An. Pediatr.* **2020**, *94*, 377–384. [CrossRef]

45. Roca, M.; Donat, E.; Masip, E.; Crespo-Escobar, P.; Cañada-Martínez, A.J.; Polo, B.; Ribes-Koninckx, C. Analysis of gluten immunogenic peptides in feces to assess adherence to the gluten-free diet in pediatric celiac patients. *Eur. J. Nutr.* **2020**, *60*, 2131–2140. [CrossRef]

46. Rodrigues, M.; Yonamine, G.H.; Fernandes-Satiro, C.A. Rate and determinants of nonadherence to a gluten-free diet and nutritional status assessment in children and adolescents with celiac disease in a tertiary Brazilian referral center: A cross-sectional and retrospective study. *BMC Gastroenterol.* **2018**, *18*, 15.

47. Zingone, F.; Massa, S.; Malamisura, B.; Pisano, P.; Ciacci, C. Coeliac disease: Factors affecting the transition and a practical tool for the transition to adult healthcare. *United Eur. Gastroenterol. J.* **2018**, *6*, 1356–1362. [CrossRef] [PubMed]

48. Johansson, K.; Norström, F.; Nordyke, K.; Myleus, A. Celiac dietary adherence test simplifies determining adherence to a gluten-free diet in swedish adolescents. *J. Pediatr. Gastroenterol. Nutr.* **2019**, *69*, 575–580. [CrossRef] [PubMed]

49. Villafuerte-Galvez, J.; Vanga, R.R.; Dennis, M.; Hansen, J.; Leffler, D.A.; Kelly, C.P.; Mukherjee, R. Factors governing long-term adherence to a gluten-free diet in adult patients with coeliac disease. *Aliment. Pharmacol. Ther.* **2015**, *42*, 753–760. [CrossRef] [PubMed]

50. Ramírez-Cervantes, K.L.; Romero-López, A.V.; Núñez-Álvarez, C.A.; Uscanga-Domínguez, L.F. Adherence to a gluten-free diet in mexican subjects with gluten-related ¡disorders: A high prevalence of inadvertent gluten intake. *Rev. Investig. Clin.* **2016**, *68*, 229–234.

51. Silvester, J.A.; Weiten, D.; Graff, L.A.; Walker, J.R.; Duerksen, D.R. Is it gluten-free? Relationship between self-reported gluten-free diet adherence and knowledge of gluten content of foods. *Nutrition* **2016**, *32*, 777–783. [CrossRef] [PubMed]

52. Muhammad, H.; Reeves, S.; Ishaq, S.; Mayberry, J.; Jeanes, Y.M. Adherence to a gluten free diet is associated with receiving gluten free foods on prescription and understanding food labelling. *Nutrients* **2017**, *9*, 705. [CrossRef] [PubMed]

53. Halmos, E.P.; Deng, M.; Knowles, S.R.; Sainsbury, K.; Mullan, B.; Tye-Din, J.A. Food knowledge and psychological state predict adherence to a gluten-free diet in a survey of 5310 Australians and New Zealanders with coeliac disease. *Aliment. Pharmacol. Ther.* **2018**, *48*, 78–86. [CrossRef]

54. Costa, A.F.; Sugai, E.; Temprano, M.P.; Niveloni, S.I.; Vázquez, H.; Moreno, M.L.; Domínguez-Flores, M.R.; Muñoz-Suano, A.; Smecuol, E.; Stefanolo, J.P.; et al. Gluten immunogenic peptide excretion detects dietary transgressions in treated celiac disease patients. *World J. Gastroenterol.* **2019**, *25*, 1409–1420. [CrossRef]

55. Paganizza, S.; Zanotti, R.; D'Odorico, A.; Scapolo, P.; Canova, C. Is adherence to a gluten-free diet by adult patients with celiac disease influenced by their knowledge of the gluten content of foods? *Gastroenterol. Nurs.* **2019**, *42*, 55–64. [CrossRef] [PubMed]
56. Fueyo-Díaz, R.; Magallón-Botaya, R.; Gascón-Santos, S.; Asensio-Martínez, Á.; Navarro, G.P.; Sebastián-Domingo, J.J. The effect of self-efficacy expectations in the adherence to a gluten free diet in celiac disease. *Psychol. Health* **2020**, *35*, 734–749. [CrossRef] [PubMed]
57. Silvester, J.A.; Comino, I.; Kelly, C.P.; Sousa, C.; Duerksen, D.R. The DOGGIE BAG study group. Most patients with celiac disease on gluten-free diets consume measurable amounts of gluten. *Gastroenterology* **2020**, *158*, 1497–1499.e1. [CrossRef] [PubMed]
58. Silvester, J.A.; Comino, I.; Rigaux, L.N.; Segura, V.; Green, K.H.; Cebolla, Á.; Weiten, D.; Dominguez, R.; Leffler, A.; Leon, F.; et al. Exposure sources, amounts and time course of gluten ingestion and excretion in patients with coeliac disease on a gluten-free diet. *Aliment. Pharmacol. Ther.* **2020**, *52*, 1469–1479. [CrossRef]
59. Tanner, G.J.; Blundell, M.J.; Colgrave, M.L.; Howitt, C.A. Creation of the first ultra-low gluten barley (*Hordeum vulgare* L.) for coeliac and gluten-intolerant populations. *Plant Biotechnol. J.* **2016**, *14*, 1139–1150. [CrossRef]
60. Sánchez-León, S.; Gil-Humanes, J.; Ozuna, C.V.; Giménez, M.J.; Sousa, C.; Voytas, D.F.; Barro, F. Low-gluten, nontransgenic wheat engineered with CRISPR/Cas9. *Plant Biotechnol. J.* **2018**, *16*, 902–910. [CrossRef]
61. Syage, J.A.; Green, P.H.R.; Khosla, C.; Adelman, D.C.; Sealey-Voyksner, J.A.; Murray, J.A. Latiglutenase treatment for celiac disease: Symptom and quality of life improvement for seropositive patients on a gluten-free diet. *GastroHep* **2019**, *1*, 293–301. [CrossRef]
62. Håkansson, Å.; Andrén-Aronsson, C.; Brundin, C.; Oscarsson, E.; Molin, G.; Agardh, D. Effects of Lactobacillus plantarum and Lactobacillus paracasei on the peripheral immune response in children with celiac disease autoimmunity: A randomized, double-blind, placebo-controlled clinical trial. *Nutrients* **2019**, *11*, 1925. [CrossRef]
63. Ruh, T.; Ohsam, J.; Pasternack, R.; Yokoyama, K.; Kumazawa, Y.; Hils, M. Microbial transglutaminase treatment in pasta-production does not affect the immunoreactivity of gliadin with celiac disease patients' sera. *J. Agric. Food Chem.* **2014**, *62*, 7604–7611. [CrossRef] [PubMed]
64. Liang, L.; Pinier, M.; Leroux, J.C.; Subirade, M. Interaction of alpha-gliadin with poly (HEMA-co-SS): Structural characterization and biological implication. *Biopolymers* **2009**, *91*, 169–178. [CrossRef]
65. Pinier, M.; Fuhrmann, G.; Galipeau, H.J.; Rivard, N.; Murray, J.A.; David, S.H.; Drasarova, H.; Tuckova, L.; Leroux, J.C.; Verdu, E. The copolymer P(HEMA-co-SS) binds gluten and reduces immune response in gluten-sensitized mice and human tissues. *Gastroenterology* **2012**, *142*, 316–325.e12. [CrossRef] [PubMed]
66. Sánchez-Valverde, F.V.; Denis, S.Z.; Etayo, V.E. Nuevas estrategias terapéuticas en la enfermedad celíaca. In *Enfermedad Celíaca: Presente y Future*; Allué, P., Ed.; Ergon: Madrid, Spain, 2013; pp. 127–134.
67. Chevallier, S.; Goeltz, P.; Thibault, P.; Banville, D.; Gagnon, J. Characterization of a prolyl endopeptidase from flavobacterium meningosepticum. complete sequence and localization of the active-site serine. *J. Biol. Chem.* **1992**, *267*, 8192–8199. [CrossRef]
68. Diefenthal, T.; Dargatz, H.; Witte, V.; Reipen, G.; Svendsen, I. Cloning of proline-specific endopeptidase gene from Flavobacterium meningosepticum: Expression in Escherichia coli and purification of the heterologous protein. *Appl. Microbiol. Biotechnol.* **1993**, *40*, 90–97. [CrossRef] [PubMed]
69. Shan, L.; Marti, T.; Sollid, L.M.; Gray, G.M.; Khosla, C. Comparative biochemical analysis of three bacterial prolyl endopeptidases: Implications for coeliac sprue. *Biochem. J.* **2004**, *383 Pt 2*, 311–318. [CrossRef]
70. Kabashima, T.; Fujii, M.; Meng, Y.; Ito, K.; Yoshimoto, T. Prolyl endopeptidase from *Sphingomonas capsulata*: Isolation and characterization of the enzyme and nucleotide sequence of the gene. *Arch. Biochem. Biophys.* **1998**, *358*, 141–148. [CrossRef]
71. Xiao, B.; Zhang, C.; Song, X.; Wu, M.; Mao, J.; Yu, R.; Zheng, Y. Rationally engineered prolyl endopeptidases from Sphingomonas capsulata with improved hydrolytic activity towards pathogenic peptides of celiac diseases. *Eur. J. Med. Chem.* **2020**, *15*, 112499. [CrossRef] [PubMed]
72. Knorr, V.; Wieser, H.; Koehler, P. Production of gluten-free beer by peptidase treatment. *Eur. Food Res. Technol.* **2016**, *242*, 1129–1140. [CrossRef]
73. Lähdeaho, M.L.; Kaukinen, K.; Laurila, K.; Vuotikka, P.; Koivurova, O.P.; Kärjä-Lahdensuu, T.; Marcantonio, A.; Adelman, D.C.; Mäki, M. Glutenase ALV003 attenuates gluten-induced mucosal injury in patients with celiac disease. *Gastroenterology* **2014**, *146*, 1649–1658. [CrossRef]
74. Osorio, C.E.; Wen, N.; Mejías, J.H.; Mitchell, S.; von Wettstein, D.; Rustgi, S. Directed-Mutagenesis of Flavobacterium meningosepticum prolyl-oligopeptidase and a glutamine-specific endopeptidase from barley. *Front. Nutr.* **2020**, *7*, 11. [CrossRef]
75. Darwish, G.; Helmerhorst, E.J.; Schuppan, D.; Oppenheim, F.G.; Wei, G. Pharmaceutically modified subtilisins withstand acidic conditions and effectively degrade gluten in vivo. *Sci. Rep.* **2019**, *9*, 7505. [CrossRef]
76. Lähdeaho, M.L.; Scheinin, M.; Vuotikka, P.; Taavela, J.; Popp, A.; Laukkarinen, J.; Koffert, J.; Koivurova, O.P.; Pesu, M.; Kivelä, L.; et al. Safety and efficacy of AMG 714 in adults with coeliac disease exposed to gluten challenge: A phase 2a, randomised, double-blind, placebo-controlled study. *Lancet Gastroenterol. Hepatol.* **2019**, *4*, 948–959. [CrossRef]
77. Kapoerchan, V.V.; Wiesner, M.; Hillaert, U.; Drijfhout, J.W.; Overhand, M.; Alard, P.; van der Marel, G.A.; Overkleeft, H.S.; Koning, F. Design, synthesis and evaluation of high-affinity binders for the celiac disease associated HLA-DQ2 molecule. *Mol. Immunol.* **2010**, *47*, 1091–1097. [CrossRef] [PubMed]
78. Tang, F.; Sally, B.; Lesko, K.; Discepolo, V.; Abadie, V.; Ciszewski, C.; Semrad, C.; Guandalini, S.; Kupfer, S.S.; Jabri, B. Cysteinyl leukotrienes mediate lymphokine killer activity induced by NKG2D and IL-15 in cytotoxic T cells during celiac disease. *J. Exp. Med.* **2015**, *212*, 1487–1495. [CrossRef] [PubMed]

79. Haghbin, M.; Rostami-Nejad, M.; Forouzesh, F.; Sadeghi, A.; Rostami, K.; Aghamohammadi, E.; Asadzadeh-Aghdaei, H.; Masotti, A.; Zali, M.R. The role of CXCR3 and its ligands CXCL10 and CXCL11 in the pathogenesis of celiac disease. *Medicine* **2019**, *98*, e15949. [CrossRef] [PubMed]

80. Cellier, C.; Bouma, G.; van Gils, T.; Khater, S.; Malamut, G.; Crespo, L.; Collin, P.; Green, P.H.R.; Crowe, S.E.; Tsuji, W.; et al. Safety and efficacy of AMG 714 in patients with type 2 refractory coeliac disease: A phase 2a, randomised, double-blind, placebo-controlled, parallel-group study. *Lancet Gastroenterol. Hepatol.* **2019**, *4*, 960–970. [CrossRef]

81. Benedetti, E.; Viscido, A.; Castelli, V.; Maggiani, C.; d'Angelo, M.; Di Giacomo, E.; Antonosante, A.; Picarelli, A.; Frieri, G. Mesalazine treatment in organotypic culture of celiac patients: Comparative study with gluten free diet. *J. Cell Physiol.* **2018**, *233*, 4383–4390. [CrossRef] [PubMed]

82. Leffler, D.A.; Kelly, C.P.; Green, P.H.; Fedorak, R.N.; DiMarin, A.; Perrow, W.; Rasmussen, H.; Wang, C.; Bercik, P.; Bachir, N.M.; et al. Larazotide acetate for persistent symptoms of celiac disease despite a gluten-free diet: A randomized controlled trial. *Gastroenterology* **2015**, *148*, 1311–1319.e6. [CrossRef]

83. Khaleghi, S.; Ju, J.M.; Lamba, A.; Murray, J.A. The potential utility of tight junction regulation in celiac disease: Focus on larazotide acetate. *Therap. Adv. Gastroenterol.* **2016**, *9*, 37–49. [CrossRef] [PubMed]

84. Daveson, A.J.M.; Ee, H.C.; Andrews, J.M.; King, T.; Goldstein, K.E.; Dzuris, J.L.; MacDougall, J.A.; Williams, L.J.; Treohan, A.; Cooreman, M.P.; et al. Epitope-specific immunotherapy targeting CD4-positive T cells in celiac disease: Safety, pharmacokinetics, and effects on intestinal histology and plasma cytokines with escalating dose regimens of NEXVAX2 in a randomized, double-blind, placebo-controlled phase 1 study. *EBioMedicine* **2017**, *26*, 78–90. [PubMed]

85. Goel, G.; King, T.; Daveson, A.J.; Andrews, J.M.; Krishnarajah, J.; Krause, R.; Brown, G.J.E.; Fogel, R.; Barish, C.F.; Epstein, R.; et al. Epitope-specific immunotherapy targeting CD4-positive T cells in coeliac disease: Two randomised, double-blind, placebo-controlled phase 1 studies. *Lancet Gastroenterol. Hepatol.* **2017**, *2*, 479–493. [CrossRef]

86. Database ClinicalTrials.gov. Available online: https://clinicaltrials.gov/ct2/show/NCT04530123 (accessed on 5 January 2021).

87. Database ClinicalTrials.gov. Available online: https://clinicaltrials.gov/ct2/show/NCT04248855 (accessed on 6 January 2021).

88. Croese, J.; Giacomin, P.; Navarro, S.; Clouston, A.; McCann, L.; Dougall, A.; Ferreira, I.; Susianto, A.; O'Rourke, P.; Howlett, M.; et al. Experimental hookworm infection and gluten microchallenge promote tolerance in celiac disease. *J. Allergy Clin. Immunol.* **2015**, *135*, 508–516. [CrossRef] [PubMed]

89. Freire, R.; Ingano, L.; Serena, G.; Cetinbas, M.; Anselmo, A.; Sapone, A.; Sadreyev, R.I.; Fasano, A.; Senger, S. Human gut derived-organoids provide model to study gluten response and effects of microbiota-derived molecules in celiac disease. *Sci. Rep.* **2019**, *9*, 7029. [CrossRef] [PubMed]

90. Cristofori, F.; Indrio, F.; Miniello, V.L.; De Angelis, M.; Francavilla, R. Probiotics in celiac disease. *Nutrients* **2018**, *10*, 1824. [CrossRef]

91. Ramedani, N.; Sharifan, A.; Gholam-Mostafaei, F.S.; Rostami-Nejad, M.; Yadegar, A.; Ehsani-Ardakani, M.J. The potentials of probiotics on gluten hydrolysis; a review study. *Gastroenterol. Hepatol. Bed Bench* **2020**, *13* (Suppl. 1), S1–S7.

92. Zanini, B.; Petroboni, B.; Not, T.; di Toro, N.; Villanacci, V.; Lanzarotto, F.; Pogna, N.; Ricci, C.; Lanzini, A. Search for atoxic cereals: A single blind, cross-over study on the safety of a single dose of Triticum monococcum, in patients with celiac disease. *BMC Gastroenterol.* **2013**, *13*, 92. [CrossRef]

93. Vaquero, L.; Comino, I.; Vivas, S.; Rodríguez-Martín, L.; Giménez, M.J.; Pastor, J.; Sousa, C.; Barro, F. Tritordeum: A novel cereal for food processing with good acceptability and significant reduction in gluten immunogenic peptides in comparison with wheat. *J. Sci. Food Agric.* **2018**, *98*, 2201–2209. [CrossRef]

94. Jouanin, A.; Gilissen, L.J.W.J.; Schaart, J.G.; Leigh, F.J.; Cockram, J.; Wallington, E.J.; Boyd, L.A.; van den Broeck, H.C.; van der Meer, I.M.; America, A.H.P.; et al. CRISPR/Cas9 gene editing of gluten in wheat to reduce gluten content and exposure-reviewing methods to screen for coeliac safety. *Front. Nutr.* **2020**, *7*, 51. [CrossRef]

95. Troncone, R.; Auricchio, R.; Granata, V. Issues related to gluten-free diet in coeliac disease. *Curr. Opin. Clin. Nutr. Metab. Care* **2008**, *11*, 329–333. [CrossRef]

96. Barro, F.; Lehisa, J.C.; Giménez, M.J.; García-Molina, M.D.; Ozun, C.V.; Comino, I.; Sousa, C.; Gil-Humanes, J. Targeting of prolamins by RNAi in bread wheat: Effectiveness of seven silencing-fragment combinations for obtaining lines devoid of coeliac disease epitopes from highly immunogenic gliadins. *Plant Biotechnol. J.* **2016**, *14*, 986–996. [CrossRef]

97. Haro, C.; Villatoro, M.; Vaquero, L.; Pastor, J.; Giménez, M.J.; Ozuna, C.V.; Sánchez-León, S.; García-Molina, M.D.; Segura, V.; Comino, I.; et al. The dietary intervention of transgenic low-gliadin wheat bread in patients with non-celiac gluten sensitivity (NCGS) showed no differences with gluten free diet (GFD) but provides better gut microbiota profile. *Nutrients* **2018**, *10*, 1964. [CrossRef]

98. Picozzi, C.; Mariotti, M.; Cappa, C.; Tedesco, B.; Vigentini, I.; Foschino, R.; Lucisano, M. Development of a Type I gluten-free sourdough. *Lett. Appl. Microbiol.* **2016**, *62*, 119–125. [CrossRef]

99. Kõiv, V.; Tenson, T. Gluten-degrading bacteria: Availability and applications. *Appl. Microbiol. Biotechnol.* **2021**, *105*, 3045–3059. [CrossRef]

100. Francavilla, R.; Piccolo, M.; Francavilla, A.; Polimeno, L.; Semeraro, F.; Cristofori, F.; Castellaneta, S.; Barone, M.; Indrio, F.; Gobbetti, M.; et al. Clinical and microbiological effect of a multispecies probiotic supplementation in celiac patients with persistent ibs-type symptoms: A randomized, double-blind, placebo-controlled, multicenter trial. *J. Clin. Gastroenterol.* **2019**, *53*, e117–e125. [CrossRef] [PubMed]

101. De Angelis, M.; Rizzello, C.G.; Fasano, A.; Clemente, M.G.; De Simone, C.; Silano, M.; De Vincenzi, M.; Losito, I.; Gobbetti, M. VSL#3 probiotic preparation has the capacity to hydrolyze gliadin polypeptides responsible for Celiac Sprue. *Biochim. Biophys. Acta* **2006**, *1762*, 80–93.

102. Harnett, J.; Myers, S.P.; Rolfe, M. Probiotics and the microbiome in celiac disease: A Randomised Controlled Trial. *Evid. Based Complement. Alternat. Med.* **2016**, *2016*, 9048574. [CrossRef]

103. Vaquero, L.; Rodríguez-Martín, L.; León, F.; Jorquera, F.; Vivas, S. Nuevas terapias en la enfermedad celiaca y sus complicaciones. *Gastroenterol. Hepatol.* **2018**, *41*, 191–204. [CrossRef]

104. Aaron, L.; Torsten, M. Microbial transglutaminase: A new potential player in celiac disease. *Clin. Immunol.* **2019**, *199*, 37–43. [CrossRef]

105. Zhou, G.; Sprengers, D.; Boor, P.P.C.; Doukas, M.; Schutz, H.; Mancham, S.; Pedroza-Gonzalez, A.; Polak, W.G.; de Jonge, J.; Gaspersz, M.; et al. Antibodies Against Immune Checkpoint Molecules Restore Functions of Tumor-Infiltrating T Cells in Hepatocellular Carcinomas. *Gastroenterology* **2017**, *153*, 1107–1119.e10. [CrossRef]

106. McCarville, J.L.; Nisemblat, Y.; Galipeau, H.J.; Jury, J.; Tabakman, R.; Cohen, A.; Naftali, E.; Neiman, B.; Halbfinger, E.; Murray, J.A.; et al. BL-7010 demonstrates specific binding to gliadin and reduces gluten-associated pathology in a chronic mouse model of gliadin sensitivity. *PLoS ONE* **2014**, *9*, e109972. [CrossRef]

107. Database ClinicalTrials.gov. Available online: https://clinicaltrials.gov/ct2/show/NCT03707730 (accessed on 10 March 2021).

108. Stadlmann, V.; Harant, H.; Korschineck, I.; Hermann, M.; Forster, F.; Missbichler, A. Novel avian single-chain fragment variable (scFv) targets dietary gluten and related natural grain prolamins, toxic entities of celiac disease. *BMC Biotechnol.* **2015**, *15*, 109. [CrossRef] [PubMed]

109. Moreno, M.L.; Arévalo-Rodríguez, M.; Mellado, E.; Martínez-Reyes, J.C.; Sousa, C. A new microbial gluten-degrading prolyl endopeptidase: Potential application in celiac disease to reduce gluten immunogenic peptides. *PLoS ONE* **2019**, *14*, e0218346. [CrossRef] [PubMed]

110. Krishnareddy, S.; Stier, K.; Recanati, M.; Lebwohl, B.; Green, P.H. Commercially available glutenases: A potential hazard in coeliac disease. *Therap. Adv. Gastroenterol.* **2017**, *10*, 473–481. [CrossRef]

111. Moreno, M.L.; Sánchez-Muñoz, D.; Sanders, D.; Rodríguez-Herrera, A.; Sousa, C. Verifying diagnosis of refractory celiac disease with urine gluten immunogenic peptides as biomarker. *Front. Med.* **2021**, *7*, 601854. [CrossRef]

112. Tian, N.; Wei, G.; Schuppan, D.; Helmerhorst, E.J. Effect of Rothia mucilaginosa enzymes on gliadin (gluten) structure, deamidation, and immunogenic epitopes relevant to celiac disease. *Am. J. Physiol. Gastrointest. Liver Physiol.* **2014**, *307*, G769–G776. [CrossRef]

113. Socha, P.; Mickowska, B.; Urminská, D.; Kacmárová, K. The use of different proteases to hydrolyze gliadins. *J. Microbiol. Biotechnol. Food Sci.* **2015**, *4*, 101–104. [CrossRef]

114. Wei, G.; Tian, N.; Valery, A.C.; Zhong, Y.; Schuppan, D.; Helmerhorst, E.J. Identification of pseudolysin (lasB) as an aciduric gluten-degrading enzyme with high therapeutic potential for celiac disease. *Am. J. Gastroenterol.* **2015**, *110*, 899–908. [CrossRef]

115. Wagh, S.K.; Gadge, P.P.; Padul, M.V. Significant hydrolysis of wheat gliadin by bacillus tequilensis (10bT/HQ223107): A Pilot Study. *Probiotics Antimicrob. Proteins.* **2018**, *10*, 662–667. [CrossRef]

116. Rashmi, B.S.; Gayathri, D.; Vasudha, M.; Prashantkumar, C.S.; Swamy, C.T., Sunil, K.S.; Somaraja, P.K.; Prakash, P. Gluten hydrolyzing activity of Bacillus spp isolated from sourdough. *Microb. Cell Factories* **2020**, *19*, 130. [CrossRef]

117. Cavaletti, L.; Taravella, A.; Carrano, L.; Carenzi, G.; Sigurta, A.; Solinas, N.; De Caro, S.; Di Stasio, L.; Piscascia, S.; Laezza, M.; et al. E40, a novel microbial protease efficiently detoxifying gluten proteins, for the dietary management of gluten intolerance. *Sci. Rep.* **2019**, *9*, 13147. [CrossRef]

118. Ehren, J.; Morón, B.; Martin, E.; Bethune, M.T.; Gray, G.M.; Khosla, C. A foodgrade enzyme preparation with modest gluten detoxification properties. *PLoS ONE* **2009**, *4*, e6313. [CrossRef]

119. Janssen, G.; Christis, C.; Kooy-Winkelaar, Y.; Edens, L.; Smith, D.; van Veelen, P.; Koning, F. Ineffective degradation of immunogenic gluten epitopes by currently available enzyme supplements. *PLoS ONE* **2015**, *10*, e0128065. [CrossRef] [PubMed]

120. Bethune, M.T.; Ribka, E.; Khosla, C.; Sestak, K. Transepithelial transport and enzymatic detoxification of gluten in gluten-sensitive rhesus macaques. *PLoS ONE* **2008**, *3*, e1857. [CrossRef] [PubMed]

121. Cornell, H.J.; Doherty, W.; Stelmasiak, T. Papaya latex enzymes capable of detoxification of gliadin. *Amino Acids* **2009**, *38*, 155–165. [CrossRef]

122. Savvateeva, L.V.; Gorokhovets, N.V.; Makarov, V.A.; Serebryakova, M.V.; Solovyev, A.G.; Morozov, S.Y.; Reddy, V.P.; Zernii, E.Y.; Zamyatnin, A.A., Jr.; Aliev, G. Glutenase and collagenase activities of wheat cysteine protease Triticain-α: Feasibility for enzymatic therapy assays. *Int. J. Biochem. Cell Biol.* **2015**, *62*, 115–124. [CrossRef]

123. Mika, N.; Zorn, H.; Rühl, M. Prolyl-specific peptidases for applications in food protein hydrolysis. *Appl. Microbiol. Biotechnol.* **2015**, *99*, 7837–7846. [CrossRef]

124. Tereshchenkova, V.F.; Goptar, I.A.; Zhuzhikov, D.P.; Belozersky, M.A.; Dunaevsky, Y.E.; Oppert, B.; Filippova, I.Y.; Elpidina, E.N. Prolidase is a critical enzyme for complete gliadin digestion in Tenebrio molitor larvae. *Arch. Insect Biochem. Physiol.* **2017**, *95*, e21395. [CrossRef]

125. Murray, J.A.; Kelly, C.P.; Green, P.H.R.; Marcantonio, A.; Wu, T.T.; Mäki, M.; Adelman, D.C.; CeliAction Study Group of Investigators. No difference between latiglutenase and placebo in reducing villous atrophy or improving symptoms in patients with symptomatic celiac disease. *Gastroenterology* **2017**, *152*, 787–798.e2. [CrossRef]

126. Syage, J.A.; Murray, J.A.; Green, P.H.R.; Khosla, C. Latiglutenase improves symptoms in seropositive celiac disease patients while on a gluten-free diet. *Dig. Dis. Sci.* **2017**, *62*, 2428–2432. [CrossRef]

127. Database ClinicalTrials.gov. Available online: https://clinicaltrials.gov/ct2/show/NCT03585478 (accessed on 13 March 2021).

128. Mitea, C.; Havenaar, R.; Drijfhout, J.W.; Edens, L.; Dekking, L.; Koning, F. Efficient degradation of gluten by a prolyl endoprotease in a gastrointestinal model: Implications for coeliac disease. *Gut* **2008**, *57*, 25–32. [CrossRef]

129. Xia, J.; Bergseng, E.; Fleckenstein, B.; Siegel, M.; Kim, C.Y.; Khosla, C.; Sollid, L.M. Cyclic and dimeric gluten peptide analogues inhibiting DQ2-mediated antigen presentation in celiac disease. *Bioorg. Med. Chem.* **2007**, *15*, 6565–6573. [CrossRef]

130. Esposito, C.; Caputo, I.; Troncone, R. New therapeutic strategies for coeliac disease: Tissue transglutaminase as a target. *Curr. Med. Chem.* **2007**, *14*, 2572–2580. [CrossRef]

131. Alhassan, E.; Yadav, A.; Kelly, C.P.; Mukherjee, R. Novel nondietary therapies for celiac disease. *Cell Mol. Gastroenterol. Hepatol.* **2019**, *8*, 335–345. [CrossRef]

132. Palanski, B.A.; Khosla, C. Cystamine and disulfiram inhibit human transglutaminase 2 via an oxidative mechanism. *Biochemistry* **2018**, *57*, 3359–3363. [CrossRef]

133. Database ClinicalTrials.gov. Available online: https://grantome.com/grant/NIH/R01-DK100619-01A1 (accessed on 1 April 2021).

134. Sollid, L.M.; Khosla, C. Novel therapies for coeliac disease. *J. Intern. Med.* **2011**, *269*, 604–613. [CrossRef]

135. Database ClinicalTrials.gov. Available online: https://clinicaltrials.gov/ct2/show/NCT04593251 (accessed on 1 April 2021).

136. Reinisch, W. How to manage loss of response to anti-TNF in Crohn's disease? *Curr. Drug Targets* **2010**, *11*, 152–155. [CrossRef] [PubMed]

137. Phan-Ba, R.; Lambinet, N.; Louis, E.; Delvenne, P.; Tshibanda, L.; Boverie, J.; Moonen, G.; Belachew, S. Natalizumab to kill two birds with one stone: A case of celiac disease and multiple sclerosis. *Inflamm. Bowel Dis.* **2011**, *17*, E62–E63. [CrossRef] [PubMed]

138. Rath, T.; Billmeier, U.; Ferrazzi, F.; Vieth, M.; Ekici, A.; Neurath, F.M.; Atreya, R. Effects of anti-integrin treatment with vedolizumab on immune pathways and cytokines in inflammatory bowel diseases. *Front. Immunol.* **2018**, *9*, 1700. [CrossRef] [PubMed]

139. Saruta, M.; Yu, Q.T.; Avanesyan, A.; Fleshner, P.R.; Targan, S.R.; Papadakis, K.A. Phenotype and effector function of CC chemokine receptor 9-expressing lymphocytes in small intestinal Crohn's disease. *J. Immunol.* **2007**, *178*, 3293–3300. [CrossRef] [PubMed]

140. Walters, M.J.; Wang, Y.; Lai, N.; Baumgart, T.; Zhao, B.N.; Dairaghi, D.J.; Bekker, P.; Ertl, L.S.; Penfold, M.E.; Jaen, J.C.; et al. Characterization of CCX282-B, an orally bioavailable antagonist of the CCR9 chemokine receptor, for treatment of inflammatory bowel disease. *J. Pharmacol. Exp. Ther.* **2010**, *335*, 61–69. [CrossRef]

141. Kivelä, L.; Caminero, A.; Leffler, D.A.; Pinto-Sanchez, M.I.; Tye-Din, J.A.; Lindfors, K. Current and emerging therapies for coeliac disease. *Nat. Rev. Gastroenterol. Hepatol.* **2021**, *18*, 181–195. [CrossRef] [PubMed]

142. Gopalakrishnan, S.; Tripathi, A.; Tamiz, A.P.; Alkan, S.S.; Pandey, N.B. Larazotide acetate promotes tight junction assembly in epithelial cells. *Peptides* **2012**, *35*, 95–101. [CrossRef] [PubMed]

143. Gopalakrishnan, S.; Durai, M.; Kitchens, K.; Tamiz, A.P.; Somerville, R.; Ginski, M.; Paterson, B.M.; Murray, J.A.; Verdu, E.F.; Alkan, S.S.; et al. Larazotide acetate regulates epithelial tight junctions in vitro and in vivo. *Peptides* **2012**, *35*, 86–94. [CrossRef] [PubMed]

144. Serena, G.; Kelly, C.P.; Fasano, A. Nondietary therapies for celiac disease. *Gastroenterol. Clin. N. Am.* **2019**, *48*, 145–163. [CrossRef] [PubMed]

145. Tye-Din, J.A.; Stewart, J.A.; Dromey, J.A.; Beissbarth, T.; van Heel, D.A.; Tatham, A.; Henderson, K.; Mannering, S.I.; Gianfrani, C.; Jewell, D.P.; et al. Comprehensive, quantitative mapping of T cell epitopes in gluten in celiac disease. *Sci. Transl. Med.* **2010**, *2*, 41ra51. [CrossRef] [PubMed]

146. Database ClinicalTrials.gov. Available online: https://clinicaltrials.gov/ct2/show/NCT03738475 (accessed on 10 April 2021).

147. Daveson, A.J.; Jones, D.M.; Gaze, S.; McSorley, H.; Clouston, A.; Pascoe, A.; Cooke, S.; Speare, R.; Macdonald, G.A.; Anderson, R.; et al. Effect of hookworm infection on wheat challenge in celiac disease-a randomised double-blinded placebo-controlled trial. *PLoS ONE* **2011**, *6*, e17366. [CrossRef] [PubMed]

148. Chibbar, R.; Dieleman, L.A. The gut microbiota in celiac disease and probiotics. *Nutrients* **2019**, *11*, 2375. [CrossRef]

149. Caminero, A.; Verdu, E.F. Metabolism of wheat proteins by intestinal microbes: Implications for wheat related disorders. *Gastroenterol. Hepatol.* **2019**, *42*, 449–457. [CrossRef]

150. Palma, M.L.; Garcia-Bates, T.M.; Martins, F.S.; Douradinha, B. Genetically engineered probiotic Saccharomyces cerevisiae strains mature human dendritic cells and stimulate Gag-specific memory CD8+ T cells ex vivo. *Appl. Microbiol. Biotechnol.* **2019**, *103*, 5183–5192. [CrossRef]

151. Francavilla, R.; Cristofori, F.; Vacca, M.; Barone, M.; De Angelis, M. Advances in understanding the potential therapeutic applications of gut microbiota and probiotic mediated therapies in celiac disease. *Expert. Rev. Gastroenterol. Hepatol.* **2020**, *14*, 323–333. [CrossRef] [PubMed]

152. Strimbu, K.; Tavel, J.A. What are biomarkers? *Curr. Opin. HIV AIDS* **2010**, *5*, 463–466. [CrossRef]

153. Gottlieb, K.; Dawson, J.; Hussain, F.; Murray, J.A. Development of drugs for celiac disease: Review of endpoints for Phase 2 and 3 trials. *Gastroenterol. Rep.* **2015**, *3*, 91–102. [CrossRef] [PubMed]

Article

Long-Term Effect of a Gluten-Free Diet on Diarrhoea- or Bloating-Predominant Functional Bowel Disease: Role of the 'Low-Grade Coeliac Score' and the 'Coeliac Lymphogram' in the Response Rate to the Diet

Fernando Fernández-Bañares [1,2,*], Beatriz Arau [1,2], Agnès Raga [1], Montserrat Aceituno [1,2], Eva Tristán [1,2], Anna Carrasco [1,2], Laura Ruiz [1,2], Albert Martín-Cardona [1,2], Pablo Ruiz-Ramírez [1] and Maria Esteve [1,2]

[1] Department of Gastroenterology, Hospital Universitari Mutua Terrassa, 08221 Terrassa, Spain; beatrizarau@mutuaterrassa.es (B.A.); araga@mutuaterrassa.cat (A.R.); maceituno@mutuaterrassa.es (M.A.); etristan@mutuaterrassa.cat (E.T.); anna.carrasco.garcia@gmail.com (A.C.); lruiz@mutuaterrassa.es (L.R.); albertmartin@mutuaterrassa.cat (A.M.-C.); pruiz@mutuaterrassa.es (P.R.-R.); mariaesteve@mutuaterrassa.cat (M.E.)

[2] Centro de Investigación Biomédica en Red de Enfermedades Hepáticas y Digestivas (CIBERehd), Instituto de Salud Carlos III, 28029 Madrid, Spain

* Correspondence: ffbanares@mutuaterrassa.es

Citation: Fernández-Bañares, F.; Arau, B.; Raga, A.; Aceituno, M.; Tristán, E.; Carrasco, A.; Ruiz, L.; Martín-Cardona, A.; Ruiz-Ramírez, P.; Esteve, M. Long-Term Effect of a Gluten-Free Diet on Diarrhoea- or Bloating-Predominant Functional Bowel Disease: Role of the 'Low-Grade Coeliac Score' and the 'Coeliac Lymphogram' in the Response Rate to the Diet. *Nutrients* **2021**, *13*, 1812. https://doi.org/10.3390/nu13061812

Academic Editors: David S. Sanders and Giacomo Caio

Received: 30 March 2021
Accepted: 25 May 2021
Published: 26 May 2021

Publisher's Note: MDPI stays neutral with regard to jurisdictional claims in published maps and institutional affiliations.

Abstract: 1. Background: The long-term effect of a gluten-free diet (GFD) on functional bowel disorders (FBDs) has been scarcely studied. The aim was to assess the effect of a GFD on FBD patients, and to assess the role of both the low-grade coeliac score and coeliac lymphogram in the probability of response to a GFD. 2. Methods: 116 adult patients with either predominant diarrhoea or abdominal bloating, fulfilling Rome IV criteria of FBD, were treated with a GFD. Duodenum biopsies were performed for both pathology studies and intraepithelial lymphocyte subpopulation patterns. Coeliac lymphogram was defined as an increase in $TCR\gamma\delta^+$ cells plus a decrease in $CD3^-$ cells. A low-grade coeliac score >10 was considered positive. 3. Results: Sustained response to GFD was observed in 72 patients (62%) after a median of 21 months of follow-up, who presented more often with coeliac lymphogram (37.5 vs. 11.4%; $p = 0.02$) and a score >10 (32 vs. 11.4%; $p = 0.027$) compared to non-responders. The frequency of low-grade coeliac enteropathy was 19.8%. 4. Conclusion: A GFD is effective in the long-term treatment of patients with previously unexplained chronic watery diarrhoea- or bloating-predominant symptoms fulfilling the criteria of FBD. The response rate was much higher in the subgroup of patients defined by the presence of both a positive low-grade coeliac score and coeliac lymphogram.

Keywords: functional bowel disease; gluten-free diet; coeliac disease; tissue biomarkers; non-coeliac gluten sensitivity; FODMAP diet

1. Introduction

Functional bowel disorders (FBDs) are a subset of a larger family of functional gastrointestinal disorders and are associated with chronic symptoms such as abdominal pain, bloating, diarrhoea, and constipation [1]. Similar to other functional disorders, FBDs have no identifiable structural or biochemical abnormalities that can account for their defining symptoms. Diagnosis is, therefore, based on reported symptoms and physical examination, in accordance with the Rome IV criteria, which are the most widely accepted standard for such symptom-based diagnoses. FBDs include, among others, irritable bowel syn drome (IBS), functional diarrhoea, and functional abdominal bloating/distention. In addition, IBS subtypes are defined in Rome IV based on the typical type of stool consistency abnormality (diarrhoea, constipation, and mixed).

Non-coeliac gluten sensitivity (NCGS), which is sometimes referred to as gluten sensitivity, gluten intolerance, or non-coeliac wheat sensitivity, is characterized by intestinal and extra-intestinal symptoms related to the ingestion of gluten-containing food in subjects that are not affected by either coeliac disease (CD) or wheat allergy. This is the original definition based on the Salerno Experts' Criteria [2]. At present, however, it is recognized that symptoms occur due not just to the ingestion of gluten proteins but potentially other wheat-related components, such as fructans [3,4]. Patients with NCGS have clinical symptoms that are indistinguishable from an IBS-like clinical picture. Conversely, recent studies support the hypothesis that gluten and other wheat components may trigger IBS symptoms [5]. In fact, diet has always played a significant role in IBS, with approximately two thirds of patients developing symptoms soon after the ingestion of food [6,7]. Recent research on diet therapy in IBS has focused on the role of a diet low in fermentable oligo-, di-, and mono-saccharides and polyols (FODMAPs) and wheat-free and gluten-free diets (GFD) [3,5]. Low-FODMAP diets are characterized by the elimination of wheat, barley, spelt, rye, and all other gluten containing cereals, as these cereals also contain fructans, which, as mentioned above, may be responsible for triggering IBS-related symptoms. In fact, a GFD has been proposed as a 'bottom up' approach to reducing fructan intake in a low-FODMAP diet [8]. The effect of GFD on patients with symptoms suggestive of IBS has been studied in a number of trials that were limited by small sample sizes and a short study duration, with an overall efficacy ranging from 34 to 71% [9,10]. Double-blind placebo-controlled trials evaluating the role of gluten reintroduction in patients with IBS and symptoms controlled on a GFD have recently been reviewed [3].

It is also known that CD patients may present with IBS-like symptoms. Performing a differential diagnosis between CD and NCGS is sometimes difficult and is especially challenging in cases with low-grade coeliac enteropathy in which CD serology is generally negative. Low-grade coeliac enteropathy lies in the milder range of the CD spectrum and was previously referred to with several different terms, including 'coeliac-light', 'coeliac-lite', 'coeliac trait', 'mild enteropathy coeliac disease', and 'low-grade gluten sensitive enteropathy' [11–16]. We have shown that a blinded gluten challenge in these patients was associated with a significantly higher clinical relapse rate and a deterioration in quality of life as compared with placebo, reinforcing the role of gluten in the pathogenesis of this mild enteropathy [17]. In addition, we derived a scoring system that identifies patients with coeliac characteristics likely to respond to a GFD and to be diagnosed with low-grade coeliac enteropathy with an AUC value of 0.91 [18]. This score uses data on coeliac serology, coeliac genetics (HLA-DQ2/8), and the number of intraepithelial lymphocytes (IEL) and CD3$^+$ T-cell receptor gamma-delta$^+$ cells (TCR$\gamma\delta^+$ cells) in duodenal mucosa. In addition, coeliac lymphogram, which is defined as an increase in TCR$\gamma\delta^+$ IEL plus the additional concomitant decrease in CD3$^-$ cells, adds specificity to the IEL assay [18–22]. It has been described that the number of TCR$\gamma\delta^+$ IELs is only elevated in CD subjects, while in NCGS patients, the number of TCR$\gamma\delta^+$ IELs is similar to that in controls [23].

The aim of the present study was to evaluate the long-term response rate to a GFD in patients with symptoms suggestive of either diarrhoea- or bloating-predominant FBD and to assess whether or not a low-grade coeliac score value >10 and the presence of coeliac lymphogram increases the probability of response to the diet.

2. Materials and Methods

From April 2010 to December 2017, all patients from whom duodenal biopsies were taken to rule out CD were prospectively recorded. The indications for duodenal biopsy sampling were long-standing gastro-intestinal or extra-intestinal symptoms suggestive of CD and/or positive coeliac serology. In addition, most patients were referred for duodenal biopsies on the additional basis of positive HLA-DQ2.5/8.

In the present study, we included consecutive patients recorded in that database based on the following inclusion criteria: (1) age 18 years or over; (2) fulfilling Rome IV criteria of FBD (IBS-D, functional diarrhoea, or functional abdominal bloating); (3) undergoing duodenal biopsies performed while on a gluten-containing diet for both pathology and flow cytometry studies; (4) starting a GFD for FBD symptom control; (5) a follow-up after starting GFD longer than six months to reduce the possibility of a placebo response. Patients were excluded if they had: (1) coeliac disease with atrophy; (2) positive coeliac serology (IgA anti-tissular transglutaminase antibodies—anti-tTG-), even those with anti-tTG borderline titres, defined as those detectable but below the manufacturer cut-off, who had positive IgA anti-endomysial antibodies (EmA); (3) inflammatory bowel disease; (4) microscopic colitis; (5) other enteropathies (olmesartan, giardiasis, etc.).

Demographic data, clinical presentation, coeliac serology (anti-tTG and EmA if indicated), coeliac genetics (HLA-DQ2.5/2.2/8), duodenal histology, IEL count, percentage of TCRγδ$^+$ and CD3$^-$ cells, and low-grade coeliac score were recorded for all included patients. A retrospective review of the medical records of all these patients to assess the response rate to a GFD was performed. A GFD was administered on the criteria of the physician in charge. Assessment of diet compliance was performed by a dietician when in the 3-month follow-up visit, there was a suspicion of non-adherence following a direct clinical interview with the patient. Afterwards, visits were at 6 months and after that every year during follow-up.

2.1. Clinical Response to a GFD

Response to GFD was defined as the sustained complete resolution of symptoms for more than six months, and renewed symptom relapse with inadvertent exposures to gluten-containing foods. In patients with chronic watery diarrhoea, defined as three or more liquid stools per day at least three days in a week, response to a GFD was considered as the complete resolution of diarrhoea. In the case of abdominal bloating, defined as symptoms of bloating and/or distention occurring either daily or at least 3 days a week, being the predominant symptoms in the past 3 months, response to the diet was defined as the sustained complete disappearance of bloating and/or distension. Non-responders to the diet were defined as those with persisting symptoms after a six-week GFD. Partial clinical responses to a GFD were considered as failures considering the retrospective nature of the study and the impossibility to quantitatively measure the response. Therefore, we considered as response only the absence of symptoms, i.e., a clear a meaningful clinical improvement. This response should be maintained at least for 6 months to consider response to a GFD.

2.2. Coeliac Serology

Serum IgA anti-tTG (or IgG anti-tTG in IgA deficient patients) was analysed using homologated commercial quantitative automated ELISAs, while the patients were on a gluten-containing diet. As mentioned, patients with anti-tTG titres that were detectable but below the cut-off suggested by the manufacturer were tested for EmA and included only if negative. Serum EmA was performed by indirect immunofluorescence assay in serum samples at 1:5 dilution (commercial sections of monkey distal oesophagus; BioMedical Diagnostics, Marne-la-Vallée, France). Total serum IgA was measured using rate nephelometry (BN II, Siemens Healthcare Diagnostics SL, Marburg, Germany).

2.3. Histological Studies

Two endoscopic biopsies from the bulb and four from the second portion of the duodenum were obtained and placed in separate vials in the index endoscopy for standard histological studies while patients were on a gluten-containing diet. Duodenal samples were processed using haematoxylin/eosin staining and CD3 immunophenotyping. Lymphocytic enteritis was considered as an IEL count of >25 IELs per 100 epithelial nuclei and normal villous architecture.

2.4. Flow Cytometry

For IEL flow cytometry, one single duodenal biopsy from the second portion of the duodenum was obtained in the index endoscopy and processed immediately as previously described [17,22]. The results of the flow cytometry were obtained in four hours. Coeliac lymphogram was then defined as an increase in TCR$\gamma\delta^+$ cells >8.5% plus a concomitant decrease in CD3$^-$ cells <10%. There were four intraepithelial lymphocyte patterns: a normal pattern, an isolated decrease in CD3$^-$, an isolated increase in TCR$\gamma\delta^+$, and the coeliac lymphogram (an increase in TCR$\gamma\delta^+$ plus a decrease in CD3$^-$). A brief methodological description of the procedures is provided in Appendix A.

2.5. Coeliac Genetics

Methods of assessment of coeliac genetics are described in Appendix A.

2.6. Low-Grade Coeliac Score and Definition of Low-Grade Coeliac Enteropathy

The low-grade coeliac score was calculated as described previously (Table 1) [18]. We use a cut-off >10 points for positive scores. In the present study, in which all included patients had negative coeliac serology, the score ranged from −2 to 17 points. Low-grade coeliac enteropathy was defined as both a score >10 and a long-term clinical response to a GFD.

Table 1. The low-grade coeliac scoring system (−2 to 25 points): a score >10 points is considered positive [18].

Predictors	Points
Serum anti-tTG2	
>20 U/mL	10
>8–20 U/mL or >2–8 U/mL plus EmA+	6
>2 to 8 U/mL plus EmA-	2
2 U/mL	0
IEL cytometry pattern	
↑TCR$\gamma\delta^+$ cells	7
Histology (IEL count)	
>25%	5
19–25%	0
<19%	−1
Coeliac genetics:	
DQ2.5+	3
DQ8+/DQ2.2+/Allele DQB1 of haplotype DQ2.5+	0
2 alleles DQ2.5- and DQ8-	−1

Serum anti-tTG2: IgA anti-transglutaminase antibodies; EmA: IgA anti-endomysium antibodies; IEL: intraepithelial lymphocytes.

The low-grade coeliac score includes among its items the increase in TCR$\gamma\delta^+$ cells, either isolated or with the concomitant decrease in CD3$^-$. In the present study, we analyse the GFD response rate in patients with a positive score comparing both IEL cytometry patterns.

2.7. Statistical Analysis

Results are expressed as mean ± SEM and as proportions. Chi-square statistics were used to compare qualitative variables, and either the Student t test or an analysis of variance was used to compare quantitative variables. Statistical calculations were performed using the SPSS for Windows statistical package (SPSS Inc., Chicago, IL, USA). Statistical significance was predetermined as $p < 0.05$. The study SPSS database can be found as Supplementary Material.

2.8. Ethical Issues

The study was conducted in accordance with the Declaration of Helsinki, and the protocol for the prospective registry was approved by the Ethics Committee of the Hospital Universitari MútuaTerrassa at the start of the registry in 2010 (Code: EO/1011; date: 25-03-2010). All participants provided informed consent for that. Since the assessment of GFD response was a retrospective, non-interventional medical record review, informed consent was not requested from patients. Researchers guaranteed strict measures for preserving patient confidentiality. The Ethics and Research Committee of the Hospital Universitari Mútua Terrassa was informed of the conduct of the medical record review.

3. Results

During the study period, a duodenal biopsy to rule out CD was performed in 260 patients with FBD, of whom 116 had been treated with a GFD. Eighty-four per cent were HLA-DQ2.5/DQ8/DQ2.2 positive, 44% presented with an IEL coeliac pattern, and 25% presented with a low-grade coeliac score >10. Three (2.6%) patients had detectable anti-tTG titres with negative EmA. As compared to the total sample of 260 patients, the frequency of an IEL count > 25%, an IEL coeliac pattern, and a score >10 was significantly higher in the subsample of patients on a GFD (see Appendix B: Table A1).

3.1. Response to Gluten-Free Diet

Clinical response to a GFD was observed in 72 of the 116 patients (62%; 95% CI, 53 to 70%), which was sustained after a median follow-up of 21 months (IQR, 12 to 36). These patients presented more often with the coeliac lymphogram pattern (37.5 vs. 11%; $p = 0.02$) and/or a score >10 (32 vs. 14%; $p = 0.027$) as compared to non-responders (Table 2). Response to GFD increased according to the presence of analytical parameters related to CD. In this sense, patients with a low-grade coeliac score $\leq$10 had the lowest GFD response rate (55.7%), which progressively increased to 86% in patients with a score >10 and positive coeliac lymphogram ($p = 0.011$) (Figure 1). The response rate to the diet was significantly different in terms of the type of IEL coeliac pattern observed (Figure 1). Those patients presenting with an isolated increase in TCR$\gamma\delta^+$ cells ($n = 20$) had a response rate of 55%, whereas for those with coeliac lymphogram ($n = 32$), the response rate was 84.4% ($p = 0.02$). In fact, seven out of the 20 (35%) patients with an isolated increase in TCR$\gamma\delta^+$ cells and 21 out of the 32 (65.6%) patients with coeliac lymphogram had a score >10 ($p = 0.046$).

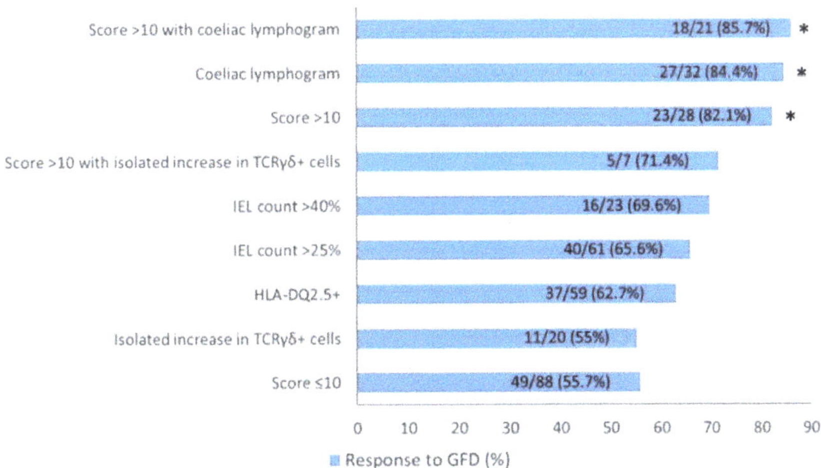

Figure 1. Response rate to a GFD in function of the presence of the different coeliac parameters and the low-grade coeliac score (* $p < 0.012$ vs. score $\leq$ 10). The different parameters are interrelated, the score integrates individual parameters, and it is not possible to separate the scoring system from the coeliac lymphogram: most patients with a positive score (>10) had a coeliac lymphogram and vice versa.

Table 2. Description of patients receiving GFD (n = 116): comparison of patients in terms of their response to the diet.

	Total (n = 116)	Response (n = 72)	Non-Response (n = 44)	p Value
Type of FBD symptoms:				
-IBS-D or functional diarrhoea	68 (58.6%)	39 (54.2%)	29 (65.9%)	0.21
-Functional bloating	48 (41.4%)	33 (45.8%)	15 (34.1%)	
Age (mean ± SEM)	42.4 ± 1.24	41.7 ± 1.6	43.7 ± 2	0.44
Sex (% female)	90 (77.6%)	55 (76.4%)	35 (79.5%)	0.69
Coeliac genetics:				
-HLA-DQ2.5	59 (51.8%)	37 (52.9%)	22 (50%)	
-HLA-DQ8	32 (27.6%)	22 (30.5)	10 (22.7%)	
-HLA-DQ2.2	11 (5%)	6 (8.3%)	5 (11.3%)	0.77
-1 allele DQ2.5	7 (3.2%)	2 (2.8%)	5 (11.3%)	
-Negative	7 (3.2%)	5 (6.9%)	2 (4.5%)	
Serology:				
-Detectable anti-tTG2 titers (EmA neg)	3 (2.6%)	2 (2.8%)	1 (2.3%)	0.87
Histology (IEL count):				
>25%	63 (54.8%)	41 (56.9%)	22 (50%)	
19–25%	18 (15.7%)	11 (15.3%)	7 (15.9%)	0.67
<19%	35 (30.2%)	19 (26.4%)	15 (34.1%)	
Coeliac IEL cytometry pattern:				
-Non-coeliac	64 (55.2%)	34 (47.2%)	30 (68.2%)	
-Isolated increase in TCRγδ$^+$ cells	20 (17.2%)	11 (15.3%)	9 (20.5)	0.019
-Coeliac lymphogram	32 (27.6%)	27 (37.5%)	5 (11.4%)	
Low-grade coeliac score > 10	28 (24.1%)	23 (31.9%)	5 (11.4%)	0.027
Score > 10 and coeliac lymphogram	21 (18.1%)	18 (25%)	3 (6.8%)	0.023
Score > 10 and isolated increase in TCRγδ$^+$ cells	7 (6.0%)	5 (6.9%)	2 (4.5%)	0.71

FBD: functional bowel disease; anti-tTG2: IgA anti-transglutaminase antibodies; EmA: IgA anti-endomysium antibodies; IEL: intraepithelial lymphocytes.

3.2. Frequency of Low-Grade Coeliac Enteropathy

Among the 72 GFD responders, there were 23 patients with a low-grade coeliac score >10 and 49 with a score ≤10. Thus, 23 out of 116 (19.8%) patients were diagnosed with low-grade coeliac enteropathy. Three patients among those with a score ≤10 presented with an HLA-DQ2.5+ and had a low-grade coeliac score equal to 10 points, because they had an IEL count between 19 and 25%, which scores 0 points. We considered that these patients had an inconclusive diagnosis [18]. Besides, 46 out of the 90 remaining patients (51.1%) had a sustained long-term clinical response to a GFD despite a negative score. Table 3 describes the clinical characteristics of these two groups of GFD responders as compared to non-responders. There were no significant differences in demographic variables, type of FBD symptoms, or the presence of HLA-DQ2.5+. Patients with low-grade coeliac enteropathy had significantly higher IEL counts and, as expected by the criteria used for diagnosis, more often coeliac lymphogram and a score >10 than the other groups. Noteworthy, there were no significant differences between non-coeliac GFD responders and non-responders.

Table 3. Comparison of the study variables among patients with low-grade coeliac enteropathy (LGCE), functional bowel disease GFD responders (FBD-R), and non-responders (FBD-NR) *.

Variable	LGCE (*n* = 23)	FBD-R (*n* = 46)	FBD-NR (*n* = 44)	*p* Value
Age (years) (mean ± SEM)	44.6 ± 2.8	39.8 ± 1.8	43.7 ± 2.0	0.24
Sex (% women)	15 (65.2%)	38 (82.6%)	35 (79.5%)	0.25
Type of FBD:				
-IBS-D/functional diarrhoea	12 (52.2%)	26 (56.5%)	29 (65.9%)	0.49
-Functional bloating	11 (47.8%)	20 (43.5%)	15 (34.1%)	
HLA-DQ2.5+	14 (63.6%)	20 (44.4%)	22 (50%)	0.34
LE (IEL > 25%) (%)	23 (100%)	19 (41.3%)	22 (50%)	<0.001
IEL count (mean ± SEM)	38.4 ± 3.4	24.5 ± 1.9	26.1 ± 2.3	0.001
Low-grade coeliac score >10	23 (100%)	0	5 (11.4%)	<0.001
Low-grade coeliac score (mean ± SEM)	13.9 ± 0.4	4.7 ± 0.5	5.7 ± 0.4	<0.001
Coeliac IEL cytometry pattern:				
-Isolated increase in TCR$\gamma\delta^+$ cells	5 (21.7%)	6 (14%)	9 (20.5%)	<0.001
-Coeliac lymphogram	18 (78.3%)	7 (15.2%)	5 (11.4%)	
TCR$\gamma\delta^+$ cells (%) (mean ± SEM)	20.6 ± 2.3	7.7 ± 1.4	6.9 ± 1.1	<0.001
CD3$^-$ cells (%) (mean ± SEM)	6.6 ± 1.1	15.9 ± 1.9	18.3 ± 1.9	<0.001

LE, lymphocytic enteritis; IEL: intraepithelial lymphocyte; * Three patients with a response to the GFD were excluded from this evaluation, since it was not possible to differentiate between LGCE and FBD-R. Two of them had coeliac lymphogram, and one had an isolated increase in TCR$\gamma\delta^+$ cells (see text).

4. Discussion

The current study presents a large series of patients fulfilling Rome IV criteria for FBD treated with a GFD. The results disclose that 62% of subjects with either diarrhoea or abdominal bloating clinical presentation show long-term clinical response to a GFD. In addition, the data support the acceptability of a GFD, since diet observance was maintained in the long term with sustained improvement. There were no differences in the frequency of HLA-DQ2/8+ between GFD responders and non-responders. However, responders more often present with a positive low-grade coeliac score and/or with coeliac lymphogram. In fact, the response rate of those patients with both a positive score and coeliac lymphogram was 86%, which is significantly higher than the 56% recorded for patients with a negative score.

The low-grade coeliac score was derived statistically to identify patients likely to respond to a GFD and be diagnosed with low-grade coeliac enteropathy with a sensitivity of 86% and a specificity of 85.2% [18]. Sensitivity is lower for patients with negative coeliac serology (77%), maintaining the same specificity (85%). Low-grade coeliac enteropathy is a term that was proposed to describe those patients characterized by lymphocytic enteritis (Marsh 1 enteropathy), positive coeliac genetics, and clinical and histological remission after a GFD [18]. Most of these patients had negative coeliac serology and present with an increased intraepithelial TCR$\gamma\delta^+$ cells count. As quoted above, several authors have considered that these patients present a mild form of CD [11–16], but despite that, they are frequently not treated as coeliacs with a GFD, and this is troubling, since both our own and other previous studies have shown that these patients may present with intestinal and extraintestinal symptoms compatible with the CD clinical spectrum, which improve after a GFD [15,16,24–26]. In this setting, the low-grade coeliac score represents a quantitative measure of the 'coeliac trait' described by Popp and Mäki [15]. Using dermatitis herpetiformis as a model disease in which there are gluten-related symptoms despite a non-atrophic enteropathy, even with negative coeliac serology in 60% of patients [27], these authors argue about the existence of a 'coeliac trait', consisting of a Marsh 1 lesion, positive coeliac genetics, and increase in TCR$\gamma\delta^+$ cells, which should be identified and treated.

While a high density of TCR$\gamma\delta^+$ intraepithelial lymphocytes in patients with non-atrophic enteropathy who also carry the susceptibility genes for CD seems to be a prerequisite for developing CD [28,29], this is not pathognomonic for the disease [19,22,29]. The low-grade coeliac score uses the TCR$\gamma\delta^+$ count, and in seronegative patients, this is the parameter that scores higher. However, results of the present study clearly show that the increase in TCR$\gamma\delta^+$ cells only has diagnostic value in seronegative Marsh 1 patients if there is a concomitant decrease in CD3$^-$ cells, i.e., when coeliac lymphogram is present. In fact, patients with a positive score presented significantly more often with the coeliac lymphogram than with an isolated increase in TCR$\gamma\delta^+$. Previous studies have shown a higher specificity in CD diagnosis for the coeliac lymphogram than for the isolated increase in TCR$\gamma\delta^+$ [19,30,31]. Since an isolated increase in TCR$\gamma\delta^+$ cells is not a useful biomarker of response to a GFD and, thus, of low-grade coeliac enteropathy, methods such as immuno-histochemistry, which only measure this parameter, are not useful in this setting. Therefore, coeliac lymphogram assessed by flow cytometry should be used instead, since it allows for the concomitant determination of CD3$^-$ cells, thereby increasing the diagnostic accuracy of the assay [18,19,21,22]. Taking an additional duodenal biopsy for flow cytometric analysis can provide useful information for decision making. Most laboratories in tertiary and even secondary hospitals dispose of a flow cytometer for diagnostic purposes, and analysing the lymphocyte subpopulations in the duodenal mucosa is an affordable technique.

Additionally, our results confirm that a cut-off of 25% IEL significantly increases the probability of low-grade coeliac enteropathy. However, as previously shown [18,31], there were a number of patients with lower cut-offs (between 19 and 25%) who were also likely to be diagnosed with low-grade coeliac enteropathy.

The response rate to GFD observed was within the range reported by previous studies. A prospective study of 41 patients with IBS-D showed clinically significant improvements in the IBS symptom severity score after six weeks on a GFD, without significant differences between HLA-DQ2/8-positive and -negative subjects. Twenty-nine of the 41 patients (71%) with clinical response were followed up for 18 months, and 21 were still on a GFD with sustained clinical response [9]. In another study, 12 out of 35 IBS-D or IBS-M patients (34%) clinically improved after a four-month period on a GFD. Additionally, the expression of HLA-DQ2/8 was not useful as diagnostic marker for GFD response [10]. As mentioned above, there are also other studies showing the effect of gluten exposure in IBS-D patients, which have recently been reviewed [3].

Independently of the presence or not of CD tissue biomarkers, the response rate to a GFD was very high in patients with symptoms suggestive of diarrhoea- or abdominal bloating-predominating FBD. In these patients, a GFD may be useful for treating patients with a low-grade coeliac enteropathy, as well as those with NCGS. In this sense, the most probable diagnosis of the non-coeliac GFD responders in the present study was NCGS. A formal diagnosis would require performing a gluten vs. placebo-controlled oral provocation [2]. However, this is controversial as the culprit triggering NCGS is currently unknown [32]. In this sense, results of a recent controlled double-blind crossover challenge study suggest that fructans rather than gluten seem to be the cause of symptoms in patients considering themselves as 'gluten-sensitive' [33].

Since a GFD may lead to a reduction in fructan intake that is sufficient to achieve sustained clinical improvement in non-coeliac individuals and may also be effective when treating those with a low-grade coeliac enteropathy, gluten restriction seems to be an effective initial approach for patients presenting with previously unexplained diarrhoea and/or abdominal bloating of presumably functional origin. In fact, it has been suggested that a GFD may be the easiest way of achieving fructan reduction [4], since fructans are a key component to be reduced in a long-term adapted low-FODMAP diet, as demonstrated in a prospective study of 103 patients [34]. In this sense, it has been suggested that a GFD may be administered as a 'bottom-up' approach in the FODMAP diet for patients with

IBS. This 'bottom-up' approach has been advocated as a way to avoid prolonged dietary restrictions in a low-FODMAP diet, potentially avoiding disruption to the gut microbiota and to nutritional status [35]. In addition, patients have rated a GFD as more acceptable than a low-FODMAP diet [36], and only 40% of patients have been shown to follow the low-FODMAP diet correctly [37].

The present study has a number of drawbacks. Firstly, the retrospective nature of the evaluation of dietary response is one limitation of the study; however, we considered response to GFD only if a complete and sustained resolution of symptoms was observed after at least 6 months of follow-up. This fact together with symptom relapse with inadvertent gluten exposure and long-term maintained observance to diet suggest a true response to gluten restriction. Secondly, the study was non-controlled, although a systematic meta-analysis of randomized controlled trials in IBS has demonstrated a pooled placebo response rate of 37.5%, with lower responses seen in those patients who fulfil the Rome criteria on study entry and who received eight weeks or more of therapy [38]. This suggests that in our study, the 62% response rate to a GFD is unlikely to be a placebo effect particularly because improvement was maintained at a median of 21 months. Third, the present study, unlike previous ones, was performed mostly in individuals having positive coeliac genetics (79% HLA-DQ2.5 and/or DQ8+ plus 5% HLA-DQ2.2+). However, this isolated parameter is not a good biomarker of response to a GFD, as has been shown both in several previous studies discussed above and in the present study, probably because of the high prevalence of these genes in the general population. Finally, the frequency of a positive low-grade coeliac score and coeliac lymphogram was higher in the sample of patients treated with a GFD than in the entire sample of 260 patients with FBD. This suggests that the actual rate of low-grade coeliac enteropathy is probably somewhat lower than the observed rate.

In conclusion, a GFD is effective in the long-term treatment of patients with previously unexplained chronic watery diarrhoea- or bloating-predominant symptoms fulfilling the criteria of FBD. The response rate is much higher in a subgroup of patients defined by the presence of both a positive low-grade coeliac score and coeliac lymphogram who may be diagnosed with low-grade coeliac enteropathy. It is mainly the presence of coeliac lymphogram and not the increase in TCR$\gamma\delta^+$ cells that is useful as a tissue biomarker of low-grade coeliac enteropathy. The results support the recommendation of administering a GFD as a 'bottom-up' approach in the FODMAP diet for patients with IBS.

Supplementary Materials: The following are available online at https://www.mdpi.com/article/10.3390/nu13061812/s1 S1. File 1.xls.

Author Contributions: Guarantor of the article: F.F.-B.; Conceptualization, F.F.-B., M.A., A.C. and M.E.; data curation, F.F.-B., B.A., A.R., E.T., A.C., L.R., A.M.-C., P.R.-R. and M.E.; formal analysis, F.F.-B., M.A., E.T., A.C., L.R. and M.E.; investigation, F.F.-B. and B.A.; methodology, F.F.-B., B.A., E.T., A.C. and M.E.; project administration, F.F.-B.; resources, A.M.-C.; software, A.M.-C.; supervision, F.F.-B.; writing—original draft, F.F.-B. and M.E.; writing—review and editing, F.F.-B., B.A., A.R., M.A., E.T., A.C., L.R., A.M.-C., P.R.-R. and M.E. All authors have read and agreed to the published version of the manuscript.

Funding: This research received no external funding.

Institutional Review Board Statement: The following statement is included on page 5 of the manuscript "The study was conducted according to the guidelines of the Declaration of Helsinki and was approved by the Ethics Committee of the Hospital Universitari Mutua Terrassa". Code: EO/1011; date: 25 March 2010.

Informed Consent Statement: Informed consent was obtained from all subjects involved in the study as noted in the manuscript.

Data Availability Statement: Database of the study supporting the reported results can be found as a Supplementary File.

Acknowledgments: The Centro de Investigación Biomédica en Red de Enfermedades Hepáticas y Digestivas (CIBERehd) is an initiative of the Instituto de Salud Carlos III, Madrid, Spain. This institution played no role in the study design, the acquisition, analysis, or interpretation of the data or the report writing.

Conflicts of Interest: The authors declare no conflict of interest.

Writing Assistance: David Bridgewater, Anglocatalan Services, Barcelona.

Appendix A.

Appendix A.1. Flow Cytometry

One single duodenal biopsy was obtained using a 2.8 mm biopsy forceps (Radial Jaw 4, Boston Scientific, USA) and immediately processed. Preparations of IEL suspensions were performed by incubation with 1 mM EDTA, 1 mM DTT in HBSS for 90 min with continuous rotation at 12 rpm in a vertical shaker at room temperature. This procedure achieved the total removal of villous epithelium and the partial removal of crypt epithelium. The proper separation of epithelial compartment was confirmed by an immunohistochemical analysis of the remaining tissue during the protocol validation. The obtained suspension, a mixture of IEL and epithelial cells, was washed once in fresh HBSS at 1500 rpm for 10 min, and IEL were immediately stained with previously titrated amounts of directly labelled antibodies for 15 min at room temperature. The antibodies used to define the different IEL subsets were anti-CD45-APC (clone 2D1), anti-CD3-PerCP (clone SK7), anti-CD103-FITC (clone Ber-ACT8), and anti-TCR $\gamma\delta$-PE (clone 11F2) (all from BD Biosciences, Franklin Lakes, NJ, USA). The intraepithelial origin of the IEL suspension was verified with CD103$^+$ staining, and it was always $\geq$85%. Cells were immediately analysed on a standard 4-color FACSCalibur instrument (BD Biosciences, Franklin Lakes, NJ, USA). Cell counts of the recovered cell number for biopsy were performed with a haemocytometer and trypan blue exclusion.

Results were obtained 3 to 4 h after biopsy sampling and expressed as percentages of bright CD45 staining and a low sideward scatter gate. The normal cut-off values for the IEL cytometric pattern in our laboratory are CD3$^+$TCR$\gamma\delta^+$ IEL $\leq$8.5% ($\leq$mean + 2SD) and CD3$^-$ IEL $\geq$10% (10th percentile). These cut-offs were calculated in a sample of 65 non-coeliac subjects. The intra-assay coefficient of variation was 5.5% (two replicates of each sample processed one immediately after the other), and the inter-sample coefficient of variation was 7.7% (two different samples from each patient obtained in the same procedure).

Appendix A.2. HLA Genotyping

Genomic DNA from whole blood was purified using the commercial Qiamp DNA Blood Mini kit (Qiagen, Düsseldorf, Germany). A commercial reverse hybridization kit for the detection of CD heterodimers HLA-DQ2.5 (A1*0501/*0505, B1*0201/*0202) and HLA-DQ8 (A1*0301, B1*0302) was used (GenID, GMBH, Strasburg, Germany). HLA-DQ2.5 haplotype was present in 24% of healthy controls and 90% of CD patients in our geographical area.

Appendix B.

Table A1. Comparison of study variables between patients with functional bowel disease on or not on a GFD.

Variable	GFD (*n* = 116)	No Diet (*n* = 144)	*p* Value
Age (years)	42.4 ± 1.2	41.2 ± 1.1	0.48
Sex (% female)	77.6%	64.6%	0.022
FBD type			

Table A1. *Cont.*

Variable	GFD (*n* = 116)	No Diet (*n* = 144)	*p* Value
SII-D/diarrhoea (%)	58.6%	58.3%	0.96
Abdominal bloating (%)	41.4%	41.7%	
HLA-DQ2.5+ (%)	51.8%	58.3%	0.29
Histology (IEL > 25%) (%)	54.8%	36.9%	0.016
Cytometry pattern			
IEL coeliac pattern (%)	43.8%	11.2%	<0.0005
Coeliac lymphogram (%)	27.6%	4.9%	<0.0005
Low-grade coeliac score >10 (%)	25%	6.3%	<0.0005
Score > 10 and coeliac lymphogram (%)	18.1%	2%	<0.0001

References

1. Mearin, F.; Lacy, B.E.; Chang, L.; Lembo, A.J.; Simren, M.; Spiller, R. Bowel disorders. *Gastroenterology* **2016**, *150*, 1393–1407.
2. Catassi, C.; Elli, L.; Bonaz, B.; Bouma, G.; Carroccio, A.; Castillejo, G.; Cellier, C.; Cristofori, F.; de Magistris, L.; Dolinsek, J.; et al. Diagnosis of non-celiac gluten sensitivity (NCGS): The Salerno experts' criteria. *Nutrients* **2015**, *7*, 4966–4977. [CrossRef] [PubMed]
3. Rej, A.; Aziz, I.; Tornblom, H.; Sanders, D.S.; Simrén, M. The role of diet in irritable bowel syndrome: Implications for dietary advice. *J. Intern. Med.* **2019**, *286*, 490–502. [CrossRef]
4. Rej, A.; Aziz, I.; Sanders, D.S. A gluten-free diet: The express route to fructan reduction. *Am. J. Gastroenterol.* **2019**, *114*, 1553. [CrossRef] [PubMed]
5. Catassi, C.; Alaedini, A.; Bojarski, C.; Bonaz, B.; Bouma, G.; Carroccio, A.; Carroccio, A.; Castillejo, G.; De Magistris, L.; Dieterich, W.; et al. The overlapping area of non-celiac gluten sensitivity (NCGS) and wheat-sensitive irritable bowel syndrome (IBS): An update. *Nutrients* **2017**, *9*, 1268. [CrossRef]
6. Bohn, L.; Storsrud, S.; Tornblom, H.; Bengtsson, U.; Simren, M. Self-reported food-related gastrointestinal symptoms in IBS are common and associated with more severe symptoms and reduced quality of life. *Am. J. Gastroenterol.* **2013**, *108*, 634–641. [CrossRef]
7. Morcos, A.; Dinan, T.; Quigley, E.M. Irritable bowel syndrome: Role of food in pathogenesis and management. *J. Dig. Dis.* **2009**, *10*, 237–246. [CrossRef] [PubMed]
8. Rej, A.; Buckle, R.L.; Shaw, C.C.; Trott, N.; Aziz, I.; Sanders, D.S. Letter: The gluten-free diet as a bottom-up approach for irritable bowel syndrome. *Aliment. Pharmacol. Ther.* **2020**, *51*, 184–195. [CrossRef]
9. Aziz, I.; Trott, N.; Briggs, R.; North, J.R.; Hadjivassiliou, M.; Sanders, D.S. Efficacy of a gluten-free diet in subjects with irritable bowel syndrome-diarrhea unaware of their HLA-DQ2/8 genotype. *Clin. Gastroenterol. Hepatol.* **2016**, *14*, 696–703. [CrossRef]
10. Barmeyer, C.; Schumann, M.; Meyer, T.; Zielinski, C.; Zuberbier, T.; Siegmund, B.; Schulzke, J.D.; Daum, S.; Ullrich, R. Long-term response to gluten-free diet as evidence for non-celiac wheat sensitivity in one third of patients with diarrhea-dominant and mixed-type irritable bowel syndrome. *Int. J. Colorectal. Dis.* **2017**, *32*, 29–39. [CrossRef]
11. Husby, S.; Koletzko, S.; Korponay-Szabó, I.R.; Mearin, M.L.; Phillips, A.; Shamir, R.; Troncone, R.; Giersiepen, K.; Branski, D.; Catassi, C.; et al. ESPGHAN Working Group on Coeliac Disease Diagnosis; European Society for Pediatric Gastroenterology, Hepatology, and Nutrition. European Society for Pediatric Gastroenterology, Hepatology, and Nutrition guidelines for the diagnosis of coeliac disease. *J. Pediatr. Gastroenterol. Nutr.* **2012**, *54*, 136–160. [CrossRef]
12. Husby, S.; Murray, J.A. Gluten sensitivity: Celiac lite versus celiac like. *J. Pediatr.* **2014**, *164*, 436–438. [CrossRef] [PubMed]
13. Ferch, C.C.; Chey, W.D. Irritable bowel syndrome and gluten sensitivity without celiac disease: Separating the wheat from the chaff. *Gastroenterology* **2012**, *142*, 664–666. [CrossRef]
14. Mooney, P.D.; Aziz, I.; Sanders, S. Non-celiac gluten sensitivity: Clinical relevance and recommendations for future research. *Neurogastroenterol. Motil.* **2013**, *25*, 864–871. [CrossRef] [PubMed]
15. Popp, A.; Mäki, M. Gluten-Induced Extra-Intestinal Manifestations in Potential Celiac Disease—Celiac Trait. *Nutrients* **2019**, *11*, 320. [CrossRef]
16. Kurppa, K.; Collin, P.; Viljamaa, M.; Haimila, K.; Saavalainen, P.; Partanen, J.; Laurila, K.; Huhtala, H.; Paasikivi, K.; Mäki, M.; et al. Diagnosing mild enteropathy celiac disease: A randomized, controlled clinical study. *Gastroenterology* **2009**, *136*, 816–823. [CrossRef]
17. Rosinach, M.; Fernández-Bañares, F.; Carrasco, A.; Ibarra, M.; Temiño, R.; Salas, A.; Esteve, M. Double-blind randomized clinical trial: Gluten versus placebo rechallenge in patients with lymphocytic enteritis and suspected celiac disease. *PLoS ONE* **2016**, *11*, e0157879. [CrossRef]

18. Fernández-Bañares, F.; Carrasco, A.; Rosinach, M.; Arau, B.; García-Puig, R.; González, C.; Tristán, E.; Zabana, Y.; Esteve, M. A scoring system for identifying patients likely to be diagnosed with low-grade coeliac enteropathy. *Nutrients* **2019**, *11*, 1050. [CrossRef]

19. Leon, F. Flow cytometry of intestinal intraepithelial lymphocytes in celiac disease. *J. Immunol. Methods* **2011**, *363*, 177–186. [CrossRef]

20. Leon, F.; Eiras, P.; Roy, G.; Camarero, C. Intestinal intraepithelial lymphocytes and anti-transglutaminase in a screening algorithm for coeliac disease. *Gut* **2002**, *50*, 740–741. [CrossRef]

21. Fernández-Bañares, F.; Carrasco, A.; García-Puig, R.; Rosinach, M.; González, C.; Alsina, M.; Loras, C.; Salas, A.; Viver, J.M.; Esteve, M.; et al. Intestinal intraepithelial lymphocyte cytometric pattern is more accurate than subepithelial deposits of anti-tissue transglutaminase IgA for the diagnosis of coeliac disease in lymphocytic enteritis. *PLoS ONE* **2014**, *9*, e101249. [CrossRef] [PubMed]

22. Fernández-Bañares, F.; Carrasco, A.; Martín, A.; Esteve, M. Systematic Review and Meta-Analysis: Accuracy of Both Gamma Delta+ Intraepithelial Lymphocytes and Coeliac Lymphogram Evaluated by Flow Cytometry for Coeliac Disease Diagnosis. *Nutrients* **2019**, *11*, 1992. [CrossRef]

23. Sapone, A.; Lammers, K.M.; Casolaro, V.; Cammarota, M.; Giuliano, M.T.; De Rosa, M.; Stefanile, R.; Mazzarella, G.; Tolone, C.; Russo, M.I.; et al. Divergence of gut permeability and mucosal immune gene expression in two gluten-associated conditions: Celiac disease and gluten sensitivity. *BMC Med.* **2011**, *9*, 23. [CrossRef]

24. Esteve, M.; Rosinach, M.; Fernández-Bañares, F.; Farré, C.; Salas, A.; Alsina, M.; Vilar, P.; Abad-Lacruz, A.; Forné, M.; Santaolalla, R.; et al. Spectrum of gluten-sensitive enteropathy in first-degree relatives of patients with coeliac disease: Clinical relevance of lymphocytic enteritis. *Gut* **2006**, *55*, 1739–1745. [CrossRef]

25. Rosinach, M.; Esteve, M.; González, C.; Temiño, R.; Mariné, M.; Monzón, H.; Sainz, E.; Loras, C.; Espinós, J.C.; Forné, M.; et al. Lymphocytic duodenosis: Aetiology and long-term response to specific treatment. *Dig. Liver Dis.* **2012**, *44*, 643–648. [CrossRef]

26. Aziz, I.; Evans, K.E.; Hopper, A.D.; Smillie, D.M.; Sanders, D.S. A prospective study into the aetiology of lymphocytic duodenosis. *Aliment. Pharmacol. Ther.* **2010**, *32*, 1392–1397. [CrossRef]

27. Mansikka, E.; Hervonen, K.; Kaukinen, K.; Collin, P.; Huhtala, H.; Reunala, T.; Salmi, T. Prognosis of dermatitis herpetiformis patients with and without villous atrophy at diagnosis. *Nutrients* **2018**, *10*, 641. [CrossRef] [PubMed]

28. Mäki, M.; Holm, K.; Collin, P.; Savilahti, E. Increase in γ/δ T cell receptor bearing lymphocytes in normal small bowel mucosa in latent celiac disease. *Gut* **1991**, *32*, 1412–1414. [CrossRef]

29. Iltanen, S.; Holm, K.; Partanen, J.; Laippala, P.; Mäki, M. Increased density of jejunal $\gamma\delta$+ T cells in patients having normal mucosa—Marker of operative autoimmune mechanisms? *Autoimmunity* **1999**, *29*, 1787–1791. [CrossRef]

30. Fernández-Bañares, F.; Crespo, L.; Núñez, C.; López-Palacios, N.; Tristán, E.; Vivas, S.; Farrais, S.; Arau, B.; Vidal, J.; Roy, G.; et al. Gamma delta+ intraepithelial lymphocytes and coeliac lymphogram in a diagnostic approach to coeliac disease in patients with seronegative villous atrophy. *Aliment. Pharmacol. Ther.* **2020**, *51*, 699–705. [CrossRef]

31. Pellegrino, S.; Villanacci, V.; Sansotta, N.; Scarfi, R.; Bassotti, G.; Vieni, G.; Princiotta, A.; Sferlazzas, C.; Magazzù, G.; Tuccari, G. Redefining the intraepithelial lymphocytes threshold to diagnose gluten sensitivity in patients with architecturally normal duodenal histology. *Aliment. Pharmacol. Ther.* **2011**, *33*, 697–706. [CrossRef] [PubMed]

32. Sergi, C.; Villanacci, V.; Carroccio, A. Non-celiac wheat sensitivity: Rationality and irrationality of a gluten-free diet in individuals affected with non-celiac disease: A review. *BMC Gastroenterol.* **2021**, *21*, 5. [CrossRef] [PubMed]

33. Skodje, G.I.; Sarna, V.K.; Minelle, I.H.; Rolfsen, K.L.; Muir, J.G.; Gibson, P.R.; Veierød, M.B.; Henriksen, C.; Lundin, K.E.A. Fructan Rather Than Gluten, Induces Symptoms in Patients with Self-Reported Non-Celiac Gluten Sensitivity. *Gastroenterology* **2018**, *154*, 529–539. [CrossRef]

34. O'Keefe, M.; Jansen, C.; Martin, L.; Williams, M.; Seamark, L.; Staudacher, H.M.; Irving, P.M.; Whelan, K.; Lomer, M.C. Long-term impact of the low-FODMAP diet on gastrointestinal symptoms, dietary intake, patient acceptability, and health care utilization in irritable bowel syndrome. *Neurogastroenterol. Motil.* **2018**, *30*, e13154. [CrossRef] [PubMed]

35. Wang, X.J.; Camilleri, M.; Vanner, S.; Tuck, C. Review article: Biological mechanisms for symptom causation by individual FODMAP subgroups—the case for a more personalised approach to dietary restriction. *Aliment. Pharmacol. Ther.* **2019**, *50*, 517–529. [CrossRef]

36. Paduano, D.; Cingolani, A.; Tanda, E.; Usai, P. Effect of Three Diets (Low-FODMAP, Gluten-free and Balanced) on Irritable Bowel Syndrome Symptoms and Health-Related Quality of Life. *Nutrients* **2019**, *11*, 1566. [CrossRef]

37. Tuck, C.J.; Reed, D.E.; Muir, J.G.; Vanner, S.J. Implementation of the low FODMAP diet in functional gastrointestinal symptoms: A real-world experience. *Neurogastroenterol. Motil.* **2020**, *32*, e13730. [CrossRef] [PubMed]

38. Ford, A.C.; Moayyedi, P. Meta-analysis: Factors affecting placebo response rate in the irritable bowel syndrome. *Aliment. Pharmacol. Ther.* **2010**, *32*, 144–158. [CrossRef] [PubMed]

MDPI
St. Alban-Anlage 66
4052 Basel
Switzerland
Tel. +41 61 683 77 34
Fax +41 61 302 89 18
www.mdpi.com

Nutrients Editorial Office
E-mail: nutrients@mdpi.com
www.mdpi.com/journal/nutrients